Putting AI in the Critical Loop

Putting AI in the Critical Loop
Assured Trust and Autonomy in Human-Machine Teams

Edited by

Prithviraj Dasgupta
Distributed Intelligent Systems Section, Naval Research Laboratory,
Washington, DC, United States

James Llinas
University at Buffalo, New York, NY, United States

Tony Gillespie
Electronic & Electrical Engineering, University College London,
London, United Kingdom

Scott Fouse
Fouse Consulting Services, Scottsdale, AZ, United States

William Lawless
Department of Mathematics, Sciences and Technology, and Department of Social
Sciences, School of Arts and Sciences, Paine College, Augusta, GA, United States

Ranjeev Mittu
Information and Decision Sciences Branch, Information Technology Division,
U.S. Naval Research Laboratory, Washington, DC, United States

Donald Sofge
Navy Center for Applied Research in Artificial Intelligence, United States Naval
Research Laboratory, Washington, DC, United States

ACADEMIC PRESS
An imprint of Elsevier

ELSEVIER

Academic Press is an imprint of Elsevier
125 London Wall, London EC2Y 5AS, United Kingdom
525 B Street, Suite 1650, San Diego, CA 92101, United States
50 Hampshire Street, 5th Floor, Cambridge, MA 02139, United States

Notices
Knowledge and best practice in this field are constantly changing. As new research and experience broaden our understanding, changes in research methods, professional practices, or medical treatment may become necessary.

Practitioners and researchers must always rely on their own experience and knowledge in evaluating and using any information, methods, compounds, or experiments described herein. In using such information or methods they should be mindful of their own safety and the safety of others, including parties for whom they have a professional responsibility.

To the fullest extent of the law, neither the Publisher nor the authors, contributors, or editors, assume any liability for any injury and/or damage to persons or property as a matter of products liability, negligence or otherwise, or from any use or operation of any methods, products, instructions, or ideas contained in the material herein.

ISBN 978-0-443-15988-6

For information on all Academic Press publications
visit our website at https://www.elsevier.com/books-and-journals

Publisher: Mara Conner
Editorial Project Manager: Toni Louise Jackson
Production Project Manager: Jayadivya Saiprasad
Cover Designer: Matthew Limbert

Typeset by STRAIVE, India

Working together
to grow libraries in
developing countries

www.elsevier.com • www.bookaid.org

Contents

12. A framework of human factors methods for safe,
ethical, and usable artificial intelligence in defense 245

Paul M. Salmon, Brandon J. King, Scott McLean, Gemma J.M. Read,
Christopher Shanahan, and Kate Devitt

13. A schema for harms-sensitive reasoning, and an
approach to populate its ontology by human annotation 265

Ariel M. Greenberg

Contributors

Shahira Ali University of Virginia, Charlottesville, VA, United States

Anthony L. Baker DEVCOM Army Research Laboratory, Aberdeen Proving Ground, Adelphi, MD, United States

Oliver Broadrick Department of Computer Science, The George Washington University, Washington, DC, United States

Carolina Centeio Jorge Delft University of Technology, Delft, The Netherlands

Erin Chiou Arizona State University, Tempe, AZ, United States

Jediah Clark School of Electronics and Computer Science, University of Southampton, Southampton, United Kingdom

Kate Devitt Human-Centred Computing, School of Electrical Engineering and Computer Science, University of Queensland; Data and Information Services, Queensland Government Customer and Digital Group, Queensland Government, Brisbane, QLD, Australia

Sean M. Fitzhugh DEVCOM Army Research Laboratory, Aberdeen Proving Ground, Adelphi, MD, United States

Daniel E. Forster DEVCOM Army Research Laboratory, Aberdeen Proving Ground, Adelphi, MD, United States

Hesham Y. Fouad Information Technology Division, US Naval Research Laboratory, Washington, DC, United States

Scott Fouse Fouse Consulting Services, Scottsdale, AZ, United States

Tony Gillespie Electronic & Electrical Engineering, University College London, London, United Kingdom

Maria Goldshtein Arizona State University, Tempe, AZ, United States

Haley N. Green University of Virginia, Charlottesville, VA, United States

Ariel M. Greenberg Johns Hopkins University Applied Physics Laboratory, Laurel, MD, United States

Benjamin Harvey Department of Computer Science, The George Washington University, Washington, DC, United States

William Hunt School of Electronics and Computer Science, University of Southampton, Southampton, United Kingdom

Tariq Iqbal University of Virginia, Charlottesville, VA, United States

Md Mofijul Islam University of Virginia, Charlottesville, VA, United States

Catholijn M. Jonker Delft University of Technology, Delft; Leiden University, Leiden, The Netherlands

Brandon J. King Centre for Human Factors and Sociotechnical Systems, University of the Sunshine Coast, Sunshine Coast, QLD, Australia

Andrea Krausman DEVCOM Army Research Laboratory, Aberdeen Proving Ground, Adelphi, MD, United States

Per Ola Kristensson Department of Engineering, University of Cambridge, Cambridge, Cambridgeshire, United Kingdom

W.F. Lawless Department of Mathematics, Sciences and Technology, and Department of Social Sciences, School of Arts and Sciences, Paine College, Augusta, GA, United States

James Llinas University at Buffalo, New York, NY, United States

Scott McLean Centre for Human Factors and Sociotechnical Systems, University of the Sunshine Coast, Sunshine Coast, QLD, Australia

Siddharth Mehrotra Delft University of Technology, Delft, The Netherlands

Michael Mylrea University of Miami, Institute of Data Science and Computing, Coral Gables, FL, United States

Mohammad Naiseh Department of Computing and Informatics, Bournemouth University, Poole, United Kingdom

Bhagi Narahari Department of Computer Science, The George Washington University, Washington, DC, United States

Catherine E. Neubauer DEVCOM Army Research Laboratory, Aberdeen Proving Ground, Adelphi, MD, United States

Max Nicosia Department of Engineering, University of Cambridge, Cambridge, Cambridgeshire, United Kingdom

Katie Parnell School of Electronics and Computer Science, University of Southampton, Southampton, United Kingdom

Charles Peeke Department of Computer Science, The George Washington University, Washington, DC, United States

Sarvapali D. Ramchurn School of Electronics and Computer Science, University of Southampton, Southampton, United Kingdom

Gemma J.M. Read Centre for Human Factors and Sociotechnical Systems, University of the Sunshine Coast, Sunshine Coast, QLD, Australia

Ray E. Reichenberg Educational Assessment Research Unit, College of Education, University of Otago, Dunedin, New Zealand

Paul M. Salmon Centre for Human Factors and Sociotechnical Systems, University of the Sunshine Coast, Sunshine Coast, QLD, Australia

Jordan Richard Schoenherr Concordia University, Montreal; Carleton University, Ottawa, ON, Canada

Christopher Shanahan Defence Science Technology Group, Canberra, NSW, Australia

Mohammad D. Soorati School of Electronics and Computer Science, University of Southampton, Southampton, United Kingdom

Myrthe L. Tielman Delft University of Technology, Delft, The Netherlands

Emma M. van Zoelen Delft University of Technology, Delft; TNO, Soesterberg, The Netherlands

Ruben Verhagen Delft University of Technology, Delft, The Netherlands

About the editors

Prithviraj Dasgupta is a computer engineer with the Distributed Intelligent Systems Section at the US Naval Research Laboratory, Washington, DC, United States. His research interests are in the areas of machine learning, AI-based game playing, game theory, and multiagent systems. He received his PhD in 2001 from the University of California, Santa Barbara. From 2001 to 2019, he was a full professor with the computer science department at the University of Nebraska in Omaha, where he established and directed the CMANTIC Robotics Laboratory. He has authored over 150 publications for leading journals and conferences in his research area. He is a senior member of IEEE.

James Llinas is an emeritus professor at the University at Buffalo, NY, United States. He established and directed the Center for Multisource Information Fusion at the university, the only academic systems-centered information fusion center in the United States, leading it to carry out well-funded multidisciplinary research for over 20 years. He was a coauthor of the first book on data fusion and has coedited and coauthored several additional books on data and information fusion. In 1998, he helped establish and was the first president of the International Society for Information Fusion.

Tony Gillespie is a visiting professor at University College London, United Kingdom and a Fellow of the Royal Academy of Engineering. His career includes academic, industrial, and government research and research management. His work on ensuring highly automated weapons meet legal requirements has been extended to other autonomous systems in recent years, and he has written a book and academic papers. He has acted as a technical adviser to the United Nations and other meetings, discussing potential bans on autonomous weapon systems.

Scott Fouse, Fouse Consulting Services, AZ, United States, had a 42-year career in aerospace R&D, mostly focused on exploring applications of AI to military applications. He was vice president of the Advanced Technology Center at Lockheed Martin Space, where he led approximately 500 scientists and engineers performing research and development in space science and a variety of space systems-related technologies and capabilities. In prior appointments, Scott served as president and CEO of ISX Corporation and was a member of the Air Force Scientific Advisory Board, where he supported a number of studies, directorate reviews, and chaired a study on experimentation to support disruptive innovation. Scott has a BS in physics from the University of Central Florida and an MS in electrical engineering from the University of Southern California.

William Lawless is professor of mathematics and psychology at Paine College, GA, United States. For his PhD topic on group dynamics, he theorized about the causes of tragic mistakes made by large organizations with world-class scientists and engineers. After his PhD in 1992, the DOE invited him to join its Citizens Advisory Board (CAB) at DOE's Savannah River Site (SRS), Aiken, SC. As a founding member, he coauthored numerous recommendations on environmental remediation of radioactive wastes (e.g., the regulated closure in 1997 of the first two high-level radioactive waste tanks in the United States). He is a member of INCOSE, IEEE, AAAI, and AAAS. His research today is on autonomous human-machine teams (A-HMT). He is the lead editor of seven published books on artificial intelligence. He was lead organizer of a special issue of *AI Magazine* on human-machine teams and explainable AI (2019). He has authored over 85 articles and book chapters and over 175 peer-reviewed proceedings. He was the lead organizer of 12 AAAI symposia at Stanford (2020). Since 2018, he has also served on the Office of Naval Research's Advisory Boards for the Science of Artificial Intelligence and Command Decision Making.

Ranjeev Mittu is the branch head for the Information Management and Decision Architectures Branch within the Information Technology Division at the US Naval Research Laboratory (NRL), Washington, DC, United States. He leads a multidisciplinary group of scientists and engineers that conduct research and advanced development in visual analytics, human performance assessment, decision support systems, and enterprise systems. Mr. Mittu's research expertise is in multiagent systems, human-systems integration, artificial intelligence (AI), machine learning, data mining, and pattern recognition; and he has authored and/or coedited nine books on the topic of AI in collaboration with national and international scientific communities spanning academia and defense. Mr. Mittu received an MS in electrical engineering in 1995 from The Johns Hopkins University in Baltimore, Maryland.

Donald Sofge is a computer scientist and roboticist at the US Naval Research Laboratory (NRL), Washington, DC, United States with 35 years of experience in artificial intelligence, machine learning, and control systems R&D. He leads the Distributed Autonomous Systems Section in the Navy Center for Applied Research in Artificial Intelligence (NCARAI), where he develops nature-inspired computing paradigms for challenging problems in sensing, artificial intelligence, and control of autonomous robotic systems. He has more than 200 refereed publications, including 11 books on robotics, artificial intelligence, machine learning, planning, sensing, control, and related disciplines.

1

Introduction

James Llinas

UNIVERSITY AT BUFFALO, NEW YORK, NY, UNITED STATES

This book is an expanded Proceedings of the March 2022 Association for the Advancement of Artificial Intelligence (AAAI) Spring Symposium titled "Putting AI in the Critical Loop: Assured Trust and Autonomy in Human-Machine Teams." That symposium was planned to start a series of symposia that will address specialized topics in system engineering processes for complex systems incorporating AI technologies and human roles. This 2022 symposium follows on the 2020 and 2021 AAAI Spring Symposia that addressed *overarching systemic aspects* of engineering such systems (2021: Leveraging systems engineering to realize synergistic AI/machine learning capabilities, https://aaai.org/Symposia/Spring/sss21symposia.php#ss07, 2020: AI welcomes systems engineering: toward the science of interdependence for autonomous human-machine teams, https://aaai.org/Symposia/Spring/sss20symposia.php#ss03); the 2022 plan was to focus on the *particular topic of human-machine teaming* and the related, specific engineering challenges associated with that theme.

1 Theme of the symposium

Close interactions between machines and humans with a high level of autonomy can lead to effective work sharing between them. Any such partnership must have underpinning, implicit assumptions about behavior and mutual trust. Performance will be maximized when the partnership aims to provide mutual benefits. Designing systems that include human-machine partnerships requires an understanding of the rationale of any such relationship, the balance of control, and the nature of human and machine autonomy. To design such systems, it is essential to understand the nature of human-machine cooperation, synergy, interdependence, and the meaning and nature of "collective intelligence." Machines and humans have distinctively different characteristics and features. These make team formation difficult but, conversely, they offer the potential to work well together, ideally overcoming each other's weaknesses, toward the cocreation of value in any application setting. The meaning of assured operation of a human-machine system also needs considerable specification; assurance has been approached historically through design processes by following rigorous safety standards in development, and by demonstrating compliance through system testing, but largely in systems of bounded capability and where human roles were similarly bounded. Across the widest range of applications, these topics remain persistent as a major concern of system design and development. Intimately related to these topics are the issues of *human-machine trust* and *"assured" performance and operation* of these complex systems, the focal topics of the 2022 symposium. Recent discussions on trust emphasize that, with regard to human-machine systems, trust is bidirectional and two-sided (as it is in humans); humans need to trust AI technology, but future AI technology at least may need to trust human inputs and guidance as well. (Once a machine has been trained to perform its role in a team, it then knows when the human(s) has become dysfunctional.) In the absence of an adequately high level of autonomy that can be relied upon, substantial operator involvement is required, which not only severely limits operational gains, but also creates significant new challenges in the areas of human-machine interaction and mixed initiative control. These intersecting themes of collective intelligence, bidirectional trust, and continual assurance formed the challenging and extraordinarily interesting themes of the symposium.

The 2½-day 2022 symposium addressed a range of topics related to the human-machine engineering theme, including the understanding and verification of autonomy and teamwork, risk determination versus risk perception, and the evaluation of the perceptions of humor in an automated partner, a topic that proved surprisingly interesting and quite innovative. Attendees included several European participants, primarily from the United Kingdom but also from the Netherlands; as a result, the symposium was run on Eastern US time to minimize time-difference issues for these colleagues. A distinguished speaker from Australia presented the last talk on Day 2, providing a kick-off paper that would set the stage for our Day 3 morning panel session on "Alternative Paths to Developing Engineering Solutions for Human-machine Teams that Meet Human Values, Laws, and Ethics."

A panel of distinguished speakers, again including European participants (for example, from the UK Royal Academy of Engineering), addressed critical themes at the core of trust in human-machine teams with a primary focus on human values, laws, and ethics. The agenda for the panel session allowed for constructive interaction between the panel and the audience that yielded a range of concrete outputs and offered ideas for new research programs.

The speakers, affiliations, and paper topics are shown in Tables 1.1 and 1.2. They illustrate what was truly an international symposium, with speakers from the United States, United Kingdom, Netherlands, and Australia.

2 Teams and teamwork

As can be seen in Table 1.1, the symposium started off with a few papers involving the notion of teams and teamwork. It would certainly be expected that the symposium contributions would all revolve around the many interrelated topics of teamwork, given the symposium theme. There is a fairly large literature on teamwork, addressing it from conceptual to engineering points of view. Some views offered here are a bit different, drawn from the field of information fusion, a field that studies how multisensor data, a priori knowledge, contextual, and other sources of data and information can be combined/fused

Table 1.1 Speakers and papers of day 1.

Speaker	Affiliation	Paper title
Michael Fisher	Univ of Manchester, UK	Understanding and Verifying Autonomy and Teamwork
Tony Gillespie	University College London, UK	Building Trust and Responsibility into Dynamic Human-Machine Teams
Carolina Centeio Jorge	Delft University of Technology, Delft, Netherlands	Toward Modeling Appropriate Mutual Trust in Human-Agent Teams
Yi Zhu	Univ at Buffalo, USA	Addressing Vulnerability of Sensors for Autonomous Driving
Alina Vereshchaka	Univ at Buffalo, USA	Human-machine Interactions in Multi-agent Reinforcement Learning
William Lawless	Paine College, USA	Risk Determination vs. Risk Perception
Shareef Uddin Mohammed	Vanguard Systems, USA	Embracing Advanced AI/ML to Help Investors Achieve Success: Vanguard's Reinforcement Learning for Financial Goal Planning
Mauricio Castillo-Effen	LMCO, USA	A Vision for Human-Autonomy Trust
Anthony Baker	US Army, USA	Toward a Causal Modeling Approach for Trust-based Interventions in Human-Autonomy Teams

Table 1.2 Speakers and papers of day 2.

Speaker	Affiliation	Paper title
Max Nicosia	Univ of Cambridge, UK	Risk Management in Human-in-the-Loop AI-Assisted, Attention Aware Systems
Sarvapali Ramchurn	Southampton, UK	Industry-led Use Case Development for Human-swarm Operations
Erin Chiou	Arizona State Univ, USA	Deferring Decisions: Effects of Varying Interaction Structures on Human-AI Performance
Haley Green	University of Virginia, USA	iSpy; a Humorous Robot: Evaluating the Perceptions of Humor Types in a Robot Partner
Enrique Dunn	Stevens Institute of Technology, USA	Augmented Reality Apparatus with Grounded Physical Mapping for Guiding Humans in Tasks
Hesham Faoud	Naval Research Laboratory, USA	Moving from the Planning Domain to the Execution Domain; Real-time AI
Michael Mylrea	Resilience, Inc., USA	A Framework for Measuring Trust in Blockchain Distributed Ledger Technology Applied in Autonomous Cyber-physical Systems
Frank Wolf	Independent Consultant, USA	For AI Ethics, Process Beats Rules
Ariel Greenberg	Johns Hopkins University	Enabling machines to reason about potential harms to humans
Kate Devitt	Trusted Autonomous Systems, Australia	Meta-Issues: Policy, Law, Ethics, Values

to support the formation of "situation awareness (SA)," a state considered to be critical to human-agent team operations. Definitions of situations are slippery, but we offer one abstraction: "a set of entities in a set of relations," where "entity" is writ large, to include physical entities, behaviors, actions, grouped entities, etc.; the key term is "relations." It can be argued that teamwork cannot begin until the team agents have a common view of a situation or problem that they are expected to address. From this point of view, the first task of a team is achieving common SA. (We remark that "awareness" and "understanding" are different. Awareness relates to a current, particular, at-the-moment situation but understanding (generalizing from the particular) implies an ability to identify a particular SA as part of a class of larger behaviors. In some cases, this distinction can be very important.) The processes involved in developing SA have also been addressed extensively in the data fusion, human factors, and cognitive communities, but Endsley's three-level [1] model has undoubtedly received the most attention of all the models presented within the literature. The three-level model describes SA as an internally held information and cognitive product (i.e., human-based) comprising three hierarchical levels that are separate from the processes (termed situation assessment) used to achieve it. Our variant of Endsley's model is presented in Fig. 1.1. The model depicts SA as a component of the information processing chain that follows perception and leads to decision-making and action execution. While Endsley's process model was formulated in the context of human-centered SA, it is also serving to guide agent-based/technology-based approaches to SA, argued here as the first operation of a team-based agent.

Pursuing this point a bit more, consider some typical sensor systems such as radar and imaging systems. Radar data, over time, provide a basis for knowing where certain objects are and how they move over time; the data provide the attributes of location and velocity. It may be possible to infer the object type from some radar data that suggest the object's length or some other feature from which the data can be correlated to an object-type database, in a classification-type operation. The same description can be asserted for imaging sensors. From the preceding discussion, these sensors provide "relata" or *entity properties* that would support reasoning from which interentity relations could be asserted, but *do not* provide "observations" of relations; those need to be inferred from the relata. This inference is an important point and suggests that the inferencing machinery for asserting an SA state will require a multiplicity of relation models and models to aggregate the inferred relations into aggregate SA states of interest. Further, these models also need to be "common" across the agent-set so that a common framework for SA development underpins the agent and team SA process. Such commonality can be realized through shared team ontologies for relations and SA component states. Ontologies for SA processes have been actively studied in the information fusion community because

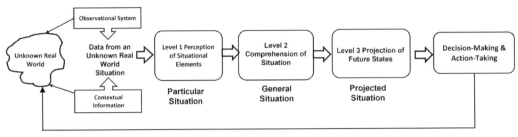

FIG. 1.1 Situation assessment processes.

multisensor and multisource data and information fusion processes to estimate situations are an inherent part of standard fusion processes; examples of such works are in Refs. [2,3]. One particular approach toward development of a computer-based ontology for SA is in Ref. [4], which builds on the situational theory of Barwise [5]. Summarizing, the design and development of computational and AI-based SA processes for human-agent team environments will require processes that link observations to entity relata, the specification of entity-relata relational models, and the specifying of SA ontology structures that model SA states of interest based on such models.

3 Team situation awareness (TSA)

Presuming a team comprises a group of agents each having local/agent-specific observational and inferencing capabilities, the combined SA of the whole team is the result of various team processes involving communication, coordination, and collaboration, leading to what most call a "shared understanding" of the same situation. (Or shared awareness, commenting on our prior distinction among these terms.) Most situations in the world are dynamic, and so there is a temporal dimension to the processes involved to yield this shared understanding; notice that Endsley's model depicted in Fig. 1.1 includes a Level 3 Projection process. The time dimension is both an important and complicating factor. Dissimilarities in agent-specific observational abilities or rates, different knowledge bases, etc. will yield local dynamics in agent-specific SA and further complicate multiagent processes to *maintain* TSA over time. But whether TSA as an all-agent or common picture requirement is truly needed in practical applications can depend on the roles and responsibilities of the agent team. If those roles and responsibilities do not require a team-wide common picture, then the question is how to achieve "shared SA" sufficient to support each agent's individual roles. Successful team performance requires that individual team members have good SA on their specific elements and also the same SA for those elements that are shared. SA, as with other systems level cognitive processes, therefore is taken to be an *emergent property* of collaborative systems; it resides in the activities among agents rather than in any one agent alone. Thus SA may be associated with agents, but it does not reside within them as it is in the emergent result of the interactions between them.

Various aspects of these sharing processes are discussed in the literature. Among the key concepts thought to be critical to TSA is the notion of shared mental models. Mental models are essentially internal representations of a system or process and have been defined as "knowledge structures, cognitive representations which humans use to organize new information, to describe, explain and predict events as well as to guide their interactions with others. Such mental models can be defined and integrated into a multiagent team using AI technologies, and thereby color the nature of human-agent and agent-agent interactions and interdependencies," from Ref. [6].

Thus, following [7], "a tripartite composition of team SA is apparent: individual team member SA (some of which may be common or "shared" with other team members), SA of other team members, and SA of the overall team." These three are closely linked, and TSA

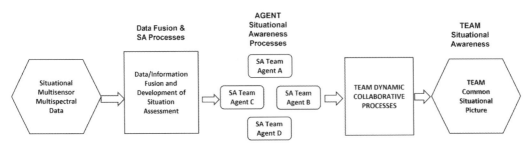

FIG. 1.2 Agent and team situation-forming processes.

is made up of individual team member SA, SA of other team members (task-specific SA), and SA of the overall team (TSA). Each of these forms of SA interact with each other, as shown in Fig. 1.2. Endsley has a recently published paper [8] that addresses the overall topic of TSA in human-agent environments, suggesting that TSA is critical to effective human-agent team performance in that it affects the key processes of decision-making, teamwork, coordination, and trust.

4 The trust dimension

Teamwork requires trust between team members. Interpersonal trust and trust in human teams is a heavily researched topic, e.g., Refs. [9,10]. In addition, over the past few years, intensive work has been carried out on how humans can trust "artificially intelligent agents," e.g., Refs. [11,12]; the works of this symposium of course are included in this research base. However, in the case of human-machine teams, there are new facets of trust to be considered, owing to the underlying peer-to-peer relationships in teams. Understanding and modeling the reciprocal nature of these trust relations is proving to be a central research topic for all AI-human interactions, but it is further complicated in the case of human-agent (AI-based) teams. First, human agents must trust the machine/AI/ML-based agents. Second, and reciprocally, the machine-based agents must trust the human team members. While there is a large body of work of discipline-specific models and methods that might be applicable, there is a lack of research on the systematic engineering of human-machine teams about how mutually trusted collaboration can be realized.

In some prior work of the author addressing two-sided adversarial decision-making [13], the works of [14] showed that levels of interagent trust over extended interactions can exhibit *hysteresis effects* wherein levels of trust can degrade over time if incidents causing distrust occur in dynamic interactions. This effect imputes the issue of *dynamics* in interagent trust exchanges, and, more importantly, those dynamics show that when incidents of distrust occur, the recovery of trust among agents is slow to occur. In Ref. [15], they found that it was more difficult to recover trust after a trust failure, given as a wrong advice, than to build trust initially. The level of performance, measured by the level of confidence in the decision the subjects made, deteriorated after the occurrence of wrong advice, and never returned to the prior level of performance. Thus incidents of distrust can modify or throttle the overall level of interagent trust across a range of dynamic interactions. (It can be seen from these older citations that interhuman, intersystem trust is a topic that has been studied for a long time, but that research base may still have implications for current-day, newer applications.)

In a multiagent system, there will likely be combinatorics of trust relations. There are human-agent trust relations (for each human-agent tuple), interagent levels of trust (for each computational agent pair), and human-human combinations of trust (for each human-human pair or tuple). This combinatoric "mapping" depends on the dynamic interaction model for the system. The interesting issue of "trust repair" (addressing the dynamics and hysteresis issue) is addressed in Ref. [16], describing four trust repair strategies, composed of the apology components *explanation* and *expression of regret* in isolation or in combination. Results indicated that expressing regret was crucial for effective trust repair. After trust declined due to some action of an agent, trust of the interacting agent only significantly recovered when an expression of regret was included in the apology. Trust management, therefore, may require types of apologies to be incorporated in multiagent teams.

5 Summary remarks

The theme of this symposium and its predecessors has been the systemic aspects of achieving advances in the design and engineering of systems comprising AI and human components. Such emphasis is on the "how" of realizing such engineering processes, importantly to include how to test and evaluate such prototypes. But developing the how knowledge also requires developing the "what" knowledge from which the how knowledge requirements evolve. It can be argued that engineering processes begin with asserted specifications or requirements that express the expected qualities of system and system components. However, developing realizable requirements demands levels of understanding about "what" is to be built, and knowledge of plausible bounds of current-day and evolving capabilities toward achieving those system goals. For specifying system requirements for a system that embodies a dynamic process, it is also important to identify the starting and ending points for that process. Importantly, such requirements must be testable and evaluated according to agreed-upon metrics and testing processes. In this chapter, the focus has been on the "what" of human-agent teams involving technological, computational agents and humans. One focus of this chapter was on the "starting point" of a dynamic human-agent interactive process, suggesting that humans and agents need to address the SA and TSA subprocesses that will govern/throttle their meta-behavior. The other focus was on some of the features involved in human-agent trust dynamics and the associated subprocess characteristics. Both of these subprocesses *underlie* the meta-processes—i.e., the purposes for which such human-agent team systems are being built—but will likely set the quality bounds that are specified for those purpose-built intentions.

References

[1] M.R. Endsley, Measurement of situation awareness in dynamic systems, Hum. Factors 37 (1) (1995) 65–84.

[2] E.P. Blasch, S. Plano, Ontological issues in higher levels of information fusion: User refinement of the fusion process, in: Proceedings of the Sixth International Conference on Information Fusion, 2003, pp. 634–641.

[3] C.J. Matheus, M.M. Kokar, K. Baclawski, A core ontology for situation awareness, in: Proceedings of the Sixth International Conference on Information Fusion, 2003, pp. 545–552.

[4] M.M. Kokar, C.J. Matheus, K. Baclawski, Ontology-based situation awareness, Inf. Fusion 10 (1) (2009) 83–98.

[5] J. Barwise, Scenes and other situations, J. Philos. 78 (7) (1981) 369–397.

[6] C.R. Paris, E. Salas, J.A. Cannon-Bowers, Teamwork in multi-person systems: a review and analysis, Ergonomics 43 (8) (2000) 1052–1075.

[7] P.M. Salmon, N.A. Stanton, G.H. Walker, D.P. Jenkins, Distributed Situation Awareness: Theory, Measurement, and Application to Teamwork, CRC Press, 2017.

[8] M.R. Endsley, Supporting human-AI teams: transparency, explainability, and situation awareness, Comput. Hum. Behav. 140 (2023) 107574.

[9] L.S. Wrightsman, Interpersonal Trust and Attitudes toward Human Nature, 1991.

[10] P. Madhavan, D.A. Wiegmann, Similarities and differences between human–human and human–automation trust: an integrative review, Theor. Issues Ergon. Sci. 8 (4) (2007) 277–301.

[11] E. Glikson, A.W. Woolley, Human trust in artificial intelligence: review of empirical research, Acad. Manag. Ann. 14 (2) (2020) 627–660.

[12] S.D. Ramchurn, D. Huynh, N.R. Jennings, Trust in multi-agent systems, Knowl. Eng. Rev. 19 (1) (2004) 1–25.

[13] J. Llinas, A. Bisantz, C. Drury, Y. Seong, J.Y. Jian, Studies and Analyses of Aided Adversarial Decision Making. Phase 2: Research on Human Trust in Automation, State Univ of New York at Buffalo Center of Multisource Information Fusion, 1998.

[14] J.D. Lee, N. Moray, Trust, control strategies and allocation of function in human-machine systems, Ergonomics 35 (10) (1992) 1243–1270.

[15] F.J. Lerch, M.J. Prietula, How do we trust machine advice? in: G. Salvendy (Ed.), Proceedings of the Third International Conference on Human-Computer Interaction on Designing and Using Human-Computer Interfaces and Knowledge Based Systems, second ed., 1989, pp. 410–419.

[16] E.S. Kox, J.H. Kerstholt, T.F. Hueting, P.W. de Vries, Trust repair in human-agent teams: the effectiveness of explanations and expressing regret, Auton. Agent. Multi-Agent Syst. 35 (2) (2021) 1–20.

2

Alternative paths to developing engineering solutions for human-machine teams

Tony Gillespie[a] and Scott Fouse[b]

[a]ELECTRONIC & ELECTRICAL ENGINEERING, UNIVERSITY COLLEGE LONDON, LONDON, UNITED KINGDOM [b]FOUSE CONSULTING SERVICES, SCOTTSDALE, AZ, UNITED STATES

1 Introduction

The final morning of the AAAI 2022 Spring Symposium was devoted to an interactive virtual panel session. The aim was to generate discussion and identify *alternative pa*ths to developing engineering solutions for human-machine teams (HMTs) that meet human values, laws, morals, and ethics. Humans, both in the team and in wider society, must trust that the machine team members will always behave so that the HMT always meets these requirements.

This chapter summarizes issues raised during the session that must be addressed before usable engineering solutions can be found for producing an HMT that is acceptable to society and its users. Specific areas requiring development work for the successful production and use of HMTs are also given.

Concepts and opinions are not attributed to any one person except those given in the speaker summaries.

2 Panel organization

The panel had four members (Andrzej Banaszuk, Bill Casebeer, Michael Fisher, and Jean-Charles Ledé) with two moderators (Tony Gillespie and Scott Fouse). Short biographies are given in the last section of this chapter.

Three test vignettes had been agreed on beforehand to focus attention on problems that affect HMTs with a high degree of autonomy in operation. These are given in Section 3.

The available time was split into two parts. The first started with brief viewpoints given by the four panelists, followed by the audience raising points, concerns, and questions arising from these. The viewpoints are summarized in the penultimate section in this chapter. After a break, the moderators summarized the main issues raised during the first part. The panelists responded, followed by further focused audience participation around the key issues and unresolved problems, refining and expanding them. These, and areas for further research and development, are given in Section 4.

3 Test vignettes

The term HMT is broad, with a wide range of possible applications and varying levels of trust required both within the team and by society during its use. To understand trust in an HMT it must be recognized that it depends not only on the human and the highly automated machine, but also on the interactions and interdependencies between them. The vignettes were chosen as ones that require high levels of human interaction with the machine and where trust is essential. They are:

3.1 Urgent recommendations

The autonomous system makes recommendations to a human for their immediate action in critical scenarios, examples being: firefighters dealing with a large fire with multiple ad hoc assets available; and military targeting in complex scenarios.

   These are life-threatening scenarios, so the human must have complete trust in the system when it takes action based on its own inputs and without reference to a human. There will probably not be time for human oversight of most decisions, but the machine must need to know when to refer its decisions to a human for approval or override. The machine trust is that the human will respond when requested to make critical decisions and also to take care of many background tasks. There is an overarching trust that the HMT is configured to deal with time-critical decisions that cannot be resolved in a safe way, that is, it adopts a fail-safe or minimum-harm mode.

3.2 Robotic support

An autonomous robot helps a person, examples being: carebots looking after people needing some form of assistance during normal life; robot first responders finding an injured person in a difficult environment and giving first-responder help.

   Both human and machine must have mutual trust; the human must trust in the carebot's judgment about their well-being and safety with regard to environmental and medical issues as these arise; the carebot has to trust that the human will accept their help and work with them to achieve predetermined long-term aims.

3.3 Periodic teaming

There will be HMTs where the nonhuman agents operate alone for long periods and then the HMT reforms, possibly with a different human member. One example was given: a robotic explorer on a nearby planet that is visited periodically by humans who rely on it for support while they are on the planet.

Trust here is almost completely one way. The machine may have learned new behaviors since the last human visit, based on its remote, autonomous operations, and possibly changed planetary conditions. The human is reliant on the operation of the machine and trusts that the reformed HMT can solve any problems that arise during the visit.

4 Major issues raised

The following issues were generally seen to be important and to underpin operations of HMTs where there is a synergistic use of team members' capabilities.

4.1 Trust and trustworthiness

The definition of trust given by Mayer et al. [3] was generally accepted as a basis for discussions about trust in an HMT. The definition is:

> *The willingness of a party to be vulnerable to the actions of another party based on the expectation that the other will perform a particular action important to the trustor, irrespective of the ability to monitor or control that other party.*

A consequence of vulnerability is that the trustor puts themselves at some risk arising from the actions of the trustee, but trust means that they see this risk as minimal. Mayer et al. point out that cooperation between agents does not require trust, but it does require confidence and predictability in their actions.

An alternative, more pragmatic definition of trust is whether a human would employ an autonomous machine in a difficult situation, especially if it will take irreversible actions. Whichever definition is used, human trust in a machine has two principal aspects, which we can call engineering trust and user trust.

Engineering trust comes from confidence that the system will respond to human commands when given and carry out some actions or monitoring activities without specific instructions to do so. It is effectively the sum of the technical properties of a cyber-physical system and its components, giving reliability, predictability, and integrity. The user could be expected to understand what these are but would not have an in-depth knowledge of any of them. Their trust would come from well-based assurances from the machine's supplier and knowledge that all regulatory requirements have been met before the user begins any operations with the machine. Supplier reputation will play a part here, as well as formal regulatory approval.

User trust is based on the machine behaving in a way that the human user can understand or accept even if the behaviors are not immediately understandable. Many behavioral factors in an autonomous system are inherently difficult if not impossible to verify, making user trust in the machine parts of the team a difficult but necessary requirement for a successful HMT.

User trust is not only subjective and context dependent, but usually increases over time. However, it is easily destroyed if the machine exhibits failures such as intermittent responses or having failures in components without informing the user. When this type of failure occurs, it is very difficult to rebuild the original trust. There are two subsets of users who need to trust an autonomous machine:

- There are a limited number of users, for example in the firefighting scenario, who all have training, experience, and some knowledge of the system's operation, all of which generate bilateral trust between the human and the machine;
- There are an unlimited number of users, many of whom will have little or no knowledge of how the system operates, and only want to use it. The wide range of users may also lead to its use in applications that were not foreseen during its design. Trust here relies heavily on the engineering trust described earlier, with user trust developing through experiencing how the machine works in an HMT rather than developing a deeper understanding of its internal operation.

In both these subsets, the user develops trust both by their own use of the system and by its reputation from other users. Both of these routes to trust rely on human understanding of the system's behavior and how this will help the human achieve their aim, i.e., they find that the machine is both reliable and beneficial in solving the human's problem. A system that can explain its intentions to its user before taking any action, and any other options considered, was considered to be more likely to be trusted than a black box. The design problem is to decide which actions or types of action need to be explained to the human and how to explain them in a comprehensible way.

Predictability is an important part of trust; the user will have expectations of what the machine will do and not be surprised if expectations are met. If the response is surprising, the user may accept it occasionally if they can understand why the machine did respond that way or see an obvious benefit from it. Too many surprises may destroy user trust if the user sees no benefit or a loss.

The human must always have an override facility, which they should only use if they doubt the machine's actions are moral, legal, or safe. Potential override actions, and their restriction to only essential times, require that the machine explain its actions to the human, or at least give them information about their intent and the factors it considers important. This helps build trust as the human can better understand the machine's likely actions. However, these facilities and the sometimes necessary wait for human approval of an action will be at the expense of machine performance, which generally has a much faster response time than a human. This trade-off between trust and performance was considered an essential aspect of HMT design.

Trustworthiness is a less-subjective concept than trust. Although not precisely defined in this context, it describes the ability of a machine to carry out a particular, definable task in a way that will be beneficial to the trustor (beneficiality). It includes all the parameters included in engineering trust plus verified limits on actions that may not benefit, or may be to the detriment of, the user. Verification of autonomous systems is already a known problem; see for example the IEEE RAS Technical Committee for Verification of Autonomous Systems website [1].

Given that trustworthiness includes many measurable parameters and can be specified in a way that builds in limitations in behavior, trustworthiness offers an engineering approach to solve one problem in establishing user trust in an HMT. However, it must be recognized that the behavior of complex and/or autonomous systems can only be determined using probabilities.

Underpinning all aspects of user trust in a machine is the trustworthiness of the models it uses and whether they can be trusted in the task being performed by the HMT.

4.2 Human-machine dialogue

The fundamental issue is that of establishing a dialogue between human and machine when the machine is a learning system such as a deep neural network (DNN). A dialogue implies understanding of both your own context and intent and that of the other party. The human will have a mental model of the machine but will not necessarily know how it has been trained or its limitations. It follows that the machine must have a model of a human's mind and its expectations so it can provide comprehensible responses to human inputs and explain decisions that the human may not have expected. Both the models will need to be refined for specific applications, the carebot being the most extreme case due to its close interaction with one user having complex needs. Despite this, there is a risk that both will act on the belief that they each understand the other when, in fact, there is limited understanding or, even worse, a condition of unrecognized mutual misunderstanding.

There are considerable variations in one human's comprehension of their environment, their interpretation of required actions, and their ability to perform them, with even wider variations between different people. Reliable modeling of the user's mental processes will necessitate new mathematical developments in order to deal with the associated uncertainties and biases. (It is a well-known problem in interactions between humans.)

A serious problem is that human-centric words such as altruism, integrity, and understanding are ported over to describe machine attributes or subroutines, but the machine does not have the human motivations that are the basis of the words' meanings for humans. The risk is that when a machine is described using one of these interpersonal words, the human user will give them a human-based interpretation that may not apply to the machine. A motive is not the same as a goal that can be clearly defined and implemented, so a machine will follow goals, but regulators and users must look at the motives of the machine's designers and their intent for their product. One important aspect is the incentives used in training a learning system and their influence on any decision that has moral implications. This is a known problem with some facial recognition algorithms that are applied in wider social systems.

The human system interface (HSI) must have a way of verifying that it understands the explicit and implicit intent of the user in the context of the current dialogue. These include wider factors affecting the human's motivation and that may limit their response.

Part of the HMT dialogue is for the machine to question or override the human's decisions. This is straightforward in the case of collision avoidance systems – no override leads to collision and harm. It should be possible to incrementally develop algorithms that can make more complex moral judgments, steadily increasing the range of criteria that the machine can consider. This could be based on an increasing order of harm to people, property, and society. There must be robust conflict-resolution processes if the human still wants to override the system response.

Every HSI must be based on actual human responses and inputs, not simulations of them. An example was given of training deep learning methods in HSI applications using humans in virtual environments. This was to ensure the training used real and not simplified virtual humans. This must be followed by training with humans using the learning system in actual scenarios.

It was noted during the panel session that trust can develop between a human and an animal, for example a dog. This system of human and an autonomous animal requires only a limited oral vocabulary, gestures, and cues for mutual understanding and trust. However, animals are motivational systems, with training by rewards, and take action in response to anticipated rewards of some kind. An inanimate machine will take actions based on its design, not the effect of the action.

4.3 Agency

The term "agent" is often used to describe the autonomous members of an HMT, with "agency" describing an agent's properties. Its use was discussed with general consensus that human and robot agents cannot be treated in the same way. The human will have a broad education, experience, and motivation, whereas the machine will have goals and training drawn from a variety of sources, not necessarily in a controlled way. One example is beneficiality; we expect that a teammate will always act to our benefit, but if they do not, they will probably warn us. Identifying benefits for another teammate and aligning actions with the benefits relies on subjective judgments, even if the benefits can be quantified: identifying that the outcome of an action will be detrimental to another is possibly more complex. Expecting a robot teammate to be able to behave in this way is a challenging problem but one that needs to be solved in order to provide trusted autonomous nonhuman agents in an HMT.

There may be a problem of loss of human agency as machine team members become more sophisticated and able to take on more tasks previously performed by a human. The problem was illustrated by the example of the loss of the skill to use a slide rule or log tables being acceptable, whereas some losses, like mariners losing the ability to navigate without GPS, may not be. On a larger scale, examples were also given of companies using large design software who found their engineers had lost their own design capabilities due to fixed processes in the software. This is also true of radio and radar component design using the dominant software package. A major danger with this trend to loss of agency in complex problems is that algorithms designed to make more moral decisions may remove the human ability to make moral judgments during design and use of a system. This trend can be seen in debates about, for example, autonomous cars.

4.4 Regulators

Questions about the role of regulators in the development and use of all aspects of HMTs were not explicitly discussed, but regulators were mentioned several times. Regulations have the force of law and play a crucial role in the transition from design to use and are distinct from standards, which may be referred to in regulations but do not have the same legal standing.

Verification of autonomous systems may need to be performed using probabilistic methods, but current methods are inadequate; this is an active research area in current approaches and developing new ones. It is well-known that verification will cause problems for the regulation of the machine parts of HMTs when mission and safety-critical decisions are made by the machine. One possible approach has been adopted by a car company that uses tiers of criticality with deterministic systems for a limited number of capabilities and human or autonomous systems carrying out the higher-level functions.

5 Future work

The panel session did not identify one "silver bullet" that could be developed to solve many of the problems in the HMT field. Instead, the discussions highlighted several major problems that are not necessarily being given the attention they deserve. They are, in no particular order:

- Human-machine dialogue requires a two-way information flow. Increasingly, more autonomous machines will interpret their instructions in the context of their understanding of the current situation and act accordingly. However, there will always be a need for the machine to comply directly with an instruction without any contextual interpretation, i.e., the machine turns the instruction into commands for its subsystems without any intermediate processing. An example is human override of autonomous actions. There is a need to develop human-machine dialogues that clearly separate instructions that assume that there will be contextual interpretation from those that are direct commands. Implicitly, the machine's feedback to the user will need to reflect this difference. This may require two interface languages and data flows running in parallel, a new concept;
- Mutual trust requires processes to avoid, or at least minimize, misunderstandings by either the human or the machine of the other's intent and wider understanding of relative priorities. This shows a need to develop algorithms to model both human and

machine cognition of their environment. Almost certainly this will need new mathematical developments for the quantitative and qualitative issues. Verification methods will also need to be developed to show that the human and machine models are mutually compatible and can interact;

- Trustworthiness depends on many factors. It needs to be defined so that it can be measured and HMT models assessed for their applicability for the HMT's tasks;
- Trust is a subjective concept, but there must be some means to calibrate it so that comparisons can be made between methods, systems, and its evolution over time;
- There will be limits to the applicability of a model in any given environment. These need to be established and/or approaches developed for a model to assess its environment and how its limits apply for the HMT task;
- Autonomous systems are being used in applications with different levels of criticality. Probabilistic verification methods will need to be developed covering different levels of criticality;
- Autonomous systems are starting to make decisions that have moral and ethical implications. There is scope for methodologies to identify when such decisions are made and to ensure that the level of moral decision-making by the system can increase without losing human moral agency.

6 Summary of presentations by panel members

6.1 Andrzej Banaszuk

Andrzej Banaszuk gave a viewpoint based on work that Lockheed-Martin and NVIDIA are undertaking looking at forest fires in general and more specifically how to predict their spread. The firefighters will have a range of assets available, but their best disposition is difficult to assess for a fire that presents a complex and dynamic scenario. The key challenges are:

1. Uncertainty management: AI accessing current state and making predictions about fire and affected population dynamics. Machine learning will need to be augmented with symbolic machine reasoning to inject human expert knowledge, enable human-AI dialogue, reduce training data required, and enable use of formal verification of the AI algorithms.
2. Complexity management: AI managing autonomous sensors to reduce uncertainty in knowledge of fire and population dynamics, and assigning firefighting assets (some autonomous) to extinguish fires.
3. Building human-AI trust: how to build firefighter trust in AI recommendations. Immersive virtual environments for joint human-AI training will be needed to facilitate AI agents and humans learning from each other and building trust over time.

A system making decisions that affect life and property will need a regulator sitting between the supplier and the user. One of their roles will be to develop user confidence that the system can be trusted when making decisions under complexity.

6.2 Bill Casebeer

Bill Casebeer argued that it is essential to understand trust in order to engineer it into products. Choosing a definition of trust is actually choosing the theory to be applied for that application. His preference is to take a behavioral/economic approach, and a dispositional theory using Mayer et al.'s definition [3]: the willingness of a party to be vulnerable to the actions of another party based on the expectation that the other will perform a particular action important to the trustor, irrespective of the ability to monitor or control that other party.

Although this definition appears simple, its implementation can be complex, as our cognitive architecture is based on causes, not reasons. The level of trust depends on measurable causes such as workload, but also on nonintuitive variables such as experience. Also, external factors such as room temperature, the recent consumption of warm drinks, and others will play a part. We need measurable determinants for trust.

One practical example from the three problem areas was given: after an automated target recognition (ATR) algorithm delivers target class judgment along with a confidence assessment; pilots are asked to give permission to an unmanned aerial vehicle teammate to prosecute a target (rules of engagement), while varying the amount and complexity of other tasks the pilot must also tackle. The basic findings were that in situations of high workload, pilots will overrely on the ATR algorithm even when confidence assessment is low. Conversely, pilots are prone to second-guess the algorithm in low workload conditions even when confidence assessment is high (underreliance). The theoretical explanation given was that workload management can be used as a heuristic to help us deal with what is important when we need to act quickly.

Measurement, whether quantitative or qualitative, requires calibration. Calibrating trust across the moral domain will almost certainly require algorithms that can assess the environment of action, the actions giving benefits, and the cognitive and contextual influences on judgment, and provide a nudge or corrective prompt to produce better judgment. The example given was prospect theory, explained as [2]:

> *Prospect theory assumes that losses and gains are valued differently, and thus individuals make decisions based on perceived gains instead of perceived losses. Also known as the "loss-aversion" theory, the general concept is that if two choices are put before an individual, both equal, with one presented in terms of potential gains and the other in terms of possible losses, the former option will be chosen.*

This is effectively an artificial conscience, or an artificial moral judgment/decision-making/action aide. This is what we must engineer.

6.3 Michael Fisher

Michael Fisher highlighted the difference between trust and trustworthiness, the former being subjective, but the latter being more definable. Trustworthiness in a deterministic cyber-physical system (CPS) is the same as reliability, but for an autonomous system, trustworthiness is reliability plus the benefits it brings to other team members (beneficiality—the degree to which something is beneficial; Macmillan dictionary, Accessed May 27, 2022).

For autonomous systems it is vital that, as well as reliability, we have:

1. Transparency of behavior/intention, and so explainability and (ideally) understanding;
2. Strong verifiability, specifically formal verification, and certainty about decisions/intentions. For individual robots/vehicles, we strongly verify key components (e.g., decision-making)—we never *rely* on anything that has not been formally verified! Similarly with teams—would you trust an opaque teammate?
3. Transparency, the explainability of decisions so the human understands the system's intent and reasons.

Ethical concerns can be verified when simple, such as ordering priorities of outcomes for alternative decisions, but verifying the operation of a full ethical code will be difficult.

Applying these conclusions to the three problem areas gives:

1. Advice: providing legal/ethical advice—e.g., firefighting • Explain laws/ethics, based on situation • Guarantees of correctness concerning advice given;
2. Social/Domestic Robotics: e.g., care robots, social robots • Reliability/predictability, especially physical interaction • Exposure of intention → gain trust • Verification of truthfulness → maintain trust • Match human ethics, e.g., "don't tell anyone when I do this";
3. Space Robotics, e.g., planetary habitat checking and remote/infrequent human interaction, need autonomy, as issues cannot all be predicted. It follows that there is a need for: verification of the decision-making processes; resilience and reconfigurability; and clear explanation of pertinent issues to operators/astronauts.

6.4 Jean-Charles Ledé

Jean-Charles Ledé defined autonomy to be the freedom to select a course of action required to achieve a higher authority's objective(s). It is built on an open environment with the three pillars of algorithms, the human system interface (HSI), and trust.

Trust is built on the basis of the underpinning concerns in the transition of an autonomous system from concept into service:

1. Technology, based on: experience, background, architecture, design
2. Requirements, based on: test standards, validation, verification, information assurance
3. Users, based on: transparency, control, liability, effectiveness, training
4. Society, based on: ethics, morality, visibility, understanding
5. Transition, based on: confidence, acceptance, deployment, use, peace

His concerns center around whether greediness, brittleness, opacity, and ulterior motives underpin AI decisions.

7 Biographies of panelists and moderators

Andrzej Banaszuk is Lockheed-Martin Advanced Technology Laboratory's Director of Strategy. Andrzej is an expert in artificial intelligence, with years of experience running the Autonomy and Artificial Intelligence group at United Technologies Research Center.

Bill Casebeer is Director of Artificial Intelligence and Machine Learning in Riverside Research's Open Innovation Center. He has decades of experience leading interdisciplinary teams of scientists and engineers to find creative solutions for pressing national security problems. He has a deep interest in the concept of an artificial conscience.

Michael Fisher is a Royal Academy of Engineering Chair in Emerging Technologies at the University of Manchester, United Kingdom, with research on the ethics, safety, trustworthiness, and verification of autonomous systems [see https://web.cs.manchester.ac.uk/~michael for details].

Jean-Charles Ledé is a strategic advisor in autonomy and AI. He was the Autonomy Technical Advisor to the Air Force Research Laboratory (AFRL) Commander overseeing the entire laboratory Autonomy and AI portfolio, and making recommendations on new programs leveraging internal and external research. Other experiences include Chief Technology Officer at RapidFlight, Program Manager at DARPA, Director of unmanned/autonomous systems at Raytheon, and Vice-President Advanced Concepts at Aurora Flight Sciences.

Tony Gillespie is a Visiting Professor at University College London and a Fellow of the Royal Academy of Engineering. His career includes academic, industrial, and government research and research management. His work on ensuring that highly automated weapons meet legal requirements has been extended to other autonomous systems in recent years; he has authored a book and academic papers. He has also acted as a technical adviser to the UN and other organizational meetings discussing potential bans on autonomous weapon systems.

Scott Fouse had a 42-year career in Aerospace R&D, mostly focused on exploring applications of AI to military applications. He ran two laboratories at Lockheed Martin, the Advanced Technology Labs based in Cherry Hill, NJ, and the Advanced Technology Center based in Palo Alto, CA, and he was president of a small company, ISX. He also served on the Air Force Scientific Advisory Board from 2003 to 2007.

References

[1] IEEE Robotics and Automation Society, https://www.ieee-ras.org/verification-of-autonomous-systems. Accessed June 2, 2022.

[2] Chen James, Prospect Theory. Investopedia, last updated July 22,2022. Available at https://www.investopedia.com/terms/p/prospecttheory.asp. Accessed Dec 9, 2022.

[3] R.C. Mayer, J.H. Davis, F.D. Schoorman, An integrative model of organizational trust, Acad. Manag. Rev. 20 (3) (1995) 709–734.

3

Risk determination vs risk perception: From hate speech, an erroneous drone attack, and military nuclear wastes to human-machine autonomy

W.F. Lawless

DEPARTMENT OF MATHEMATICS, SCIENCES AND TECHNOLOGY, AND DEPARTMENT OF SOCIAL SCIENCES, SCHOOL OF ARTS AND SCIENCES, PAINE COLLEGE, AUGUSTA, GA, UNITED STATES

1 Introduction

For this long-term work in progress, with case studies, we consider the effectiveness in the commercial sector of using artificial intelligence (AI) algorithms by humans to reduce the risks from "hate speech" attempted by Facebook using machine intelligence. We attempt to generalize that finding with a recent case study of a military drone attack by the Department of Defense (DoD) in Afghanistan, and then briefly with discovery by the public of the Department of Energy's (DOE) mismanagement of its own military nuclear wastes. Based on the lessons learned from these three case studies, we consider how to improve risk determinations and counter the adverse effects of risk perceptions by exploiting them to produce better decisions. We also briefly consider the effect of deception on risk determination and risk perception; the difficulty of adopting rational approaches to mitigate risks; and the similarity of a process that suppresses risk perceptions but also impedes innovation. We make a major discovery regarding suppression of information about the risks determined, and we end with conclusions.

2 Situation

The number of robots and drones in use around the world is increasing dramatically. In 2019, there were about 373,000 industrial robots sold, with a prevalence ranging from about 1 per 20 workers in Singapore to about 2 per 100 workers in the United States (https://www.statista.com/topics/1476/industrial-robots/). Compare that with the *World*

Robotics 2020 Industrial Robots report that found a total of 2.7 million robots already working across the world, averaging about 3 robots for every 10,000 humans.

In the United States, there are an estimated 866,000 drones, with about 40% registered for commercial use (https://www.faa.gov/uas/resources/by_the_numbers/); the number estimated in 2021 was approaching almost 2 million (https://seedscientific.com/drone-statistics/). In 2019, almost 100 countries had military drones [1]. Overall, however, none of the drones and robots presently in use are autonomous, a problem we are studying [2,3].

3 Case studies

1. *A commercial case study.* Categorization of risks with AI's machine intelligence is a difficult problem. It is the reason why self-driving cars rely on machine learning by operating repeatedly over closed courses to produce a closed context. From Mantica [4], "Self-driving cars are powered by machine learning algorithms that require vast amounts of driving data in order to function safely." As reported by *The Wall Street Journal* [5], Facebook, probably the operator of one of the most sophisticated machine learning algorithms, if not the most, has had limited success detecting "hate speech" across its system of learning machines. For example, its classifiers have labeled the video of a car wash as a first-person shooter event, and the video of a shooting as a car crash. Yet, Sheryl Sandberg, Facebook's Chief Operating Officer, responded to queries by *The Wall Street Journal* that its algorithms detected 91% of its 1.5 million posts removed for violating its company's hate policies. Casting doubt on Sandberg's claim, however, the software engineers and scientists at Facebook had earlier reported to *BuzzFeed* [6] as follows:

> *Using internal Facebook data and projections to support their points, the data scientists said in their post that roughly 1 of every 1000 pieces of content — or 5 million of the 5 billion pieces of content posted to the social network daily — violates the company's rules on hate speech. More stunning, they estimated using the company's own figures that, even with artificial intelligence and third-party moderators, the company was "deleting less than 5% of all of the hate speech posted to Facebook. ... We might just be the very best in the world at it," he wrote, "but the best in the world isn't good enough to find a fraction of it."*

One of the confounding factors is the use of deception to lure unsuspecting users. We discuss deception later (for details, see Ref. [7]).

2. *A military case study.* A categorization problem similar to Facebook's exists in determining the risk posed by potential adversaries in combat theaters. For example, the last US missile thought to have been fired by a US drone in Afghanistan occurred after a lengthy surveillance of a car on Aug. 29, leading the US military to conclude erroneously that the car had contained a bomb that posed the risk of an imminent threat to US troops at Kabul's

airport. In remarks subsequently by the US President, after this risk had been eliminated, he said that:

> *We will maintain the fight against terrorism in Afghanistan and other countries. We just don't need to fight a ground war to do it. We have what's called over-the-horizon capabilities, which means we can strike terrorists and targets without American boots on the ground — or very few, if needed. We've shown that capacity just in the last week. We struck ISIS-K remotely, days after they murdered 13 of our servicemembers and dozens of innocent Afghans. And to ISIS-K: We are not done with you yet. As Commander-in-Chief, I firmly believe the best path to guard our safety and our security lies in a tough, unforgiving, targeted, precise strategy that goes after terror where it is today, not where it was two decades ago. That's what's in our national interest.*

That "targeted, precise" strike was the planned result sought by DoD's risk assessment for this particular car. However, an investigation by *The New York Times* [8] began to immediately raise doubts about the DoD's risk assessment and the President's interpretation and justification of the drone attack, doubts about whether explosives were in the vehicle, doubts about whether the driver was a terrorist, and doubts about whether the missile's explosion generated secondary explosions. And after the DoD was confronted for a follow-up story 7 days later, *The New York Times* [9] began to suspect that the US military was becoming defensive:

> *Defense Secretary Lloyd J. Austin III and Gen. Mark A. Milley, the chairman of the Joint Chiefs of Staff, have said that the missile was launched because the military had intelligence suggesting a credible, imminent threat to Hamid Karzai International Airport in Kabul, where U.S. and allied troops were frantically trying to evacuate people. General Milley later called the strike "righteous."*

Moreover, subsequent news accounts (e.g., Ref. [10]) indicated that the attack was a mistake and may have killed 10 civilians, of which 7 were children. To investigate, US Air Force Lt. Gen. Sami D. Said, Inspector General of the Air Force, was tasked to conduct an investigation of the drone attack:

> *The service is asking Said to consider whether anyone in the chain of command should be held accountable for what Marine Gen. Frank McKenzie, the head of U.S. Central Command, called a tragic mistake.*

In addition to the US Air Force investigation by Lt. Gen. Said, the Department of Defense's (DoD) inspector general launched an investigation into the US drone strike in Kabul that may have inadvertently killed civilians, per DoD officials [11]. Moreover, from the AP, an apology had already been issued by the US military, senior Pentagon officials, and personally by the US Defense Secretary [12] for the Aug. 29 drone strike in

Kabul that killed 10 civilians, including seven children, calling it a "tragic mistake." Furthermore [13],

> *A senior U.S. Democrat said on Thursday that multiple congressional committees will investigate a drone strike that killed 10 Afghan civilians last month, to determine what went wrong and answer questions about future counterterrorism strategy.*

The extraordinary financial costs and loss of prestige from this erroneous risk assessment had also made news. From *Reuters* news service [14],

> *The Pentagon has offered unspecified condolence payments to the family of 10 civilians who were killed in a botched U.S. drone attack in Afghanistan in August during the final days before American troops withdrew from the country. The U.S. Defense Department said it made a commitment that included offering "ex-gratia condolence payments", in addition to working with the U.S. State Department in support of the family members who were interested in relocation to the United States. Colin Kahl, the U.S. Under Secretary of Defense for Policy, held a virtual meeting on Thursday with Steven Kwon, the founder and president of Nutrition & Education International, the aid organization that employed Zemari Ahmadi, who was killed in the Aug. 29 drone attack, Pentagon Press Secretary John Kirby said late on Friday. Ahmadi and others who were killed in the strike were innocent victims who bore no blame and were not affiliated with Islamic State Khorasan (ISIS-K) or threats to U.S. forces, Kirby said.*

3. *A brief case study of Department of Energy's (DOE) military nuclear wastes.* Until 1985, DOE had caused extraordinary damage across the United States from its mismanagement of its own military nuclear wastes. The cleanup was estimated at up to $200 billion for its two largest sites, the Hanford facility in Washington State and the Savannah River Site (SRS) in South Carolina [15]. From Slovic et al. [16], the risk perceptions, right or wrong, created by DOE's mishandling of its military nuclear wastes created a "profound state of distrust that cannot be erased quickly or easily" (p. 1603). To recover from its public debacle, to guide its risk determinations, and to assess the public's risk perceptions, DOE installed nine public committees to advise DOE; these committees made decisions by either seeking consensus (e.g., at Hanford) or majority rules (e.g., SRS). In the mismanagement of military nuclear wastes by DOE, however, by rapidly accelerating the cleanup of DOE's mismanagement at SRS compared to its slow-down at Hanford, we found that majority-rule decisions by DOE's Citizen Advisory Boards (CABs) were superior to those made by consensus-seeking CAB decisions [17]. What we found was that majority rule produced decisions that led to concrete decisions compared to consensus-seeking consensus decisions [15].

Regarding consensus-seeking, a White Paper reporting on a study of improving decision-making for Europe concluded, "The requirement for consensus in the European

Council often holds policy-making hostage to national interests in areas which Council could and should decide by a qualified majority" [18,19]. The problem with consensus-seeking is that it provides a minority, say an authoritarian leader or an obdurate minority, the power to block any undesired action, allowing us to redefine it as minority control [15]. Surprisingly, consensus-seeking rules block or impede the construction of a consensus, which requires interdependence to be built [2]. In contrast, the danger with majority rule is making decisions in haste or mob rule.

But minority control is sought even in China, which recently concluded a major plenum of the Chinese Communist Party, producing a formal resolution on party history that officially defines China's General Secretary Xi Jinping's political position within the Chinese Communist pantheon [20,21]:

> *With this resolution the party has elevated Mr. Xi and "Xi Jinping Thought" to a status that puts them beyond critique. Because both are now entrenched as objective historical truth, to criticize Mr. Xi is to attack the party and even China itself. Mr. Xi has rendered himself politically untouchable.*

Minority control is best exemplified by command economies where all economic activity is controlled by a central authority. For example (from Ref. [22]),

> *Command economies were characteristic of the Soviet Union and the communist countries of the Eastern bloc, and their inefficiencies were among the factors that contributed to the fall of communism in those regions in 1990–91.*

Extreme authoritarianism gives us insight into the problems with minority or consensus-seeking rules. By blocking alternative positions or interpretations of reality, which may threaten a leader's rule, authoritarians have striven to keep individuals independent of each other. The end result is a stable rule that blocks innovation [2,3].

4 How to fix?

A fix to this problem begins by decomposing it into two parts: an engineering risk assessment (or cost-benefit analysis), and perceived risks.

Engineering risk perspective (ERP). From an engineering risk perspective (ERP), can the machine or human-machine system perform as requested? Can the system become autonomous? Can the machine or human-machine team solve the problem it faces? How much uncertainty exists in the problem faced by the human-machine system? For a system as complicated as the one implicated in the drone attack, are handoffs between teams of humans and machines a part of the problem? (https://www.ready.gov/risk-assessment).

Perceived risk perspective (PRP). Risk perspectives are subjective estimates of a hazard constructed by intuition, emotion, and the media; these risk assessments are followed by risk communications in the attempt to persuade the members of a team (e.g., a risk assess-ment team, an investigation, a jury) or the public about the actual risks from an engineer-ing perspective [23]. Perceived risk in nuclear matters has been strongly linked to the public's trust [16]. From Brown [24], risk perceptions arise from "actual threats, sights, sounds, smells, and even words or memories associated with fear or danger," promoting an anxiety about risks, however, that may be sufficient to create "risks all by itself" (p. 3), if left unchecked.

Returning to the military case study, like DOE's resolution of the mismanagement of its military radioactive wastes [25], a critical step in fixing a problem is an open assessment by DOE of the causes of its own mismanagement of military nuclear wastes. To its credit, DoD [26] held an open briefing for the press about its report, which is classified because, according to Lt. Gen. Said, its author, "the sources and methods and tactics, techniques, and procedures used in executing such strikes are classified." Said interviewed "29 indi-viduals, 22 directly involved with this strike, and under oath."

Said described the context at the moment before the strike was launched in a process that "transpired over eight hours." He stated that "the risk to force at HKIA (Hamid Karzai International Airport) and the multiple threat streams that they were receiving of an immi-nent attack, mindful that, three days prior, such an attack took place, where we lost 13 sol-diers – or lost 13 members and a lot of Afghan civilians." Moreover, Said continued, the US military was one day away from leaving Afghanistan,

> so the ability for defense had declined. We're concentrated in one location, with a lot of threat streams indicating imminent attacks that looked similar to the attack that happened three days prior. So you can imagine the stress on the force is high and the risk to force is high, and not appreciating what I'm about to say through that lens I think would be inappropriate.

Said added that the strike "was unique in the sense that it was a self defense strike ... the norm ... [is] where you have a long time to do things like [to produce an attacker's] pattern of life. You have days to assess the intelligence and determine how you're going to execute the strike. It's a very different construct and very different execution."

In his report, Said "confirmed that the strike resulted in the death of 10 Afghan civilians, including three men and seven children. [However, US Military] Individuals involved in this strike, interviewed during his investigation, truly believed at the time that they were targeting an imminent threat to US forces on HKIA. The intended target of the strike, the vehicle, the white Corolla, its contents and occupant were genuinely assessed at the time to be a threat to U.S. forces." Said attributed DoD's erroneous risk determination overall to an "aggregate process breakdown" involving many people.

Per Said, DoD's [26] risk determination was overwhelmed by emotions for the risk per-ceptions in play. He stated that the.

> assessment was primarily driven by interpretation of intelligence and correlating that to observe movement throughout an eight hour window in which the vehicle was

tracked throughout the day before it was ultimately struck. Regrettably, the interpretation or the correlation of the intelligence to what was being perceived at the time, in real time, was inaccurate. In fact, the vehicle, its occupant and contents did not pose any risk to U.S. forces. … The investigation found no violation of law, including the law of war. It did find execution errors, confirmed by confirmation – or combined with confirmation bias and communication breakdowns that regrettably led to civilian casualties.

As part of his investigation, Said made three recommendations: First, adopt procedures in a strike cell for a similar situation where the military is time-constrained to act quickly to exercise self-defense in urban terrain and where it has to interpret or correlate intelligence rapidly, with procedures that mitigate the risk of confirmation bias. Second, enhance situational awareness by sharing information thoroughly within the confines of the strike cell but also outside of the cell to those supporting elements located elsewhere. Third, include an independent assessment in the targeting cell of the presence of civilians, specifically children, or anything that may magnify the costs of an erroneous decision, that is, the "severity" of a misjudgment.

Without stopping, Said recommended in addition a process of "red-teaming"; for whatever is the decision, an independent team should be assigned to push back against the interpretation to break "confirmation bias." To help DOE recover its lost prestige and to regain the public's trust, we drew conclusions similar to the DoD's Lt. Gen. Said's fourth recommendation that the confrontation among competing alternatives that are decided by majority rule accelerated DOE's cleanup compared to consensus-seeking rules [15,17]. In our model, because majority rule produces a concrete decision that promotes action, cleanup accelerates; in contrast, because concrete action is difficult if not impossible to reach via consensus, seeking consensus is inferior to majority rule.

5 A work-in-progress: Future autonomous systems

Autonomous systems. The importance of a process that checks risk determinations to uncover errors becomes even more important when autonomous systems are introduced to the battlefield. For these, legal limits are being considered for autonomous systems. From the Congressional Research Service [27],

Lethal Autonomous Weapon Systems (LAWS) are a class of weapon systems capable of independently identifying a target and employing an onboard weapon system to engage and destroy the target without manual human control. LAWS require computer algorithms and sensor suites to classify an object as hostile, make an engagement decision, and guide a weapon to the target. This capability would enable the system to operate in communications-degraded or -denied environments where traditional systems may not be able to operate. LAWS are not yet in widespread development, and some senior military and defense leaders have expressed concerns about the ethics of ever fielding such systems. For example, in 2017 testimony before the Senate Armed Services Committee, then-Vice Chairman of the Joint Chiefs of Staff General Paul Selva stated, "I do not think it is reasonable for us to put robots in charge of whether or not we take a human life." Currently, there are no domestic or international legal prohibitions on the development of LAWS; however, an international group of government experts has begun to discuss the issue. Approximately 30 countries have called for a preemptive ban on the systems due to ethical considerations, while others have called for formal regulation or guidelines for development and use. DOD Directive 3000.09 establishes department guidelines for the development and fielding of LAWS to ensure that they comply with "the law of war, applicable treaties, weapon system safety rules, and applicable rules of engagement."

Mayes [28] compared LAWS for autonomous military systems with self-driving autonomous vehicles (AVs):

AVs are essentially robots with the same requirements as human-driven vehicles—driving and parking skills, the ability to communicate with other cars and the infrastructure, navigation skills, and access to a source of energy. … Most new cars sold today are Level 1 with features such as automated cruise control and park assist. A number of companies including Tesla, Uber, Waymo, Audi, Volvo, Mercedes-Benz, and Cadillac have introduced Level 2 vehicles with automated acceleration and braking and are required to have a safety driver in the front seat available to take over if something goes wrong. … Waymo has a fleet of hybrid cars in Phoenix, Arizona, that it is using to test and develop Level 5 technology specifically to pick up and drop off passengers. (For a review of the levels of autonomy in autonomous vehicles, see https://www.nhtsa.gov/technology-innovation/automated-vehicles-safety.)

To delve more deeply into the risks associated with autonomy, we briefly present a mathematical physics model. Afterwards, we review the predictions and implications of the model. Our model is simple; its implications are not.

First, we assume that a major part of modeling is to simplify a model to the extent possible. The simplest group of models, such as game theory or agent-based models, however, assume closed systems. Games work well when the context is known or preset, when a situation can be clearly established, or when the payoffs are well-known, but the evidence suggests that game theory may not work when facing conflict or uncertainty, nor does any other rational model [29]. As an example from the fleet, the use of war games results in "preordained proofs," per retired General Zinni [19]; that is, choose a game for a given context to obtain the desired outcome.

Second, instead, as outlined in detail by the National Academies of Sciences in its report [30], needed are open-system models, the most challenging type of model, to address situations like the failed drone strike in Afghanistan. But for open systems, their chief characteristic is uncertainty; e.g., in economics, Rudd [20,21] captures the essence of his field's present inability to predict when facing uncertainty in the controversial first sentence of his new paper:

> *Mainstream economics is replete with ideas that "everyone knows" to be true, but that are actually arrant nonsense.*

Applied to intelligent systems, the chief characteristic in response to uncertainty is an interdependent reactivity to perceived risks to reduce uncertainty. In an open-system model of teams, we propose that a trade-off exists between uncertainty in the structure of an autonomous human-machine system, represented by Δ(structure), and uncertainty in its performance (represented by Δ(performance); for details, see Ref. [31]):

$$\Delta(\text{structure}) \ * \ \Delta(\text{performance}) \approx C \qquad (3.1)$$

In words, uncertainty in the structural arrangement of a team or system times uncertainty in its performance is about constant. Uncertainty is characterized by entropy. The predictions from Eq. (3.1) are counterintuitive. Applying it to the uncertainty between concepts and action results in a trade-off: as uncertainty in a concept reaches a minimum, the goal of social scientists, uncertainty in the behavioral actions covered by that concept increase exponentially, rendering the concept invalid, the result that has been found for multiple social science concepts, including self-esteem [32]; implicit attitudes [33]; or ego-depletion [34]. These problems with concepts have led to the widespread demand for replication [35]. But the demand for replication more or less overlooks the larger problem with the lack of generalizability [36].

Applying Eq. (3.1) first to risk determination and then risk perception, if a risk determination indicates that an autonomous structure (of say a team) is perfect, its structure should generate no information in the limit, i.e., zero entropy production [37]. Such a situation would allow, but not guarantee, an autonomous team to generate maximum entropy production to achieve its mission. Regarding risk perception, however, human witnesses can generate an infinite spectrum of possible interpretations, including nonsensical and even dangerous ones, as experienced by the staff that formed the strike cell for DoD's erroneous drone attack. Humans have developed two solutions to this quandary: suppress all but the desired perception, e.g., with consensus-seeking rules that preclude action [17] (Groupthink could be included here; however, groupthink condones an action without a challenge to the leadership's decision [38].); or battle test the risk perceptions in a competitive debate between the chosen perception and its competing alternative perception, deciding the best with majority rules [15]. In this regard, social science appears to be more interested in "changing the ingrained attitudes" associated with the suppression of undesirable attitudes [39], amplified in an editorial by the new editor of the *Journal of Personality and Social Psychology: Interpersonal Relations and Group Processes*, who seeks to publish articles to reflect that "our field is becoming a nexus for social-behavioral science on individuals in context" [40], making it of little help regarding uncertain contexts. Our problem with context is that it assumes certainty.

Success and failure experiences help to sort through misperceived risk perceptions. In business, the motivation to reduce uncertainty and the fittedness that results with business mergers combine to offer examples of the risks at play in open systems. Reducing the risk from uncertainty is the driving motivation for mergers in the marketplace; e.g., the eBay-PayPal deal is a general example [41]; a more relevant example of a merger for this chapter is FiscalNote's acquisition of Forge.AI, Inc. to obtain the technology to help it better model risk for autonomous vehicles [42]. For driverless cars, however, Aurora Innovation had acquired Uber's self-driving cars last year in the hope that the risks determined to exist for self-driving vehicles had been reduced, but concluding subsequently that the "technology isn't there yet" [43]. The failure to achieve or maintain fittedness increases the risk of breakups or spin-offs, such as has happened with General Electric, an industrial giant in the late 20th century renowned for its management prowess (led by Jack Welch from 1981 to 2001, considered by many to be the greatest leader of his time, in Ref. [44]; succeeded by Jeff Immelt, a leader who was slow to see the financial risks emerging, in Ref. [45]), but in recent years, a shadow of its former self, GE has faced increasing risks as it has struggled to survive (e.g., Ref. [46]).

Eq. (3.1) teaches a harsh lesson about the risk perceptions associated with autonomous systems that is strikingly similar to one about quantum mechanics: "There is no such thing as a "true reality" that's independent of the observer; in fact, the very act of making a measurement alters your system irrevocably" [47].

According to Milburn [48], cost-benefit analyses to determine risks may not be formally conducted and published, but formal reviews are the best way to improve risk determination processes and procedures to minimize risks. We disagree. For example, Justice Ginsburg [49] said that short-circuiting the tests from the courts at different levels inherent in the review process afforded by appeals from lower to higher courts would prevent an "informed assessment of competing interests" (p. 3). Far better, according to Justice Ginsburg, to complete all of the tests.

Counterintuitively, the costs of the process laid out by DOE to recover from the environmental-worker-public risks caused its mismanagement of military nuclear wastes, and by Justice Ginsburg to assess competing interests, saves money, equipment, embarrassment, and morale. The failure to enact such a process is highlighted following the offers of payment as compensation made by the United States to the families of the victims reported by *The Wall Street Journal* [50]:

The deaths of the civilians in the August drone strike raised questions about the ability of the U.S. military to conduct from afar "over-the-horizon" counterterrorism operations following the departure of the U.S. troops and their intelligence-gathering capacity. "There's no question that it will be more difficult to identify and engage threats that emanate from the region," Mr. Austin said last month. Former Trump administration national security adviser H.R. McMaster, who served as deputy commander for U.S.-led coalition forces in Afghanistan, told the House Foreign Affairs Committee on Oct. 5, "It is almost impossible to gain visibility of a terrorist network without partners on the ground who are helping you with human intelligence to be able to map those networks."

In trusting a human-machine system in a high-risk environment, there is always the danger of failure caused by rushing from a decision to take an action without first testing the decision. These tests must be balanced. In the event of a catastrophic failure like the case study of DoD's military drone strike, there is considerable danger that the perceptions of risk may well increase to the point that a working weapon system is even precluded from offsetting newly determined risks even when its use is warranted.

6 Rationality

Martinez and Sequoiah-Grayson [51] see the relation between logic and information as bidirectional. Information for rational decisions leads to an inference that underlies the intuitive understanding of standard logical notions (e.g., the process that makes implicit information explicit) and computation. Oppositely, logic is the formal framework to study information to achieve logical decisions. Martinez and Sequoiah-Grayson specify that:

Acquiring new information corresponds to a reduction of that range, thus reducing uncertainty about the actual configuration of affairs. ... an epistemic action is any action that facilitates the flow of information ... The information-as-correlation stance focuses on information flow as it is licensed within structured systems formed by systematically correlated components. ... Information-as-code is the syntax-like structure *of information pieces (their* encoding) *and the* inference and computation processes *that are licensed by virtue (among other things) of that structure. A most natural logical setting to study these informational aspects is the algebraic proof theory underpinned by a range of* substructural logics. ... *the correlations between the parts naturally allow for 'information flow' ... Formally speaking,* negative information *is simply the extension-via-negation of the positive fragment of any logic built around information-states. ...*

Putting their process of logic aside, what Martinez and Sequoiah-Grayson leave out, however, is the need to test information, especially the determinations of risk so eloquently raised by Lt. Gen. Said's report [26], by Lawless et al. [17], and by Justice Ginsburg [49]. From Pinker [52], goals need not be rational; however,

rationality emerges from a community of reasoners who spot each other's fallacies. ... [Rational thinking] is the ability to use knowledge to obtain goals. But we must use reason to choose among them when they conflict.

Risk assessments and risk perceptions form a core part of rational decision-making. Thagard [53] has studied whether science is rational, concluding, "A person or group is rational to the extent that its practices enable it to accomplish its legitimate goals." Applied to science, "scientific theories should make predictions about observable phenomena [that] aims for explanation as well as truth …" Testing, then, is necessary.

For Wilson [54], the unity of knowledge must be achieved through the scientific method. Wilson's goal is to reduce subjective determinations. As an evolutionary scientist, Wilson writes that we should "have the common goal of turning as much philosophy as possible into science." Respectfully, we disagree; Wilson does not perceive the value of society's contribution to its own evolution, where the evolution of technology and society coevolve [55]. To the extent that societies are suppressed, this point about coevolution is missed entirely by authoritarians and logicians.

But, and despite great effort [29], rationality is limited to nonconflictual and clearly certain contexts. Machine learning is context dependent [56]. Machine learning, like Shannon's [57] information theory, is restricted to i.i.d. data (independent, identically distributed data; in Ref. [36]). Thus whatever social event is observed and captured through the lens of Shannon cannot be reproduced [3].

7 Deception

Part of the difficulty of the problem with calculating a risk assessment of a terrorist strike is that terrorists often cloak many of their activities in deception. Deception's use in cloak and dagger work is well described by Luttwak [58]:

> human intelligence is needed: CIA or other field officers who speak local languages well enough to pass, can physically blend in, identify insurgents, uncover their gatherings and direct attacks on them.

The use of deception occurs in businesses, too, especially in monopolies that suppress alternative views [3]. From *The Wall Street Journal*, Amazon has been charged with using deception against its marketers by stealing the designs of the companies it markets [59]:

> Amazon.com Inc. AMZN 3.31% employees have used data about independent sellers on the company's platform to develop competing products, a practice at odds with the company's stated policies.

If true, deception occurs to those successful businesses that use Amazon to market their goods. The charge against Amazon is currently being investigated by Congress [60] (see also https://www.thefashionlaw.com/amazon-using-third-party-data-to-copy-product-boost-private-label-products-and-sales-per-new-report/). Amazon counters this claim in its privacy policy that includes a statement for its users: "We know that you care how information about you is used and shared, and we appreciate your trust that we will do so carefully and sensibly" (https://www.amazon.com/gp/help/customer/display.html?nodeId=GX7NJQ4ZB8MHFRNJ).

Deception is difficult to uncover. To uncover deception requires a challenge to the appearance of a normal, smooth or well-running operation, as may often occur with breaches of cybersecurity [7].

8 Innovation: A trade-off between innovation and suppression

Generalizing, briefly, competing risk misperceptions are one of the motivating drivers of innovation. It requires an interdependence between society and technology to coevolve [55]. Exemplified by an op-ed in *The Wall Street Journal*, innovation more likely occurs in small businesses, thereby increasing the risks to large businesses, which in turn seek to mitigate their increased perceived risks by asking Congress to adopt regulations that protect their dominant interests by reducing the risks that they have perceived [61]:

> *Research in recent years has demonstrated that new businesses account dispropor-tionately for the innovations that drive productivity growth, economic growth and new job creation. … The Platform Competition and Opportunity Act … would restrict and in some cases ban the acquisition of startups by larger companies. Ostensibly, the goal is to foster competition by preventing dominant online platforms from expand-ing their sway through acquisitions. But the legislation risks hurting the startups it aims to benefit.*

No surprise, this newly proposed legislation is actively supported by big business [62]:

> *The call for government action is part of a shifting ethos in Silicon Valley. In the past, the region has championed libertarian ideals and favored government's staying out of the way of its innovations. But tech leaders have begun to encourage Washington to get more involved in the tech industry as competition with China escalates, cyberat-tacks intensify and lawmakers express concerns about misinformation and censor-ship on social-media platforms.*

Similarly, China approaches the same problem as Amazon and big business by also sup-pressing outright the perceptions it considers to be undesirable but that consequently suppress the animal spirits associated with alternative risk perceptions, first associated with emotion [63] in a way that requires government to step in to "countervail the excesses that occur because of our animal spirits" [9,64]. But as Keynes and others have noted, ani-mal spirits are associated with confidence, trust, and creativity in a market; suppressing animal spirits can impede innovation (in China, see Ref. [65]).

The US military has shown a similar tendency when little attention is paid to one of DoD's mistakes. From *The New York Times* [66], now admitted by the United States [67], near the end of the fight in Syria against the Islamic State, with women and children associated with the once-fierce caliphate cornered in a field near a town called Baghuz, a US military drone was hunting for military targets but saw instead the women and children by a river bank, followed by American F-15E attack jets that dropped bombs on them, leaving no survivors.

> *The Defense Department's independent inspector general began an inquiry, but the report containing its findings was stalled and stripped of any mention of the strike. "Leadership just seemed so set on burying this ... ," said Gene Tate, an evaluator who worked on the case for the inspector general's office and agreed to discuss the aspects that were not classified. "It makes you lose faith in the system when people are trying to do what's right but no one in positions of leadership wants to hear it." Mr. Tate, a former Navy officer who had worked for years as a civilian analyst with the Defense Intelligence Agency and the National Counterterrorism Center before moving to the inspector general's office, said he criticized the lack of action and was eventually forced out of his job.*

Unlike the "red teams" recommended by Lt. Gen. Said, for this incident, instead, a report was written to describe "the shortcomings of the process ... [and that] the assessment teams at times lacked training and some did not have security clearances to even view the evidence" [66]. The *Times* reported, however, that the assessments of the failed strike were flawed because they were done by the same units involved in the strikes and they were grading their own performance. To reiterate, the United States has now admitted that the Syria strike in question was a mistake [67].

Internationally, open coercion may serve to prevent conflict [68], as it has in the past when the United States threatened military intervention against Haiti's military coup in 1994, leading to a peaceful transfer of power back to its previously deposed president, Jean-Bertrand Aristide; but similar and open threats to intervene militarily in 1998 against Iraq's rejection of UN inspections failed to achieve a peaceful outcome.

In general, minority control is one way that humans make decisions (e.g., for authoritarian Cuba, see Ref. [69]; for gangs, see Ref. [70]; for consensus-seeking, see Ref. [18]). As a major new finding with Eq. (3.1), instead of seeking the best fit to minimize structural entropy production so that an autonomous team or system can maximize its entropy production (MEP) for its mission, minority control expends copious amounts of energy to achieve a forcible fitness that significantly reduces MEP, the opposite of exploiting interdependence, and, working backwards, a way to deevolve (e.g., spin-offs, like J&J, in [71]; the human trafficking, promoted by Cuba to keep its government afloat (https://www.cia.gov/the-world-factbook/countries/cuba/#transnational-issues); the self-induced famine that regularly occurs in N. Korea (https://www.cia.gov/the-world-factbook/countries/korea-north/#transnational-issues)).

9 Conclusions

What many authors often call "AI" is little more than fancy i.i.d. machine data processors (independent identically distributed). These machine processors are easily fooled (e.g., a Tesla car crashed into the side of a semitrailer truck (https://www.ntsb.gov/news/press-releases/Pages/NR20200319.aspx)), easily manipulated, and easily deceived. Humans, in contrast, live in an interdependent social universe that, presently, machines cannot process, duplicate, nor understand [36]. Can machine intelligence do a lot of damage today? Yes. Can intelligent machines overtake humans, a worry expressed by Kissinger et al. [72]? No, not today.

From the perspective of perceived risks (PRP), should blue and red strike teams be concerned with how rigorous was the ERP determination? Was there a test of alternatives? How much uncertainty surrounds the target in the field? Is there, as was found by Said [26] in the tragic drone attack in Afghanistan, an emotional rush to judgment? Alternatively, did the failed drone strike create a barrier to future strikes?

Determining risk is already difficult, made more so with the introduction of autonomous systems, especially autonomous human-machine systems. Risk determination needs to be limited in scope to the problem at hand (Said, in Ref. [26]). In contrast, unmitigated, risk perception can compound the problem of determining risks. Risk perceptions can rise to the point that a human-machine system is no longer used. However, a strong process of confrontation (as public as possible, and when and where possible) between a chosen risk perception and its competing alternative that challenges the chosen risk perception arising in conjunction with a risk determination can help to keep risk perceptions in check, uncover deceptions, and increase innovation. When the stakes are life and death, confrontations test risk determinations and risk perceptions, allowing the best risk determination to be strengthened and the worst risk perceptions to be rejected and discarded. In contrast, suppressing alternative risk perceptions can backfire and even lead to deevolution (e.g., Syria [73]).

Minority control is one way that humans make decisions (e.g., for authoritarians, see Cuba, China, or North Korea; gangs; consensus-seeking). Based on Eq. (3.1), instead of seeking the best fit to minimize structural entropy production so that an autonomous team or system can freely choose to maximize its entropy production (MEP) to achieve its mission, we discovered that minority control expends copious amounts of energy to achieve a forcible fitness that significantly reduces MEP in the process, the opposite of exploiting interdependence, and, working backwards, a way to deevolve (e.g., spin-offs in businesses; human trafficking in Cuba; trafficking in body parts in China; famine in North Korea).

We also conclude by reiterating that consensus-seeking rules surprisingly block or impede the construction of a consensus, its stated purpose, which requires interdependence to be built, slowing the action that may need to be taken [2]. In contrast, the danger with majority rule is making decisions in haste or mob rule; this danger is reduced with checks and balances [3].

Lastly, our study points to the need for a theory of values in information, a future project. This project is thorny because values are bistable.

References

[1] R. Pickrell, Nearly 100 Countries have Military Drones, and it's Changing the Way the World Prepares for War, Business Insider, 2019. retrieved 10/10/2021 from https://www.businessinsider.com/world-rethinks-war-as-nearly-100-countries-field-military-drones-2019-9.

[2] W.F. Lawless, Risk determination versus risk perception: a new model of reality for human–machine autonomy, Informatics 9 (2) (2022) 30, https://doi.org/10.3390/informatics9020030. 2022.

[3] W.F. Lawless, Toward a physics of interdependence for autonomous human-machine systems: the case of the Uber fatal accident, 2018, Frontiers in physics, in: Section Interdisciplinary Physics, 2022, https://doi.org/10.3389/fphy.2022.879171.

[4] G. Mantica, Self-Taught, Self-Driving Cars? Like Babies Learning to Walk, Autonomous Vehicles Learn to Drive by Mimicking Others. Researchers Develop New Machine Learning Algorithm that Teaches Cars to Self-Drive by Observing Other Traffic, The Brink at Boston University, 2021. retrieved 11/8/2021 from https://www.bu.edu/articles/2021/self-taught-self-driving-cars/.

[5] D. Seetharaman, J. Horwitz, J. Scheck, Facebook says AI will clean up the platform. Its own engineers have doubts. AI has only minimal success in removing hate speech, violent images and other problem content, according to internal company reports, Wall Street J. (2021). retrieved 10/18/2021 from https://www.wsj.com/articles/facebook-ai-enforce-rules-engineers-doubtful-artificial-intelligence-11634338184.

[6] R. Mac, C. Silverman, After The US Election, Key People Are Leaving Facebook And Torching The Company In Departure Notes. A departing Facebook Employee Said the Social Network's Failure to Act on Hate Speech "Makes it Embarrassing to Work Here", BuzzFeed, 2020. retrieved 10/18/2021 from https://www.buzzfeednews.com/article/ryanmac/facebook-rules-hate-speech-employees-leaving.

[7] W.F. Lawless, R. Mittu, I.S. Moskowitz, D. Sofge, S. Russell, Cyber-(in)security, revisited: proactive cyber-defenses, interdependence and autonomous human-machine teams (A-HMTs), in: P. Dasgupta, J.B. Collins, R. Mittu (Eds.), Adversary Aware Learning Techniques and Trends in Cyber Security, Switzerland, Springer Nature, 2020.

[8] M. Aikins, 9/10; updated 10/2. Times Investigation: In U.S. Drone Strike, Evidence Suggests No ISIS Bomb. U.S. officials said a Reaper drone followed a car for hours and then fired based on evidence it was carrying explosives. But in-depth video analysis and interviews at the site cast doubt on that account, 2021, New York Times. Reported by Aikins, M., Koettl, C., Hill, E., & Schmitt, E., retrieved 10/9/2021 from https://www.nytimes.com/2021/09/10/world/asia/us-air-strike-drone-kabul-afghanistan-isis.html.

[9] H. Cooper, E. Schmitt, Pentagon Defends Deadly Drone Strike in Kabul. A New York Times Investigation, with Video Analysis and Interviews at the Site, Has Cast Doubt on the U.S. Military's Account, New York Times, 2021. , 9/13, updated 9/17. retrieved 10/9/2021 from https://www.nytimes.com/2021/09/13/us/politics/pentagon-drone-strike-kabul.html.

[10] T. Tritten, Air Force Secretary Taps Watchdog to Weigh Accountability in Botched Kabul Airstrike, Military.com, 2021. retrieved 10/9/2021 from https://www.military.com/daily-news/2021/09/21/air-force-secretary-taps-watchdog-weigh-accountability-botched-kabul-airstrike.html.

[11] C. Doornbus, DoD Inspector General Launches Investigation into Kabul Drone Strike that Killed 10 Civilians, 2021, Military.com, retrieved 10/9/2021 from https://www.military.com/daily-news/2021/09/24/dod-inspector-general-launches-investigation-kabul-drone-strike-killed-10-civilians.html. 24 Sep 2021.

[12] P. Stewart, I. Ali, U.S. says Kabul Drone Strike Killed 10 Civilians, Including Children, in 'Tragic Mistake', AP, 2021. retrieved 10/9/2021 from https://www.reuters.com/world/asia-pacific/us-military-says-10-civilians-killed-kabul-drone-strike-last-month-2021-09-17/.

[13] P. Zengerle, U.S. Fallout over Kabul Drone Strike Grows with Plans for Multiple Probes, AP, 2021. retrieved 10/9/2021 from https://www.reuters.com/world/asia-pacific/us-fallout-over-kabul-drone-strike-grows-with-plans-multiple-probes-2021-09-23/.

[14] K. Singh, U.S. Offers Payments, Relocation to Family of Afghans Killed in Botched Drone Attack, Reuters, 2021. retrieved 10/19/2021 from https://www.reuters.com/world/asia-pacific/us-offers-unspecified-payments-family-those-killed-botched-drone-attack-2021-10-16/.

[15] W.F. Lawless, M. Akiyoshi, F. Angjellari-Dajcic, J. Whitton, Public consent for the geologic disposal of highly radioactive wastes and spent nuclear fuel, Int. J. Environ. Stud. 71 (1) (2014) 41–62.

[16] P. Slovic, J. Flynn, M. Layman, Perceived risk, trust, and the politics of nuclear waste, Science 254 (1991) 1603–1607.

[17] W.F. Lawless, M. Bergman, N. Feltovich, Consensus-Seeking Versus Truth-Seeking, ASCE Practice Periodical of Hazardous, Toxic, and Radioactive Waste Management, 2005, pp. 59–70. 9(1).

[18] WP, White Paper, 2001. European governance (COM (2001) 428 final; Brussels, 25.7.2001). Brussels, Commission of the European Community.

[19] M. Augier, S.F.X. Barrett, General Anthony Zinni (ret.) On Wargaming Iraq, Millennium Challenge, and Competition, 10/18, Center for International Maritime Security, 2021. retrieved 10/21/2021 from https://cimsec.org/general-anthony-zinni-ret-on-wargaming-iraq-millennium-challenge-and-competition/.

[20] J.B. Rudd, Why Do We Think That Inflation Expectations Matter for Inflation? (And Should We?), Finance and Economics Discussion Series Divisions of Research & Statistics and Monetary Affairs, Federal Reserve Board, Washington, D.C., 2021. https://doi.org/10.17016/FEDS.2021.062.

[21] K. Rudd, Xi Jinping Thought' Makes China a Tougher Adversary. The party ratifies it as a Marxist 'breakthrough,' consolidating his power and expanding its control, Wall Street J. (2021). retrieved 11/12/2021 from https://www.wsj.com/articles/xi-jinping-thought-makes-china-a-tougher-adversary-ccp-rise-marxist-reappointment-11636750676?mod=opinion_lead_pos6.

[22] T. Britannica, Editors of Encyclopaedia, Command Economy. Encyclopedia Britannica, 2017. https://www.britannica.com/topic/command-economy.

[23] H.-J. Paek, T. Hove, Risk Perceptions and Risk Characteristics, Oxford Research Encyclopedia of Communication, 2017, https://doi.org/10.1093/acrefore/9780190228613.013.283. retrieved 10/10/2021 from.

[24] V.J. Brown, Risk perception: It's personal, environmental health, Perspective 122 (10) (2014) A276–A279, https://doi.org/10.1289/ehp.122-A276 (Retrieved 10/10/2021).

[25] M. Akiyoshi, J. Whitton, I. Charnley-Parry, W.F. Lawless, Effective Decision Rules for Systems of Public Engagement in Radioactive Waste Disposal: Evidence from the United States, the United Kingdom, and Japan, Springer, 2021, pp. 509–533, https://doi.org/10.1007/978-3-030-77283-3_24. Chapter 24.

[26] DoD, Pentagon Press Secretary John F. Kirby and Air Force Lt. Gen. Sami D. Said Hold a Press Briefing, 2021, Retrieved 11/3/2021 from https://www.defense.gov/News/Transcripts/Transcript/Article/2832634/pentagon-press-secretary-john-f-kirby-and-air-force-lt-gen-sami-d-said-hold-a-p/.

[27] CRS, Defense Primer: Emerging Technologies, Congressional Research Service, 2021, retrieved 10/19/2021 from https://crsreports.congress.gov/product/pdf/IF/IF11105.

[28] R. Mayes, Autonomous Vehicles: Hype or Reality?, Quillette, 2021. retrieved 10/20/2021 from https://quillette.com/2021/10/19/autonomous-vehicles-hype-or-reality/.

[29] R.P. Mann, Collective decision making by rational individuals, PNAS 115 (44) (2018) E10387–E10396. from https://doi.org/10.1073/pnas.1811964115.

[30] M.R. Endsley, Chair, Human-AI Teaming: State-of-the-Art and Research Needs. The National Academies of Sciences-Engineering-Medicine, National Academies Press, Washington, DC, 2021. Retrieved 12/27/2021 from https://www.nap.edu/catalog/26355/human-ai-teaming-state-of-the-art-and-research-needs.

[31] W.F. Lawless, Quantum-like interdependence theory advances autonomous human–machine teams (A-HMTs), Entropy 22 (11) (2020) 1227, https://doi.org/10.3390/e22111227.

[32] R.F. Baumeister, J.D. Campbell, J.I. Krueger, K.D. Vohs, Exploding the self-esteem myth, Sci. Am. 292 (1) (2005) 84–91. January. from https://www.uvm.edu/~wgibson/PDF/Self-Esteem%20Myth.pdf.

[33] H. Blanton, J. Klick, G. Mitchell, J. Jaccard, B. Mellers, P.E. Tetlock, Strong claims and weak evidence: reassessing the predictive validity of the IAT, J. Appl. Psychol. 94 (3) (2009) 567–582.

[34] M.S. Hagger, N.L.D. Chatzisarantis, H. Alberts, et al., A multilab preregistered replication of the Ego-depletion effect, Perspect. Psychol. Sci. 11 (4) (2016) 546–573, https://doi.org/10.1177/1745691616652873. hdl:20.500.11937/16871. PMID 27474142.

[35] B. Nosek, corresponding author from OCS, Open Collaboration of Science: Estimating the reproducibility of psychological science, Science 349 (6251) (2015) 943. supplementary: 4716–1 to 4716–9. (National Academies of Sciences, Engineering, and Medicine. 2019. Reproducibility and Replicability in Science. Washington, DC: The National Academies Press https://doi.org/10.17226/25303.

[36] B. Schölkopf, F. Locatello, S. Bauer, N.R. Ke, N. Kalchbrenner, A. Goyal, Y. Bengio, Towards Causal Representation Learning, 2021, arXiv, retrieved 7/6/2021 from https://arxiv.org/pdf/2102.11107.pdf.

[37] W.F. Lawless, Interdependence for human-machine teams, in: Foundations of Science, Springer, 2019, https://doi.org/10.1007/s10699-019-09632-5.

[38] I.L. Janis, Groupthink, Houghton, New York, 1971.

[39] C.S. Crandall, A. Eshleman, L. O'Brien, Social norms and the expression and suppression of prejudice: the struggle for internalization, J. Pers. Soc. Psychol. 82 (3) (2002) 359–378.

[40] C.W. Leach, Journal of Personality and Social Psychology: Interpersonal Relations and Group Processes, 2021, Editorial https://doi.org/10.1037/pspi0000226. retrieved 11/15/2021 from https://www.apa.org/pubs/journals/features/psp-pspi0000226.pdf.

[41] E.M. Jackson, How eBay's purchase of PayPal changed Silicon Valley, VentureBeat, 2012. retrieved 10/19/2021 from https://venturebeat.com/2012/10/27/how-ebays-purchase-of-paypal-changed-silicon-valley/.

[42] B. Dummett, J. Steinberg, CQ roll call owner FiscalNote strikes $1.3 billion SPAC Deal. The Washington, D.C.-based firm would merge with Nasdaq-listed Duddell street acquisition, Wall Street J. (2021). retrieved 11/8/2021 from https://www.wsj.com/articles/cq-roll-call-owner-fiscalnote-in-talks-to-list-through-1-3-billion-spac-deal-11636322400.

[43] S. Wilmot, Driverless 'Robotaxis' arrive at the stock market. Newly listed shares of Aurora innovation will be a key gauge of investor interest in autonomous vehicles, particularly for private peers Waymo, cruise and Argo AI, Wall Street J. (2021). retrieved 11/8/2021 from https://www.wsj.com/articles/driverless-robotaxis-arrive-at-the-stock-market-11636123007.

[44] C. Fernández-Aráoz, Jack Welch's Approach to Leadership, Harvard Business Review, 2020. retrieved 11/9/2021 from https://hbr.org/2020/03/jack-welchs-approach-to-leadership.

[45] EB, The GE Empire Breaks Up. The company that was once worth $600 billion will split itself in three parts, Wall Street Journal Editorial Board, 2021. retrieved 11/9/2021 from https://www.wsj.com/articles/the-end-of-the-ge-empire-general-electric-splitting-into-three-companies-larry-culp-11636495000?mod=opinion_lead_pos3.

[46] S. Lohr, M.J. de la Merced, General Electric Plans to Break Itself up into Three Companies, New York Times, 2021. retrieved 11/9/2021 from https://www.nytimes.com/live/2021/11/09/business/news-business-stock-market#general-electric-break-up.

[47] E. Siegel, Ask Ethan: What Should Everyone Know About Quantum Mechanics? Quantum Physics isn't Quite Magic, But it Requires an Entirely Novel Set of Rules to Make Sense of the Quantum Universe, Medium, 2021. retrieved 11/11/2021 from https://medium.com/starts-with-a-bang/ask-ethan-what-should-everyone-know-about-quantum-mechanics-99426c76a06.

[48] A. Milburn, Drone Strikes Gone Wrong: Fixing a Strategic Problem, Small Wars Journal, 2021. retrieved 10/9/2021 from https://smallwarsjournal.com/jrnl/art/drone-strikes-gone-wrong-fixing-strategic-problem.

[49] R.B. Ginsburg, 2011. American Electric Power Co., Inc., et al. v. Connecticut et al., 10-174 http://www.supremecourt.gov/opinions/10pdf/10-174.pdf. Accessed 11 May 2017.

[50] D. Nasaw, U.S. Offers Payments to Families of Afghans Killed in August Drone Strike. State Department to support slain aid worker's family's effort to relocate to U.S., Pentagon says, Wall Street J. (2021). retrieved 10/16/2021 from https://www.wsj.com/articles/u-s-offers-payments-to-families-of-afghans-killed-in-august-drone-strike-11634350706?mod=hp_listb_pos3.

[51] M. Martinez, S. Sequoiah-Grayson, Logic and information, in: E.N. Zalta (Ed.), The Stanford Encyclopedia of Philosophy, 2019. retrieved 10/20/2021 from https://plato.stanford.edu/archives/spr2019/entries/logic-information/.

[52] S. Pinker, Rationality: What It Is, Why It Seems Scarce, Why It Matters, Viking Press, 2021.

[53] P. Thagard, Rationality and science, in: A. Mele, P. Rawlings (Eds.), Handbook of Rationality, Oxford University Press, Oxford, 2004, pp. 363–379.

[54] E.O. Wilson, Consilience: The Unity of Knowledge, 1998.

[55] M.S. Ponce de León, A. Marom, S. Engel, et al., The primitive brain of early Homo, 372 (6538) (2021) 165–171, https://doi.org/10.1126/science.aaz0032.

[56] J.C. Peterson, D.D. Bourgin, M. Agrawal, D. Reichman, T.L. Griffiths, et al., Using large-scale experiments and machine learning to discover theories of human decision-making, Science 372 (6547) (2021) 1209–1214, https://doi.org/10.1126/science.abe2629.

[57] C.E. Shannon, A mathematical theory of communication, Bell Syst. Tech. J. 27 (1948) 379–423. 623–656.

[58] E.N. Luttwak, How the CIA lets America down. 'Counterinsurgency warfare' is a nullity without human intelligence, Wall Street J. (2021). retrieved 11/15/2021 from https://www.wsj.com/articles/cia-intelligence-coin-insurgency-afghanistan-war-afghan-11636866563?mod=Searchresults_pos1&page=1.

[59] D. Mattioli, Amazon scooped up data from its own sellers to launch competing products. Contrary to assertions to congress, employees often consulted sales information on third-party vendors when developing private-label merchandise, Wall Street J. (2020). retrieved 10/18/2021 from https://www.wsj.com/articles/amazon-scooped-up-data-from-its-own-sellers-to-launch-competing-products-11587650015?mod=article_inline.

[60] D. Mattioli, Members of congressional committee question whether Amazon executives misled congress. In a letter, bipartisan group of representatives asks for documents, 'exculpatory' evidence as they consider whether to recommend Justice Department investigation, Wall Street J. (2021). retrieved 10/18/2021 from https://www.wsj.com/articles/members-of-congressional-committee-question-if-amazon-executives-misled-congress-11634551201?mod=hp_lead_pos2.

[61] B. Hein, Lawmakers plan to tank the startup economy. A measure aimed at big tech would curb innovation, risk-taking and entrepreneurship by small companies, Wall Street J. (2021). retrieved 10/19/2021 from https://www.wsj.com/articles/lawmakers-plan-to-tank-the-startup-economy-acquisition-antitrust-ipo-11634592207?page=1.

[62] T. Mickle, Google CEO Sundar Pichai calls for government action on cybersecurity, innovation. Executive urges governments to adopt a Geneva convention for cybersecurity, and for the U.S. to invest

more in tech, Wall Street J. (2021). retrieved 10/19/2021 from https://www.wsj.com/articles/google-ceo-sundar-pichai-calls-for-government-action-on-cybersecurity-innovation-11634580600?mod=djem10point.

[63] J.M. Keynes, The General Theory of Employment, Interest and Money, Macmillan, London, 1936, pp. 161–162.

[64] G.A. Akerlof, R.J. Shiller, Animal Spirits : How Human Psychology Drives the Economy, and why it Matters for Global Capitalism, Princeton University Press, 2009. See https://archive.org/details/animalspiritshow00aker.

[65] S. Roach, China's Animal Spirits Deficit, Project Syndicate, 2021. retrieved 11/10/2021 from https://www.project-syndicate.org/commentary/chinese-tech-crackdown-crushes-animal-spirits-by-stephen-s-roach-2021-07.

[66] D. Philipps, E. Schmitt, How the U.S. Hid an Airstrike That Killed Dozens of Civilians in Syria? The Military Never Conducted an Independent Investigation into a 2019 Bombing on the last Bastion of the Islamic State, Despite Concerns about a Secretive Commando Force, New York Times, 2021. retrieved 11/14;2021 from https://www.nytimes.com/2021/11/13/us/us-airstrikes-civilian-deaths.html.

[67] M. Brest, US Admits to Strikes that Killed Civilians in Syria Years Ago, Washington Examiner, 2021. retrieved 11/15/2021 from https://www.washingtonexaminer.com/policy/defense-national-security/us-admits-to-strikes-that-killed-civilians-in-syria-years-ago.

[68] C. James, Victims and Bullies: Understanding the Optics of Coercion in A New Era Of Us Foreign Policy, Modern War Institute, 2021. retrieved 11/14/2021 from https://mwi.usma.edu/victims-and-bullies-understanding-the-optics-of-coercion-in-a-new-era-of-us-foreign-policy/.

[69] B. Hoffmann, The international dimension of authoritarian regime legitimation: insights from the Cuban case, J. Int. Relat. Dev. 18 (4) (2015) 556–574, https://doi.org/10.1057/jird.2014.9.

[70] H. Beech, They Warned Their Names Were on a Hit List. They Were Killed, 2021, In less than a month, assassins have killed at least eight people in the Rohingya refugee camps of Bangladesh, silencing those who have dared to speak out against the violent gangs," New York Times, retrieved 11/14/2021 from https://www.nytimes.com/2021/11/14/world/asia/rohingya-refugees-bangladesh.html.

[71] J.D. Rockoff, P. Loftus, Johnson & Johnson to Split consumer from pharmaceutical, medical-device businesses, creating two companies. Consumer business—home to band-aid and Tylenol—will be shed within 24 months, Wall Street J. (2021). retrieved 11/13/2021 from https://www.wsj.com/articles/johnson-johnson-plans-to-split-into-two-public-companies-11636715700?mod=hp_lead_pos4.

[72] H. Kissinger, E. Schmidt, D. Huttenlocher, The challenge of being human in the age of AI. Reason is our primary means of understanding the world. How does that change if machines think ? Wall Street J. (2021). retrieved 11/9/2021 from https://www.wsj.com/articles/being-human-artifical-intelligence-ai-chess-antibiotic-philosophy-ethics-bill-of-rights-11635795271?page=1.

[73] Human Rights Watch, Syria: Returning refugees face grave abuse. Struggle to survive amid devastation, property destruction (2001), Retrieved 11/2/2023 from: https://www.hrw.org/news/2021/10/20/syria-returning-refugees-face-grave-abuse.

4

Appropriate context-dependent artificial trust in human-machine teamwork*

Carolina Centeio Jorge[a], Emma M. van Zoelen[a,b], Ruben Verhagen[a], Siddharth Mehrotra[a], Catholijn M. Jonker[a,c], and Myrthe L. Tielman[a]

[a]*DELFT UNIVERSITY OF TECHNOLOGY, DELFT, THE NETHERLANDS* [b]*TNO, SOESTERBERG, THE NETHERLANDS* [c]*LEIDEN UNIVERSITY, LEIDEN, THE NETHERLANDS*

1 Introduction

Artificial agents are becoming more intelligent and able to execute relevant tasks for our daily lives, including tasks in work environments, home assistance, on the battlefield, and crisis response [1]. For some of these tasks, humans and artificial agents should learn to cooperate, coordinate, and collaborate, forming *human-machine teams*. (These teams have alternative names, such as human-AI teams, human-agent teams, and human-automation teams. We use *human-machine team* in this chapter.) A key driver for achieving effective teamwork is *mutual trust* [2], that is, teammates should trust each other. In particular, we consider that *appropriate* mutual trust is a fundamental property in effective human-machine teamwork. When there is appropriate trust, there is no undertrust (leading to underreliance) or overtrust (leading to overcompliance) [3], which minimizes negative performance outcomes [4]. As such, we take appropriate to mean that a human's trust in an agent (natural trust) should correspond to that agent's trustworthiness, and an agent's trust in a human (artificial trust) should correspond to the human's trustworthiness. In fact, assessing a teammate's trustworthiness is one of the decisive factors when a person considers whether to engage in an interdependent relationship with that teammate [5]. To achieve appropriate mutual trust, we first need to understand trust and how to implement and measure it in the artificial teammates. This chapter is meant to explore how artificial agents can appropriately trust their human teammates. We explore what trust and trustworthiness mean in the context of human-machine teams, which construct the artificial agent needs to understand to reason about trust, and how these constructs can be estimated from interaction.

*This document is the result of the research project funded by AI*MAN lab from TU Delft AI Initiative.

Artificial agents (referred to as the cognitive part of the machine) need to be able to observe, direct, and predict teammates [6] in order to make decisions and ensure effective human-machine teamwork. We argue that between observing and predicting a human teammate there is a process of assessing the human trustworthiness, which we call artificial trust. In this process, artificial agents model the krypta of the teammates (model of their internal characteristics) through the accessible manifesta (behavioral cues of the teammates) [7, 8]. The krypta can then be used to form their beliefs of artificial trust, that is, *competence* belief and *willingness* belief [9]. What the krypta should be, how it is present in the manifesta, and then how it transforms into formal beliefs for the artificial agent are not trivial, but we can explore human-human models of trust as a first step [10]. In the literature, we can find trust models in human-human teams, such as the ABI model [11], which may be suitable for krypta of human teammates in human-machine teams. Once we know what the krypta should be, we can work on manifesta to learn the dimensions of the krypta by interaction (e.g., prior task performance as manifesta of the krypta's ability). Depending on the situation, however, it may not be possible to build the krypta over extensive and frequent interactions. In this case, we can consider different ways of assessing trust, for example, with *swift trust* [12], which relies mostly on first interaction. By assessing trust, artificial agents can then decide on whether or not to trust a human for a certain task and act accordingly (by helping the human, e.g., mitigating risks and ensuring the team's goal). Engaging (or not) in a trusting action involves risks and it is also an important part of the decision-making, that is, after knowing how much I trust someone, I still have to decide whether I should engage in a trusting action. This decision, as well as the trust assessment itself, depends on the context.

Trust is then context-dependent. In human-machine teamwork, this context can be composed of task and team configuration. We followed the taxonomy presented by Parashar et al. [13] to reflect on the different characteristics of context that can affect trust (we particularly look at the teammate's krypta that the agent should reason about). This taxonomy aims at characterizing human-robot interactions in a teamwork setting and is illustrated with the examples of the urban search and rescue (USAR) domain as well as the assembly line manufacturing setting. Navigating from one to the other, we can also explore how trust models are sensitive to the context. Both manifesta and krypta are highly dependent on this situation characterization, that is, what is important to observe and reason about in USAR and in manufacturing setting may differ based on the characteristics of the context. In a USAR setting, for example, the task may require integrity from the trustee given that moral decisions may be required, whereas in an assembly line perhaps ability may be the only important aspect to consider when trusting a teammate. Moreover, much of the existing work on human-machine trust has the goal of defining one model of trust that fits any situation.

In this chapter, we argue that it could be useful not to take a "one-model-fits-all" approach, and instead see (1) how different trust models might accommodate different contexts, (2) how within one model some dimensions may be more relevant than others, and (3) how we can start formalizing trust as a belief of context-dependent trustworthiness. Thus the chapter is structured as follows: we start by presenting the definition of trust for this chapter in Section 2, and then we go through the related work and important concepts required to understand the rest of the chapter in Section 3; we present our taxonomy of context-dependent trust in Section 4, and explore a possible formalization of the beliefs of trust in Section 5; we finally discuss the main findings in Section 6 and conclude in Section 7.

2 Trust definition

Trust is a dyadic attitude or behavior between a trustor (the one who trusts) and a trustee (the entity being trusted) and it can be defined as "the willingness of a party to be vulnerable to the actions of another party based on the expectation that the other will perform a particular action important to the trustor, irrespective of the ability to monitor or control that other party" (p. 712) [11]. In a team composed of both humans and machines, we need to write this definition in a more formal way in order to implement and measure it. Thus we approach trust from a functional perspective, in which trust is a relational construct between the trustor x, the trustee y, about a defined (more or less specialized) task (τ), as in Falcone et al. [7]. Particularly, we propose that trust is one agent's *perception of the trustworthiness* of another, meaning that how much x trusts y depends on how trustworthy x believes y is. This means x appropriately trusts y when x's belief in y's trustworthiness actually corresponds to y's trustworthiness. For example, if an agent x trusts another agent y to execute a task (e.g., driving a car) that requires skills that y does not have, agent x overtrusts agent y and the consequences can be negative and even disastrous (e.g., car accident). On the other hand, if agent x does not trust agent y to execute a task (e.g., driving a car) and agent y is perfectly capable of successfully executing the task, agent x is undertrusting agent y, which can also negatively affect team effectiveness (e.g., walking instead). In particular, when x is a human and y is an artificial agent, and trust is not appropriate, this will lead to disuse or misuse of technology [1]. Thus a dyadic relationship between a human and an artificial agent in a human-machine team should be designed in such a way that it supports (1) appropriate trust from the human toward the agent and (2) appropriate trust of the agent toward the human. As such, we need the artificial agent to understand trust, how to form these beliefs and how others form these beliefs. In particular, artificial agents need to have models of trustworthiness of their teammates.

3 Trust models, Krypta and Manifesta

3.1 Models

Trust has been vastly explored in the context of human-human interaction, with well-known contributions such as the ABI model [11] (which suggests trustworthiness is based on ability, benevolence, and integrity), for organizational behavior. In particular, trust in human teams has been recently explored in contexts such as virtual teams [14], sports [15], and university group projects [16]. Furthermore, in multiagent systems (MAS) trust has been used as a security and control mechanism, to protect agents from not knowing other agents' code of conduct [17]. Among others, we can find a formalization for trust and reputation (e.g., [18]), ways of categorizing agents to explain internal qualities (krypta) with their observable signs (manifesta) in order to promote trust (e.g., [7, 19]), and, more recently, models for assessing an agent's trust based on human values (e.g., [20, 21]). Similarly, trust in human-machine interaction has been gaining increasing attention. We can consider the most consensual model of human trust in technology as being Performance, Process, and Purpose [22]. Moreover, there are works on the dynamics of human trust toward technology (e.g., [23, 24]), how agents can assess and promote appropriate trust in humans (e.g., [25–29]), and the role of (appropriate) trust in human-machine teams (e.g., [3, 30–32]). There are also contributions on artificial trust, such as how an artificial agent can detect that a situation requires trust [33, 34] and also how an artificial agent can detect whether a human is being trustworthy, based on episodic memory [35] and social cues [36]. Furthermore, Azevedo-Sa et al. [37] suggest a model of trust prediction in human-machine teams, based on capabilities and task requirements. Essentially, we can find in the literature: (1) how humans trust humans, (2) how agents can trust other agents, (3) how humans trust artificial agents, and (4) how artificial agents can calibrate this trust with certain actions. Nonetheless, we found the literature on trust from the perspective of an agent toward a human to be scarce. Making artificial agents able to detect under which situations they could use trust and when they can trust a human, based on social cues, memory, and capabilities, is of utmost importance. Enabling them to understand human trustworthiness and its dimensions can lead to a better human-machine understanding and team effectiveness.

3.2 Manifesta and Krypta

We assess trustworthiness through available cues (manifesta) [38]. Often, the manifesta are cues of certain internal qualities (krypta), which are dimensions of trustworthiness. Frequently, it is suggested that human trustworthiness has as dimensions the krypta of ability, benevolence, and integrity (ABI model [11]). In the literature, we can find instruments that follow the ABI model and measure, through questionnaires, propensity to trust [39], and perceived trustworthiness (of teammates) in military teams [40, 41]. Although we can find instruments to measure trust subjectively, it is in our interest to measure trust using objective measures that can be used in interaction. So far, studies have measured trust objectively through physiological signals, such as electroencephalography (EEG) and electrocardiography (ECG) and, sometimes, audio and electrooculography (EOG) [42], which are not ecologically valid in human-machine teamwork. Breuer et al. [14] present a taxonomy of behaviors that affect how teammates perceive each other's trustworthiness in virtual human teams, although it is not clear how this taxonomy can transfer to human-machine teams. Finally, some works have presented how the krypta can be computed from the manifesta (once we know which manifesta and krypta suit our domain), such as POMDPs [29], dynamic-Bayesian networks [43], machine theory of mind [44], and instance-based learning [45]. None of these methods have been successfully tested when interacting with humans in order to estimate their trustworthiness.

3.3 Context-dependent models and their dimensions

Most of the existing models represent trust as something that develops between people over time and is built over a series of shared experiences and interactions [12, 46]. The ABI model [11] considers trust to be, besides the trustor's propensity to trust and other external factors, the perception of trustworthiness as ability, benevolence, and integrity. Ability comprises the set of skills and competences of the trustee; benevolence has to do with the relationship between the trustor and the trustee and whether the trustee is believed to want the trustor's good; finally, integrity deals with the set of principles and moral values that the trustee adopts and whether the trustor finds them acceptable. However, there are certain situations in life, including in human-machine teams, in which time is not a given nor are some of the dynamics that allow trustors to interact and share experiences sufficiently to build their trust in such detail. Certain situations require *swift trust*. Swift trust usually happens in situations that are temporary and that may require some level of urgency [12]. This type of trust model does not build after an extensive observation, but is rather built at first (based, e.g., on imported information, propensity to trust, and surface-level cues [47]) and fine-tuned later, through interaction and observation. Although these are not the only two models of human trust in organizational settings, they show that, depending on the context, the relevant internal characteristics of teammates (krypta) may differ, as well as how they will show through behavioral cues (manifesta). Similarly, although we do not know if these models can be used for artificial trust, that is, an artificial agent trusting a human, this is the closest we have so far. In order to reason about these models and build them from interaction, an agent needs to understand both the relevant *krypta* (which would be the dimensions of trustworthiness in this case) and the *manifesta*, that is, the cues that an artificial agent can perceive in order to build the teammates' krypta. Which krypta and manifesta are important for each situation is not trivial, but we suggest we need to study them within a well-characterized context.

4 Trust as a context-dependent model

Artificial agents need to build beliefs regarding their teammates' and their own trustworthiness. We have seen before that there are models presented in the literature that represent trust in different situations and contexts. In particular, there are models of how humans trust other humans, such as the ABI model and Swift trust model, as well as how humans trust machines, such as Performance, Process, and Purpose. When we consider these models in human-machine settings, it is not trivial which of these models fits the situation best. In particular, these models are constructs, that is, they contain several different conceptual dimensions, and we might wonder whether all of these dimensions are equally important. Taking the ABI model as an example and its three antecedents as dimensions, namely, ability, benevolence, and integrity, one can wonder whether integrity is relevant (or as relevant as other antecedents) in tasks such as assembly lines in manufacturing. Moreover, the literature still does not provide trust models that artificial agents can use to assess their teammates' and their own trustworthiness in the context of teamwork. We do not know to which krypta are important for each situation. Consequently, we also do not know which manifesta the artificial agent should pay attention to build its mental models. Nevertheless, we suggest that we may not be able to find a general krypta and manifesta for all human-machine interactions. This being said, we should start by characterizing well the context in which the agent needs to assess trust in their teammates. In this section, we reflect on which characteristics of the context, including task and team configuration, may affect the krypta that the artificial agent should build to assess trustworthiness.

In the literature, we can find a taxonomy of the interactions of human-robot teams by Parashar et al. [13], which comprehends characteristics of tasks and team configuration. Departing from this chapter, and making use of the illustrative examples it provides (USAR and assembly line), we have built a taxonomy that can be used to describe a situation when an artificial agent needs to trust a human during human-machine teamwork, which can be found in Fig. 4.1. We have also included certain concepts from inspirations of other papers, such as *set of stimuli* and *time* from Farina et al. [48], *workload* from Neerincx et al. [49], *lifespan* from Haring et al. [12], and *nature* and *output* from Farina et al. [48],

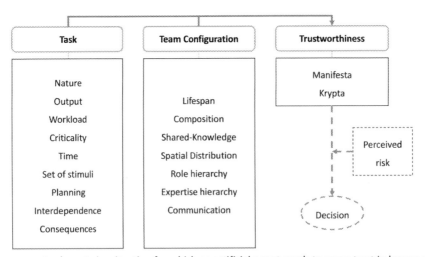

FIG. 4.1 Taxonomy to characterize situation for which an artificial agent needs to assess trust in human-machine teams. Characteristics of task and team configuration influence trustworthiness's components of Krypta and Manifesta. The assessed trustworthiness will contribute to the decision on whether to engage in a trusting action (trust decision) after a risk evaluation (perceived risk).

Wildman et al. [50], and McGrath [51]. According to our interpretation, task characteristics comprise the basic information required to distinguish one task from the other, such as type of output required, the expected time, etc. On the other hand, team configuration consists of the information regarding the team that will execute the task or the set of tasks and their dynamics, for example, the *lifespan* of the team can be 2 months for a certain project, irrespective of the tasks and their *time* that will be involved in the same project.

4.1 Task

We start by reflecting on the **Nature** of the task and its impact when assessing a teammate's trustworthiness. Although nature is quite generic, we would like to make the distinction between *cognitive* and *physical* tasks. The choice of the model of trust may depend on its nature, in particular on the manifesta expected from one another. This means that the visual cues that an artificial agent can use to, for example, conclude whether a teammate is able and/or benevolent, may vary considerably depending on the nature of the task. As nature is still a broad concept, we find it important to include the task's **Output**. In Parashar et al. [13], we can find *focus*, with examples of transit, area coverage, management, etc. Output is in that sense similar to focus but has the intention of being more measurable and implementable. For example, instead of *management* as a possibility, we intend to have a specification such as "Allocation of three tasks among the team for today." The complexity of the task as well as some of the necessary skills to be successful at it should be expressed in the output. For example, verbs of Bloom's taxonomy for educational purposes [52] could be used to better express the type of task (e.g., build, rate, choose). **Workload** classifies a task in terms of the amount of work it requires. Certainly workload varies from person to person, but it can still be useful to characterize different tasks, for example, in terms of cognitive load [49, 53].

The trust decision is highly related to the perceived risk [11]. As such, we consider **Criticality** (*low, medium, high, severe*) to be important for task characterization. This criticality concerns the risks involved with the performance of the task or the failure of such task, for example, choosing who to help in a USAR situation. In such scenarios, some constructs of trust may be more important, for example, integrity or high ability, whereas in an assembly line, if there is a task of low criticality, perhaps the ability and integrity required may be lower. When talking of criticality, we automatically think of *urgency*, which is criticality in terms of **Time**. The timeframe set for a task, that is, when it has to be done, is also important when assessing trust. We speculate that more important than assessing trust, a task's timeframe may play a major role when deciding whether to engage in the risk-taking relationship (trust decision) [5]. This means that, for example, we may decide to trust someone for an urgent task (e.g., carry a victim) that we would not trust if there was more time (perhaps we would do it ourselves instead). This characteristic also plays with the risk of not trusting being higher than trusting. Finally, we included **Set of stimuli** so that we can have a measure of motivation to successfully do a certain task or engagement while doing the task. This characteristic may be a hard one to quantify, as the present stimuli may not be obvious. However, aligned with output and workload, it can give us a better sense of the difficulty and complexity of the task. We chose stimuli to include the features of a task that can instinctively create some reaction on the agent, in particular the human. These can range from the objects involved in the task (e.g., lights, screws, robots) to social stimuli (e.g., as other people involved, a baby, and even a social robot), music, etc. [54]. We speculate that the set of stimuli can be very important to assessing overall competence and willingness of a teammate to perform a certain task, given that it may influence their engagement. Furthermore, we can even reflect on whether certain types of stimuli, for example, social stimuli, can be important to understand the relevance of benevolence in a task.

Planning (*online, offline, hybrid*) has to do with how the task is prepared and how much is ad hoc or improvised. The assessment of trust may be different when this characteristic changes. For example, one's assessment of trust in a USAR situation where we do not know what to expect and, as such, are not able to plan it beforehand, may change considerably in an assembly line situation where everything is planned from start to end. Additionally, it is the joint nature of key tasks that defines teamwork, which is only possible through the effective management of **Interdependence** [6] (*none, soft, hard*), that is, "the set of complementary relationships that two or more parties rely on to manage required (hard) or opportunistic (soft) dependencies in joint activity" (p. 3). We are unsure how and whether interdependence affects the trust assessment, but we can imagine that it affects the engagement in risk-taking relationships. For example, by knowing that someone's action will affect mine, that may lead me to engage in the risk-taking relationship even though I would not trust them if our actions were independent, or the opposite. Last but not least, we speculate that the **Consequence** of the task influences the trust decision. Consequence differs from output in the sense that it entails what happens to the system after the task is completed (or failed), that is, it is the direct consequence of the output. In video games, we can illustrate consequences in more measurable ways, such as rewards, levels, etc. In human-machine teams, these consequences may take more complex rewards, such as saving lives (in USAR), items to sell (in assembly lines), and so on. However, the consequences of a task may also be social, such as getting closer to someone by the means of helping them, or feeling we did what was right. These consequences can influence the weight of dimensions such as benevolence and/or integrity, respectively. Ultimately, consequences of a task are related to personal risk and reward. Although consequences may not be easy to quantify or measure, we believe we can categorize tasks as *high-risk-high-reward, high-risk-low-reward, low-risk-high-reward,* and *low-risk-low-reward.*

4.2 Team configuration

The **Lifespan** of a team is very important when choosing the trust model to use. It is important to note that this is not the **Time** of the task, for example, if I am in a group project at university for a semester, the lifespan of the team is a semester, but probably we will have tasks that take a different time, for example, preparing a presentation that takes us 2 days. In a team with a short lifespan, trust forms in a very different way than when one has time to actually get to know their teammates. Take the example of the USAR context; one may be in the field with people they have not met before (e.g., they came to reinforce help after a terrible catastrophe), or it could be daily life at a manufacturing assembly line, where they see their colleagues every day during their shift, over a few months. We do not expect trust to develop in the same way in both situations and one reason is the lifespan, since trust does not have the same time to develop. For example, in the case of USAR, one will probably not have enough time to form beliefs regarding another's benevolence and/or integrity. Or perhaps those dimensions of trustworthiness are just not important in this case, since the team will dissolve once the mission is accomplished. Swift trust models may be more suitable to teams with low lifespan, whereas ABI can suit long-term teams better.

On the other hand, team **Composition** (*single human to single machine, multihuman to multimachine*, etc.) may affect the type of model we use to assess a teammate's trustworthiness, given its nature and overall context of the team. Moreover, it will affect the overall team trust model, that is, the trust we have in the team, despite (not necessarily independent of) the trust we have in the teammates individually, see, for example, Ulfert et al. [30]. In human-machine teams, it is also important to consider the **Shared-Knowledge** (*independent, partially independent, overlap*), which may affect the way we assess trust regarding a certain teammate. For example, if I know everything my teammate knows by default, that is, we both have access to the same information, then the way I build my trust in them will be different than in a situation where we each have access to different information. It is also important to characterize how much, what, and among whom knowledge in a team should be shared [55]. This knowledge may comprise information regarding ontologies (e.g., domain, team member, and organization), world state (e.g., map and task sequence), and team member's models (e.g., their availability, capabilities, etc.) [56].

The **Spatial Distribution** (*proximal, remote, hybrid*) of the team can also change the way we perceive trust. As humans we know that having in-person meetings is simply different than having online meetings, whatever you may prefer. With remote work being more and more part of our lives, some people prefer working remotely, while others prefer to be physically present in the office, while there is yet another group of people that prefers a hybrid setting, which includes both spatial distributions (remote and in person) [57]. In fact, spatial distribution affects people's satisfaction, performance, and productivity [58], and it can also affect trust [59, 60]. Proximal distribution allows team members to perceive each other's characteristics differently. In particular, in human-robot teams anthropomorphic features of robots, for example, also change the way we perceive them and, consequently, how we trust them [61]. Furthermore, when we trust someone in our team, the **Role Hierarchy** (*supervisor, peer, mentor*) and **Expertise Hierarchy** (*fixed, fluid*) are important to consider. In terms of roles, we wonder whether one assesses the trustworthiness of their subordinates in the same way one would do for a superior. Or whether, if in different roles we simply expect different constructs and our trust model is bound to that. There are recent contributions suggesting structured roles in human-machine teams, including the roles of coordinator, creator, perfectionist, and doer [62]. It may be that even without a necessary hierarchy we construct our trust toward teammates in these different roles differently, for example, it may be that one expects higher integrity of a coordinator, given their authority, than a doer. It can also happen that the dynamics of trust change when there is hierarchy in terms of role or expertise. For example, in a surgery situation where a machine is used, we can imagine that the way a nurse trusts the machine may depend on how the doctor trusts the machine, or vice versa. We can argue that in that case there is trust transitivity [63] and that constructs such as integrity may play a bigger role than, for example, benevolence or ability. Finally, **Communication** (*environment-based, sensing-based, direct-partial, direct-full*) is highly related with trust in the sense that we depend on it to build our trust models [2]. The way we communicate may affect not only how we perceive the krypta but also which krypta we end up building. Communication is related to shared-knowledge as well [2], meaning that we can share knowledge through communication and build shared knowledge from communication.

4.3 Summary: A taxonomy

In this section, we have explored several characteristics of the context that may affect how an artificial agent should assess trust. The resulting taxonomy (in Fig. 4.1) proposes a set of important characteristics that influence the choice of trustworthiness model. In particular, we looked at task and team configuration and reflected on which characteristics may influence the choice of krypta (model containing the internal characteristics of the teammate that are relevant to assess trustworthiness) and the manifesta (behavioral cues of the teammate that hint to their krypta) to appropriately estimate a teammate's trustworthiness.

Certain task and team configuration characteristics may not only impact the estimation of trustworthiness but also the decision to trust, that is, engage in a trusting action. The decision on whether to engage in a trusting relationship is dependent on the risk that decision represents. It is important to note that the decision on whether to engage in a trusting relationship may have risks for both positive and negative decisions. This means that sometimes it may be riskier not to trust than to trust.

The taxonomy is proposed as a tool to choose krypta and manifesta, in order to assess trustworthiness appropriately. Once krypta and manifesta are chosen, it is important to formalize trust so that it can be implemented in the artificial agents. Trust must be formalized as a belief of trustworthiness, which is a construct dependent on task, team configuration, trustor, and trustee. In the next section, we propose a general formalization of these beliefs and reflect on what it takes to make them appropriate.

5 Trust as a belief of trustworthiness

We have seen that assessing trust is not trivial. In particular, trustworthiness is a complex concept, and following the literature it can consist of a set of dimensions that range from the trustee's competence to their intentions [64]. In the previous section, we reflected on

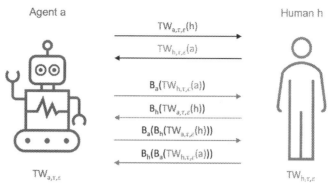

FIG. 4.2 Human-machine dyadic trust. *Modified from C. Centeio Jorge, S. Mehrotra, C.M. Jonker, M.L. Tielman, Trust should correspond to trustworthiness: a formalization of appropriate mutual trust in human-agent teams, in: D. Wang, R. Falcone, J. Zhang (Eds.), Proceedings of the 22nd International Workshop on Trust in Agent Societies (TRUST 2021) Co-located with the 20th International Conferences on Autonomous Agents and Multiagent Systems (AAMAS 2021), London, UK, May 3–7, 2021, CEUR Workshop Proceedings, vol. 3022, CEUR-WS.org, 2021, http://ceur-ws.org/ Vol-3022/paper4.pdf.*

how trustworthiness can be context-dependent, including the manifesta (behavioral cues that show trustworthiness) and krypta (the construct that defines trustworthiness in a situation). Examples of krypta are *ability, benevolence,* and *integrity* (ABI) model [11] or *willingness, competence* [65]. It is important to formalize trustworthiness as beliefs of the artificial agent so that it can use them to make decisions. The artificial agent can then form beliefs of trustworthiness regarding the adequate (in terms of context) krypta from the manifesta and other possible indirect sources of information, such as overall reputation of another teammate [17]. In this section, we propose a first step toward modeling the beliefs of trustworthiness, taking into account its possible krypta dimensions and context dependencies. As several of krypta's dimensions (e.g., Benevolence, Integrity), may relate to both trustor and trustee, we stipulate that we need to define the trust of an agent x in agent y as the belief $\mathcal{B}$ of agent x regarding the trustworthiness of y with respect to x:

$$T(x, y, \tau, c) = \mathcal{B}_x(TW_{y,\tau,c}(x)) \tag{4.1}$$

where τ is the *task* and c is the *team configuration* as explored in the previous section. Fig. 4.2 schematizes a dyadic human-machine relationship.

5.1 Forming (appropriate) artificial trust

As an example, let us consider the task of driving a car. Inspired by Mecacci and de Sio [66], let's imagine a dual-mode vehicle, which can be driven both by an artificial agent or by a human. The default setting is the human driving according to the agent's instructions, but the agent takes over when it recognizes dangerous situations. Although it may be counter-intuitive, we need the agent to *trust* the human to drive safely (their joint goal), while complying with societal ethics, so that it knows when to take over. In this example, we will have the trustworthiness of the agent a, given a human h, $TW_{a,\tau,c}(h)$, and the trustworthiness of the human h given an artificial agent a, $TW_{h,\tau,c}(a)$ in a certain context, that is, task τ and team configuration c. In practical terms, this means that the way the human is going to follow the agent's instructions may vary according to the agent that is helping (e.g., depending on whether the human relies on this particular agent's knowledge/intelligence). Moreover, we have the trust of the artificial agent in the human, meaning the agent's belief in the human's trustworthiness, $T(a, h, \tau, c) = \mathcal{B}_a(TW_{h,\tau,c}(a))$ (from Eq. 4.1), and the trust of the human in the agent, which is the human's belief on agent's trustworthiness $T(h, a, \tau, c) = \mathcal{B}_h(TW_{a,\tau,c}(h))$. The trust of the artificial agent in the human $(T(a, h, \tau, c))$ is what the agent believes that the human will do if the agent gives the human a certain instruction.

In order to estimate $\mathcal{B}_a(TW_{h,\tau,c}(a))$, we may also need the agent's belief in the human's trust in the agent, that is, $\mathcal{B}_a(\mathcal{B}_h(TW_{a,\tau,c}(h)))$, since some dimensions of trust (such as benevolence) depend both on the trustor and trustee, by definition. Following the example, for the agent to trust the human to follow an instruction, the agent needs to believe that the human trusts the agent (e.g., the human relies on this particular agent's knowledge/intelligence). Finally, when estimating whether it can trust its human teammate to follow an instruction, the *agent*'s trust in the human should correspond to the actual human's trustworthiness (e.g., to what actually the human can and/or wants to do), that is,

$$T(a, h, \tau, c) \equiv \mathcal{B}_a(TW_{h,\tau,c}(a)) \equiv TW_{h,\tau,c}(a) \qquad (4.2)$$

which requires that the agent also accurately estimates the human's trust in the agent, $T(h, a, \tau, c) \equiv \mathcal{B}_a(\mathcal{B}_h(TW_{a,\tau,c}(h)))$. The *human*'s trust in the agent, on the other hand, is the belief of the human in the agent's trustworthiness, $\mathcal{B}_h(TW_{a,\tau,c}(h))$, and should correspond to the agent's actual trustworthiness $(TW_{a,\tau,c}(h))$, that is,

$$T(h, a, \tau, c) \equiv \mathcal{B}_a(\mathcal{B}_h(TW_{a,\tau,c}(h))) \equiv \mathcal{B}_h(TW_{a,\tau,c}(h)) \equiv TW_{a,\tau,c}(h) \qquad (4.3)$$

What's more, we argue that appropriateness of trust depends on the task and team configuration, meaning that one may trust another in a certain context but not in another, for example, a may appropriately trust h to drive a car (a believes that h can drive a car) but not to pilot a plane (e.g., a believes that h can pilot a plane but h actually cannot). We do not illustrate possible different team configurations for the sake of simplification, but that is definitely to consider as well.

Lastly, since the nested concepts presented in Eq. (4.3) are based on $TW_{a,\tau,c}(h)$, this means that we may be able to calibrate the human's trust in the agent $(T(h, a, \tau, c))$, by manipulating $TW_{a,\tau,c}(h)$ through the accurate belief of the agent's own trustworthiness. This means that if the agent is aware of its own trustworthiness, meaning that if the agent's belief in the agent's trustworthiness corresponds to the actual agent's trustworthiness, that is,

$$\mathcal{B}_a(TW_{a,\tau,c}(h)) \equiv TW_{a,\tau,c}(h) \qquad (4.4)$$

the agent may be able to alter its own trustworthiness (or simply how it lets the human perceive it) and, consequently, calibrate human's trust. In our example, the agent might understand that it is not being perceived as intelligent, and start justify its instructions, possibly leading the human to trust it more.

We want to apply this formalization to existing trustworthiness models, by modeling and learning their dimensions. As discussed before, with the information regarding τ and c, we can choose the krypta that best fits the situation, as well as which dimensions should have more impact when forming a belief. For example, we can consider the ABI trustworthiness model once more. We consider the trustworthiness of the human to be the weighted sum of their ability, integrity, and benevolence toward a specified task, in a certain environment. Similarly, we consider the agent's trust in the human to be the weighted sum of the agent's beliefs in the ability, benevolence, and integrity of the human, toward a specified task (τ), in a certain team configuration (c). The belief in *ability* (*Ab*) of the human takes into account the task τ and team configuration c. The belief in human's *benevolence* (*Ben*), however, besides the task and team configuration, also takes into account the agent, given that, by definition, benevolence has to do with the relationship between both [11]. Benevolence may also, among other things, implicitly use the belief of the human's trust in the agent, $\mathcal{B}_a(T(h,a))$, which, as previously discussed, can be expressed as $\mathcal{B}_a(\mathcal{B}_h(TW_a(h)))$, since it comprises how the trustor (trustee) perceives the trustee's willingness to do good to them (trustor) [11]. Finally, the belief in the human's *integrity* (*I*) depends on the agent, task, and environment. By definition, perception of integrity deals with how the trustor finds the trustee's values and moral principles acceptable. Thus

$$\mathcal{B}_a(TW_{h,\tau,c}(a)) = W \cdot [\mathcal{B}_a(Ab_{h,\tau,c}),\ \mathcal{B}_a(Ben_{h,\tau,c}(a)),\ \mathcal{B}_a(I_{h,\tau,c}(a))] \tag{4.5}$$

where *W* is a weight vector. This weight vector is once more dependent on context, as discussed in the previous section.

5.2 Calibrating natural trust

Although we focus mostly on how to make artificial trust appropriate, that is, how an artificial agent can trust their human teammates appropriately, it is also important that the human's trust in the agent (natural trust) is also appropriate. By giving the necessary tools to the agent to reason about trust, we argue that it can also affect natural trust, once these beliefs may be nested (as illustrated in Fig. 4.2). As such, leveraging on the idea that agents reflect about their own trustworthiness, we may be able to influence humans to appropriately fine-tune their trust in them. For example, let us again consider the task of driving a car. Considering that the agent reflects about its own trustworthiness regarding its ability and willingness to drive the vehicle, the agent may then influence the human teammate to adapt to the agent's strengths and weaknesses (fine-tuning the human's trust in the agent). As such, it is important that the agent not only reflects on its own trustworthiness but that it does so considering the context, that is, the task and team configuration, since the context may influence the relevance of certain internal characteristics.

We posit that how trustworthy an agent is for a human and how a human trusts the agent (human's belief in agent's trustworthiness) in a certain context (task and team configuration) should be similar to get appropriate trust. If the belief of an agent in their own trustworthiness toward the human is different from their belief of the human's trustworthiness toward them in a certain context, then we come closer to undertrust $T(a, h, \tau, c)$ $\downarrow$ or overtrust $T(a, h, \tau, c)$ $\uparrow$, that is

$$\mathcal{B}_h(TW_{a,\tau,c}(h)) > \mathcal{B}_a(TW_{a,\tau,c}(h)) \rightarrow T(a, h, \tau, c)\uparrow \tag{4.6}$$

$$\mathcal{B}_h(TW_{a,\tau,c}(h)) < \mathcal{B}_a(TW_{a,\tau,c}(h)) \rightarrow T(a, h, \tau, c)\downarrow \tag{4.7}$$

Therefore, to avoid such situations, the agent's belief in their own trustworthiness should match with their belief about the belief of the human's trustworthiness in them. This will result in eliciting appropriate trust in a human from an agent's perspective. Most literature sees appropriate trust as the alignment between the perceived and actual performance of the agent by the human in terms of the agent's abilities [67] looking at "ability" as the core factor of estimating trust [68–70]—that is, *focusing upon the engineering aspect of trust*. However, as seen before, we propose to view trustworthiness as more than just ability, including *psychological aspects* [71] such as benevolence and integrity. What's more, we argue that to find this alignment, we need to first define the context as a task and team configuration. In this work, we aim to propose a first attempt on how several context-dependent dimensions of the krypta can be modeled (so that they can be learned) by an artificial agent, both to appropriately trust (artificial trust) and be trusted by a human teammate (natural trust).

6 Discussion and future work

In this chapter, we have argued that not only the way in which we trust our team members is context-dependent, but also whether we decide to engage in trusting actions is context-dependent as well. This is especially important when we talk about human-machine teams and more specifically artificial trust, that is, how artificial team members (such as machines) should estimate the trustworthiness of their human partners, making use of krypta and respective manifesta. We have based our reasoning on experience with past human-machine trust research, in which it proved difficult to determine exactly why people chose to trust or to be trustworthy at a given moment. Moreover, we have also experienced that it is not always obvious how and in which contexts human trust models (such as ABI) can be imported to human-machine scenarios. Our hypothesis is that analyzing task characteristics and team configuration can help to assess how trust models should be built up for a given scenario.

We have used existing task models and taxonomies to create a new human-machine task and team configuration taxonomy that can support such analysis. Given that many human-machine team tasks are used in the literature, but that there are few taxonomies to describe or categorize these tasks, our taxonomy provides a tool for task and team configuration design and analysis in the context of human-machine teaming research. While our aim is to provide pointers for what trust model can be used given a task and team configuration, we have currently only provided some suggestions of how different task and team configuration factors might influence the choice of model. We believe that, in order to draw clear conclusions on how task and team configuration characteristics can influence trust models, systematic empirical research is necessary, both in human-human as well as human-machine team settings. Additionally, our taxonomy presents a tool that may help in the process of designing experiments in human-machine teams. This taxonomy can also open the possibility of better describing experimental scenarios, tasks and teams, as to make research more reusable and organized. As such, our taxonomy can be used as a starting point for research, by

1. creating tasks that vary on specific factors in the taxonomy, such as Nature (cognitive vs. physical) or Criticality (high, medium, or low);
2. creating team configurations that vary on specific factors in the taxonomy, such as Lifespan (short [e.g., an hour] vs. long [e.g., a month]) or Shared-Knowledge (full shared knowledge vs. partial shared knowledge); and
3. setting up experiments that use different sets of krypta to model trust (and as a consequence different manifesta) in different contexts, as defined by task characteristics and team configurations.

To maintain consistency throughout such experiments, the formalization we described in Section 5 can be a starting point. Once we know what trustworthiness should look like (in terms of krypta and manifesta) in a certain human-machine teamwork scenario, we aim at implementing it on the artificial agents. The ultimate goal of exploring artificial trust is that an artificial agent can form beliefs regarding a teammate's trustworthiness and reason about these beliefs toward a better team performance, through decision-making. With such beliefs, an artificial agent can predict and direct its human teammates toward the team goal, while avoiding risks, that is, by helping the human or allocating tasks differently within a team. It is also important that the agent can reflect on its own trustworthiness in a certain context, so that it can calibrate its teammates' trust. By ensuring there is no under-trust, we aim at minimizing risks in human-machine interaction. With a set of expressions we show how this formalization can be used to implement appropriate (both artificial and natural) trust.

Our formalization provides a first step toward making it possible to implement artificial trust beliefs, which take into account the trustor, the trustee, and the context, that is, task and team configuration. In this chapter, we propose that the alignment of beliefs of trustworthiness and, consequently, the appropriateness of trust depend on the task and team configuration. This means that one may trust appropriately another in certain contexts but not in others. It is important to study how these beliefs can actually form from manifesta and how they can represent each different krypta adequately. When illustrating how this formalization can be used with a certain krypta, we have also presented an array of undefined weights. Although these weights are ultimately defined by the context, we argue that more complex structures may be appropriate to consider the different dimensions of krypta. We do not expect the learning of these beliefs, as well as their implementations, to be trivial and recognize these beliefs require further studying through simulation and human-machine experiments. The update of the beliefs of artificial trust is also something that should be addressed in future work. We find it important to mention that manifesta are not the only source of information to build the beliefs of krypta, that is, indirect information such as reputation may also have its role. However, we focus mainly on the interaction itself and we can infer krypta from observable behavior.

7 Conclusion

In this chapter, we present the concept of appropriate context-dependent artificial trust in human-machine teamwork. The goal of this chapter is twofold. First, we propose a taxonomy based on existing literature that can be used to choose the most appropriate model of human trustworthiness, when assessing artificial trust. Through this taxonomy we reflect on how, depending on several task and team configuration characteristics, the internal characteristics of a teammate (krypta) and how they show them (manifesta) can vary. We argue that we may not find one trustworthiness model that fits all the situations, but instead, a taxonomy that helps in characterizing the context and choosing the right model. This taxonomy contributes to the field of human-machine teamwork by proposing a set of characteristics that can define a certain context, which can facilitate the experimental design and definition of research questions. Second, we propose how we can formalize artificial trust as a belief of context-dependent trustworthiness. Our work provides a departure point for the implementation of artificial trust in artificial agents, which will make machines more adaptable and useful to their human teammates.

References

[1] M. Lewis, K. Sycara, P. Walker, The role of trust in human-robot interaction, in: Studies in Systems, Decision and Control, Springer International Publishing, 2018, pp. 135–159, https://doi.org/10.1007/978-3-319-64816-3_8.

[2] E. Salas, D.E. Sims, C. Burke, Is there a "Big Five" in teamwork? Small Group Res. 36 (2005) 555–599.

[3] M. Lewis, H. Li, K. Sycara, Deep learning, transparency, and trust in human robot teamwork, in: Trust in Human-Robot Interaction, Elsevier, 2020, pp. 321–352.

[4] S. Ososky, D. Schuster, E. Phillips, F. Jentsch, Building appropriate trust in human-robot teams, in: AAAI Spring Symposium: Trust and Autonomous Systems, 2013.

[5] M. Johnson, J.M. Bradshaw, Chapter 16—The role of interdependence in trust, in: C.S. Nam, J.B. Lyons (Eds.), Trust in Human-Robot Interaction, Academic Press, 2021, pp. 379–403, https://doi.org/10.1016/B978-0-12-819472-0.00016-2.

[6] M. Johnson, J.M. Bradshaw, P.J. Feltovich, C.M. Jonker, M.B. Van Riemsdijk, M. Sierhuis, Coactive design: designing support for interdependence in human-robot teamwork, J. Hum. Robot Interact. 3 (1) (2014) 43–69.

[7] R. Falcone, M. Piunti, M. Venanzi, C. Castelfranchi, From manifesta to krypta: the relevance of categories for trusting others, ACM Trans. Intell. Syst. Technol. 4 (2013), https://doi.org/10.1145/2438653.2438662.

[8] M. Bacharach, D. Gambetta, Trust as type detection, in: C. Castelfranchi, Y.-H. Tan (Eds.), Trust and Deception in Virtual Societies, Springer Netherlands, Dordrecht, 2001, pp. 1–26, https://doi.org/10.1007/978-94-017-3614-5_1.

[9] R. Falcone, C. Castelfranchi, Trust dynamics: how trust is influenced by direct experiences and by trust itself, in: AAMAS, IEEE Computer Society, 2004, pp. 740–747, https://doi.org/10.1109/AAMAS.2004.10084.

[10] C. Centeio Jorge, M.L. Tielman, C.M. Jonker, Assessing artificial trust in human-agent teams: a conceptual model, in: C. Martinho, J. Dias, J. Campos, D. Heylen (Eds.), IVA '22: ACM International Conference on Intelligent Virtual Agents, Faro, Portugal, September 6–9, 2022, ACM, 2022, pp. 24:1–24:3, https://doi.org/10.1145/3514197.3549696.

[11] R.C. Mayer, J.H. Davis, F.D. Schoorman, An integrative model of organizational trust, Acad. Manag. Rev. 20 (1995) 709–734.

[12] K.S. Haring, E. Phillips, E.H. Lazzara, D. Ullman, A.L. Baker, J.R. Keebler, Applying the swift trust model to human-robot teaming, in: Trust in Human-Robot Interaction, Elsevier, 2021, pp. 407–427.

[13] P. Parashar, L.M. Sanneman, J.A. Shah, H.I. Christensen, A taxonomy for characterizing modes of interactions in goal-driven, human-robot teams, in: 2019 IEEE/RSJ International Conference on Intelligent Robots and Systems, IROS 2019, Macau, SAR, China, November 3–8, 2019, IEEE, 2019, pp. 2213–2220, https://doi.org/10.1109/IROS40897.2019.8967974.

[14] C. Breuer, J. Hüffmeier, F. Hibben, G. Hertel, Trust in teams: a taxonomy of perceived trustworthiness factors and risk-taking behaviors in face-to-face and virtual teams, Hum. Relat. 73 (2020) 3–34.

[15] H. Huynh, C.E. Johnson, H.S. Wehe, Humble coaches and their influence on players and teams: the mediating role of affect-based (but not cognition-based) trust, Psychol. Rep. 123 (2019) 1297–1315.

[16] A.M. Naber, S.C. Payne, S.S. Webber, The relative influence of trustor and trustee individual differences on peer assessments of trust, Pers. Individ. Differ. 128 (2018) 62–68, https://doi.org/10.1016/j.paid.2018.02.022.

[17] J. Sabater-Mir, L. Vercouter, Trust and reputation in multiagent systems, in: Multiagent Systems, MIT Press, 2013, p. 381.

[18] A. Herzig, E. Lorini, J.F. Hübner, L. Vercouter, A logic of trust and reputation, Logic J. IGPL 18 (2009) 214–244, https://doi.org/10.1093/jigpal/jzp077.

[19] C. Burnett, T.J. Norman, K. Sycara, Stereotypical trust and bias in dynamic multiagent systems, ACM Trans. Intell. Syst. Technol. 4 (2013), https://doi.org/10.1145/2438653.2438661.

[20] K. Chhogyal, A.C. Nayak, A. Ghose, K.H. Dam, A value-based trust assessment model for multi-agent systems, in: 28th International Joint Conference on Artificial Intelligence (IJCAI-19), 2019.

[21] C. Cruciani, A. Moretti, P. Pellizzari, Dynamic patterns in similarity-based cooperation: an agent-based investigation, J. Econ. Interact. Coord. 12 (1) (2017) 121–141.

[22] J.D. Lee, K.A. See, Trust in automation: designing for appropriate reliance, Hum. Factors 46 (2004) 50–80.

[23] M. Winikoff, Towards trusting autonomous systems, Lect. Notes Comput. Sci. 10738 (2018) 3–20, https://doi.org/10.1007/978-3-319-91899-0_1.

[24] C. Nam, P. Walker, H. Li, M. Lewis, K. Sycara, Models of trust in human control of swarms with varied levels of autonomy, IEEE Trans. Hum.-Mach. Syst. 50 (2020) 194–204, https://doi.org/10.1109/THMS.2019.2896845.

[25] M.W. Floyd, M. Drinkwater, D.W. Aha, Learning trustworthy behaviors using an inverse trust metric, in: Robust Intelligence and Trust in Autonomous Systems, Springer, 2016.

[26] I.B. Ajenaghughrure, S.C. Sousa, I.J. Kosunen, D. Lamas, Predictive model to assess user trust: a psycho-physiological approach, in: Proceedings of the 10th Indian Conference on Human-Computer Interaction, 2019, pp. 1–10.

[27] Y. Guo, X.J. Yang, Modeling and predicting trust dynamics in human-robot teaming: a Bayesian inference approach, Int. J. Soc. Robot. (2020), https://doi.org/10.1007/s12369-020-00703-3.

[28] C. Neubauer, G. Gremillion, B.S. Perelman, C.L. Fleur, J.S. Metcalfe, K.E. Schaefer, Analysis of facial expressions explain affective state and trust-based decisions during interaction with autonomy, in: Advances in Intelligent Systems and Computing, Proceedings of the 3rd International Conference on Integrating People and Intelligent Systems, February 19–21, 2020, Modena, Italy, vol. 1131, Springer, 2020, pp. 999–1006, https://doi.org/10.1007/978-3-030-39512-4_152.

[29] M. Chen, S. Nikolaidis, H. Soh, D. Hsu, S.S. Srinivasa, Planning with trust for human-robot collaboration, in: T. Kanda, S. Sabanovic, G. Hoffman, A. Tapus (Eds.), Proceedings of the 2018 ACM/IEEE International Conference on Human-Robot Interaction, HRI 2018, Chicago, IL, USA, March 5–8, 2018, ACM, 2018, pp. 307–315, https://doi.org/10.1145/3171221.3171264.

[30] A.-S. Ulfert, E. Georganta, A model of team trust in human-agent teams, in: Companion Publication of the 2020 International Conference on Multimodal Interaction, Association for Computing Machinery, New York, NY, 2020, pp. 171–176, https://doi.org/10.1145/3395035.3425959.

[31] K.E. Schaefer, B.S. Perelman, G.M. Gremillion, A.R. Marathe, J.S. Metcalfe, A roadmap for developing team trust metrics for human-autonomy teams, in: Trust in Human-Robot Interaction, Academic Press, 2021, https://doi.org/10.1016/B978-0-12-819472-0.00012-5.

[32] E.J.D. Visser, M.M.M. Peeters, M.F. Jung, S. Kohn, T.H. Shaw, R. Pak, M.A. Neerincx, Towards a theory of longitudinal trust calibration in human-robot teams, Int. J. Soc. Robot. 12 (2020) 459–478, https://doi.org/10.1007/s12369-019-00596-x.

[33] A.R. Wagner, R.C. Arkin, Recognizing situations that demand trust, in: 2011 RO-MAN, IEEE, 2011, pp. 7–14.

[34] A.R. Wagner, P. Robinette, A. Howard, Modeling the human-robot trust phenomenon: a conceptual framework based on risk, ACM Trans. Interact. Intell. Syst. 8 (2018), https://doi.org/10.1145/3152890.

[35] S. Vinanzi, M. Patacchiola, A. Chella, A. Cangelosi, Would a robot trust you? Developmental robotics model of trust and theory of mind, Philos. Trans. R. Soc. B 374 (2019), https://doi.org/10.1098/rstb.2018.0032.

[36] V. Surendran, A. Wagner, Your robot is watching: using surface cues to evaluate the trustworthiness of human actions, in: 2019 28th IEEE International Conference on Robot and Human Interactive Communication (RO-MAN), 2019, pp. 1–8.

[37] H. Azevedo-Sa, X.J. Yang, L.P. Robert, D.M. Tilbury, A unified bi-directional model for natural and artificial trust in human-robot collaboration, IEEE Robot. Autom. Lett. 6 (3) (2021) 5913–5920, https://doi.org/10.1109/LRA.2021.3088082.

[38] N. Schlicker, M. Langer, Towards warranted trust: a model on the relation between actual and perceived system trustworthiness, in: S. Schneegass, B. Pfleging, D. Kern (Eds.), MuC '21: Mensch und Computer 2021, Ingolstadt, Germany, September 5–8, 2021, ACM, 2021, pp. 325–329, https://doi.org/10.1145/3473856.3474018.

[39] R.C. Mayer, J.H. Davis, The effect of the performance appraisal system on trust for management: a field quasi-experiment, J. Appl. Psychol. 84 (1) (1999) 123.

[40] B.D. Adams, R. Webb, Trust in Small Military Teams, Command and Control Research Program, 2002.

[41] B.D. Adams, S. Waldherr, J. Sartori, Trust in Teams Scale, Trust in Leaders Scale: Manual for Administration and Analyses, 2008.

[42] I.B. Ajenaghughrure, S. Sousa, D.J.R. Lamas, Measuring trust with psychophysiological signals: a systematic mapping study of approaches used, Multimodal Technol. Interact. 4 (2020) 63.

[43] A. Xu, G. Dudek, OPTIMo: online probabilistic trust inference model for asymmetric human-robot collaborations, in: ACM/IEEE International Conference on Human-Robot Interaction, 2015, IEEE Computer Society, 2015, pp. 221–228, https://doi.org/10.1145/2696454.2696492. vol.

[44] N.C. Rabinowitz, F. Perbet, H.F. Song, C. Zhang, S.M.A. Eslami, M.M. Botvinick, Machine theory of mind, in: J.G. Dy, A. Krause (Eds.), Proceedings of Machine Learning Research, Proceedings of the 35th International Conference on Machine Learning, ICML 2018, Stockholmsmässan, Stockholm, Sweden, July 10–15, 2018, vol. 80, PMLR, 2018, pp. 4215–4224.

[45] T.N. Nguyen, C. Gonzalez, Cognitive machine theory of mind, in: S. Denison, M. Mack, Y. Xu, B.C. Armstrong (Eds.), Proceedings of the 42th Annual Meeting of the Cognitive Science Society—Developing a Mind: Learning in Humans, Animals, and Machines, CogSci 2020, virtual, July 29 to August 1, 2020.

[46] J.K. Rempel, J.G. Holmes, M.P. Zanna, Trust in close relationships, J. Pers. Soc. Psychol. 49 (1) (1985) 95.

[47] J.L. Wildman, M.L. Shuffler, E.H. Lazzara, S.M. Fiore, C.S. Burke, E. Salas, S. Garven, Trust development in swift starting action teams: a multilevel framework, Group Org. Manag. 37 (2) (2012) 137–170.

[48] A.J. Farina, G.R. Wheaton, E.A. Fleishman, Development of a taxonomy of human performance: the task characteristics approach to performance prediction, American Institutes for Research, 1971.

[49] M.A. Neerincx, et al., Cognitive task load analysis: allocating tasks and designing support, in: Handbook of Cognitive Task Design, vol. 2003, Lawrence Erlbaum Associates, Mahwah, NJ, 2003, pp. 283–305.

[50] J.L. Wildman, A.L. Thayer, M.A. Rosen, E. Salas, J.E. Mathieu, S.R. Rayne, Task types and team-level attributes: synthesis of team classification literature, Hum. Resour. Dev. Rev. 11 (1) (2012) 97–129.

[51] J.E. McGrath, Groups: Interaction and Performance, vol. 14, Prentice-Hall, Englewood Cliffs, NJ, 1984.

[52] B.S. Bloom, Committee of College and University Examiners, Taxonomy of Educational Objectives, vol. 2, Longmans, Greene, NY, 1964.

[53] J. Sweller, Cognitive load theory, in: Psychology of Learning and Motivation, vol. 55, Elsevier, 2011, pp. 37–76.

[54] J. Cohen-Mansfield, M.S. Marx, L.S. Freedman, H. Murad, N.G. Regier, K. Thein, M. Dakheel-Ali, The comprehensive process model of engagement, Am. J. Geriatr. Psychiatry 19 (10) (2011) 859–870.

[55] M. Harbers, C.M. Jonker, M.B. van Riemsdijk, Context-sensitive sharedness criteria for teamwork, in: A.L.C. Bazzan, M.N. Huhns, A. Lomuscio, P. Scerri (Eds.), International Conference on Autonomous Agents and Multi-Agent Systems, AAMAS '14, Paris, France, May 5–9, 2014, IFAAMAS/ACM, 2014, pp. 1507–1508.

[56] C.M. Jonker, M.B. van Riemsdijk, I. van de Kieft, M.L. Gini, Compositionality of team mental models in relation to sharedness and team performance, in: H. Jiang, W. Ding, M. Ali, X. Wu (Eds.), Lecture Notes in Computer Science, Advanced Research in Applied Artificial Intelligence—25th International Conference on Industrial Engineering and Other Applications of Applied Intelligent Systems, IEA/AIE 2012, Dalian, China, June 9–12, 2012. Proceedings, vol. 7345, Springer, 2012, pp. 242–251, https://doi.org/10.1007/978-3-642-31087-4_26.

[57] Slack, Moving beyond remote: workplace transformation in the wake of Covid-19, 2020. https://slack.com/blog/collaboration/workplace-transformation-in-the-wake-of-covid-19.

[58] A. Alfaleh, A.N. Alkattan, A. Alageel, M. Salah, M.M. Almutairi, K. Sagor, K. Alabdulkareem, Onsite versus remote working: the impact on satisfaction, productivity, and performance of medical call center workers, Inquiry 58 (2021), https://doi.org/10.1177/00469580211056041.

[59] S.P. Mikawa, S.K. Cunnington, S.A. Gaskins, Removing barriers to trust in distributed teams: understanding cultural differences and strengthening social ties, in: S.R. Fussell, P.J. Hinds, T. Ishida (Eds.), Proceedings of the 2009 International Workshop on Intercultural Collaboration, IWIC '09, Palo Alto, California, USA, February 20–21, 2009, ACM, 2009, pp. 273–276, https://doi.org/10.1145/1499224.1499275.

[60] D.S. Staples, P. Ratnasingham, Trust: the panacea of virtual management? in: J.I. DeGross, R. Hirschheim, M. Newman (Eds.), Proceedings of the Nineteenth International Conference on

Information Systems, ICIS 1998, Helsinki, Finland, December 13–16, 1998, Association for Information Systems, 1998, pp. 128–144.

[61] M. Natarajan, M.C. Gombolay, Effects of anthropomorphism and accountability on trust in human robot interaction, in: T. Belpaeme, J.E. Young, H. Gunes, L.D. Riek (Eds.), HRI '20: ACM/IEEE International Conference on Human-Robot Interaction, Cambridge, United Kingdom, March 23–26, 2020, ACM, 2020, pp. 33–42, https://doi.org/10.1145/3319502.3374839.

[62] D. Siemon, Elaborating team roles for artificial intelligence-based teammates in human-AI collaboration, Group Decis. Negot. 31 (2022) 871–912.

[63] L. Huang, N.J. Cooke, R.S. Gutzwiller, S. Berman, E.K. Chiou, M. Demir, W. Zhang, Distributed dynamic team trust in human, artificial intelligence, and robot teaming, in: Trust in Human-Robot Interaction, Elsevier, 2021, pp. 301–319.

[64] N. Griffiths, Task delegation using experience-based multi-dimensional trust, in: AAMAS '05, 2005.

[65] C. Castelfranchi, R. Falcone, Trust & Self-Organising Socio-Technical Systems, Springer International Publishing, 2010, pp. 209–229, https://doi.org/10.1007/978-3-319-29201-4_8.

[66] G. Mecacci, F.S de Sio, Meaningful human control as reason-responsiveness: the case of dual-mode vehicles, Ethics Inf. Technol. 22 (2) (2020) 103–115, https://doi.org/10.1007/s10676-019-09519-w.

[67] J.M. McGuirl, N.B. Sarter, Supporting trust calibration and the effective use of decision aids by presenting dynamic system confidence information, Hum. Factors 48 (4) (2006) 656–665.

[68] F. Yang, Z. Huang, J. Scholtz, D.L. Arendt, How do visual explanations foster end users' appropriate trust in machine learning? in: Proceedings of the 25th International Conference on Intelligent User Interfaces, 2020, pp. 189–201.

[69] F. Ekman, M. Johansson, J. Sochor, Creating appropriate trust in automated vehicle systems: a framework for HMI design, IEEE Trans. Hum.-Mach. Syst. 48 (1) (2017) 95–101.

[70] S.H. Huang, K. Bhatia, P. Abbeel, A.D. Dragan, Establishing appropriate trust via critical states, in: 2018 IEEE/RSJ International Conference on Intelligent Robots and Systems (IROS), IEEE, 2018, pp. 3929–3936.

[71] R.R. Hoffman, A taxonomy of emergent trusting in the human-machine relationship, in: Cognitive Systems Engineering: The Future for a Changing World, Talyor & Francis, Boca Raton, FL, 2017, pp. 137–163.

5

Toward a causal modeling approach for trust-based interventions in human-autonomy teams*

Anthony L. Baker[a], Daniel E. Forster[a], Ray E. Reichenberg[b], Catherine E. Neubauer[a], Sean M. Fitzhugh[a], and Andrea Krausman[a]

[a]DEVCOM ARMY RESEARCH LABORATORY, ABERDEEN PROVING GROUND, ADELPHI, MD, UNITED STATES [b]EDUCATIONAL ASSESSMENT RESEARCH UNIT, COLLEGE OF EDUCATION, UNIVERSITY OF OTAGO, DUNEDIN, NEW ZEALAND

The literature on human-autonomy teams (HATs) has established the importance of calibrating human team members' knowledge and expectations to their intelligent, autonomous teammates' capabilities. In doing so, teams can utilize their autonomous resources appropriately. Ideally, the team will delegate to the autonomy the tasks and decisions for which human input is not necessary, while also directly supervising or managing the autonomy when its assignments extend beyond its abilities. This state, known as calibrated trust, is key to delivering assured performance in a human-autonomy team. To maintain appropriately calibrated team trust, three steps must be taken. First, the relevant trust constructs for a given teaming situation must be clearly and consistently defined. This reduces the ambiguity imposed by the great variety of trust definitions and concepts that exist. Second, based on the selected trust definitions, appropriate measures of trust must be identified and implemented. This can be challenging given the limitations of some trust measures, such as traditional self-report surveys, which are unable to capture real-time fluctuations in team trust dynamics. Third, when the selected measures reveal that team trust levels are not appropriately calibrated to the capabilities of the autonomous system, trust interventions can be implemented to bring the team's trust level to a more optimized state. Recent work has established the importance of integrating multiple trust measurement approaches to gain a holistic understanding of the team's trust. However, to design appropriate interventions, it is important to consider a similar holistic understanding of the team's trust-related factors (antecedents, covariates, confounds, mediators/moderators, etc.). To achieve this, we propose a causal analysis approach to understanding the effects of interventions. In this chapter, we will first review the concept

---

*The views, opinions, and/or findings contained in this chapter are those of the authors and should not be construed as an official Department of the Army position, policy, or decision unless so designated by other documentation.

Putting AI in the Critical Loop. https://doi.org/10.1016/B978-0-443-15988-6.00011-X

of trust in human-autonomy teams, as well as current research on trust measurement. Then, we will examine the process of implementing interventions in teams. These discussions will form the basis for how causal modeling techniques provide a useful framework for evaluating interventions. We then provide a model scenario for human-autonomy teaming, involving an autonomy-supported navigation and threat identification task, and introduce causal modeling and two of its common approaches: Bayesian networks and structural equation modeling. Following that, we will construct a Bayesian network based on our model scenario, highlighting the ability of the causal model to reveal the importance of various factors in the human-autonomy teaming intervention. The goal of this chapter is to relay to readers the importance of more completely characterizing and analyzing teamwork and trust-related factors when designing and implementing interventions, and to present the causal modeling approach as a useful way forward for human-autonomy teaming research.

1 Human-autonomy teams

While a vast literature exists that has examined the components of effective human teams, there is a burgeoning need to understand the processes that govern effective HAT. Human-autonomy teams can be defined as one or more human teammates coupled with one or more autonomous systems (or more general intelligent agents) that collaborate to accomplish a task or goal [1]. Behind the concept of teaming humans and autonomous systems is the supposition that, when utilized appropriately, humans and autonomy working together may facilitate better performance than either entity alone [2,3]. For example, autonomous systems can collect, process, and analyze large quantities of data quickly [3], thereby freeing up a human teammate to use the data outputs, coupled with their intuition and experience, to make well-informed, timely decisions [4]. Examples of humans and autonomy working in proximity are becoming more prevalent in a variety of settings including manufacturing, medicine, and the military, owing to advances in AI and machine learning [4,5]. With further advancements, autonomous systems will possess increased capacity for independence and interdependence, as well as skills that enable heightened collaboration with human team members [6].

In theory, the goal of HAT development is for autonomous systems to advance beyond simply serving as tools that support the team, and instead to serve as full-fledged team members [7,8]. This requires understanding how to manage interdependence and elicit teamwork behaviors that are characteristic of high-performing human teams, such as information sharing, requesting backup, observing team members, and adapting where necessary [6,9]. Successful integration of humans and autonomy into effective teams is complex and full of challenges. For example, autonomy and/or intelligent agents, whether embedded or embodied (e.g., robots), perform a variety of tasks at various levels of autonomy, at times providing little feedback to human teammates regarding their actions and decisions, making it difficult to assess and understand the agent's actions and decisions [10]. Similarly, human team member actions and decisions can be opaque to other team members including autonomous teammates, making it difficult to arrive at a shared understanding [11]. Therefore, to successfully team humans and autonomous systems, it is necessary to examine how these differences influence teamwork dynamics and the emergence of critical team processes such as trust.

2 Trust in human-autonomy teams

Human-autonomy team trust builds on the principles of interpersonal trust in the literature and can be described as "team members' positive expectations about each other's competence and motivations, as well as a shared acceptance of vulnerability based on the assumption that teammates will act in the best interest of the team" [12, p. 396], where a teammate can be a human or an autonomous system. There is strong consensus in the teams literature that trust is foundational to team function and effectiveness [13]. For example, trust influences how team members interact and work together, as well as enabling key team processes such as information sharing, monitoring behaviors, and assisting other team members when needed [12]. Without trust, team members focus on their own individual interests rather than those of the team, share less information with each other, and spend valuable time and energy excessively monitoring one another and double-checking each other's work [11,14–16]. Trust also shapes teams' decisions whether and when to use autonomous systems. Insufficient trust in an autonomous system leads to underutilization and failure to take advantage of a skillset that often exceeds those of humans, such as rapid text parsing or visual scanning capabilities [3]. Trust and appropriate trust calibration then become even more important as autonomous capabilities advance and enable systems to adapt to human teammates.

What is clear in the literature is that trust is a multifaceted and complex construct [11]. Part of the complexity is that trust is not all or none [17], but is continuous, emergent, and dynamic. Once trust is established it changes over time based on the experience, expectations, and observations of team member behaviors [18], as well as an individual's predisposition or propensity to trust [11,19], and other factors such as personality, system reliability, or workload [20]. Further, trust may differ according to context – team, task, environmental, and situational contexts can all impact the development or maintenance of trust in an agent, whether human or autonomy [21]. For instance, a driver may trust a car's navigation system to identify the best route during rush hour but may be less willing to trust the car to drive that route home in a heavy snowstorm. While this is a simple example, it highlights the need to consider context and recognize the complexity of trust as a construct that both influences and is influenced by behaviors of and interactions between humans and autonomy [17]. The complexity of the myriad constructs that influence trust's dynamic nature also underscores the importance of formalizing the causal relationships within a trust network. Doing so will help scholars appropriately examine the boundaries between trust and trust-relevant constructs, as well as the processes that are more likely to promote or hinder trust development. As mentioned previously, as autonomous capabilities continue to advance and these systems become more intelligent and adaptable, the interactions between humans and their autonomous counterparts will change, leading to an evolution in how trust evolves, and subsequently how it should be measured. In the next section, we will discuss an approach to measuring and modeling trust within the context of human autonomy teamwork.

3 Trust measurement in HAT

Defining a method of measuring trust requires a foundational causal understanding of what outcomes are directly caused by trust—indeed, this is the approach for building most measurement models [22]. Specifically, psychometricians design measurements for inherently unobservable constructs (e.g., math ability, pain, propensity to trust) that are thought to be the underlying causes of specific behaviors, which are typically responses on a test or questionnaire (e.g., correct answers on a math test, self-reported levels on a pain or propensity to trust scale). By collecting and observing several responses, researchers can infer respondents' levels on the underlying construct by examining the common variance shared across all their responses. However, to get to this level of measuring a latent construct such as trust, in this case, researchers need to clearly define what the construct is in terms of what it is expected to cause (e.g., endorsing the phrase, "I trust my teammate") and, for validation purposes, what is expected to cause it (e.g., changing a teammate's reliability should change a person's trust in that teammate).

Following Mayer et al. [23], we argue that trust is a latent state that generally represents one's willingness to be vulnerable to the actions of another party in anticipation that the other will perform some action beneficial to the trustor. Trust can be indicated by prosocial actions and feeling safe, expressing beliefs, etc. By extension, team trust is an emergent property that represents the dynamic impacts of trust on teammate interactions, which in turn affect each member's trust. Typically, trust is defined at different levels in terms of relatively stable traits (e.g., one's disposition toward trusting others) and more context-relevant states (e.g., one's newly formed yet fragile trust in a new acquaintance), in addition to distinct categories of antecedents and indicators, such as affect-based, cognitive-based, and behavioral trust [20,24]. When assessing a complex state such as trust, and how it emerges and manifests within a human-autonomy team, it is important to describe the difference between trait and state measures of trust, as trust has generally been conceptualized and recognized to be both a trait and a state [24]. Trait-based trust is relatively stable over time and relates to an individual's predisposition to trust in others, and can more commonly be referred to as dispositional, generalized, and/or propensity to trust. However, all three types of trust traits are inherent to the person, and somewhat independent of the context within which a person finds themselves. Conversely, state-based trust is very much dependent on a person's situation or context and therefore can fluctuate greatly. Hence, it is our argument that trust should be considered a multi-dimensional construct [20]. Approaches to defining trust point toward a larger network of causal interactions between constructs that influence one's willingness to be vulnerable to others, thus requiring researchers and practitioners to make clear decisions about what they mean by trust and how that construct is situated in a causal network.

How we decide to measure trust depends on how we define trust (e.g., whether it is an attitude or behavior) and what we theorize about the causal relationship between the construct and the measurement indicator (e.g., trust causes people to self-report higher trust levels). When attempting to assess trust, it is imperative to select appropriate measures and metrics to determine whether trust is indeed present and changing over time. Traditional measurement approaches of trust focus on self-report; however, given the complex, multidimensional nature of the construct as described here, practitioners need more robust measures that provide a more comprehensive picture of team dynamics over time. As such, not all measures and metrics are appropriate in all situations and contexts, nor do they offer the same robust power for assessment [25]. For example, subjective assessments of trust are commonly used in studies and have been instrumental in trust research advancements; however, subjective assessments are limited by temporal constraints and subjective bias [26]. While self-report measures have shown great benefits for understanding certain constructs that are more difficult to capture through other means, such as attitudes and emotions, self-report measures can suffer from certain biases, such as primacy and recency effects. To avoid subject fatigue and response patterns, researchers often err on the side of administering surveys less frequently and therefore may miss details necessary for inferring team states at a sufficient temporal resolution.

Because trust is emergent and dynamic, adequate trust measurement requires a multimethod, multimodal approach [27] that considers the strengths, weaknesses, and causal relevance of each measure, and can capture changes in trust that occur over time. In so doing, we can arrive at a more comprehensive understanding of how trust develops and what impacts trust as the team evolves and responds to various situations [12]. Rather than rely on self-report measures alone, we propose a multimodal approach to trust assessment that allows us to track trust levels over time (e.g., over the course of a mission or specific task). Further, synchronizing mission and/or task data with different trust measures can provide a more comprehensive picture of team dynamics and what led to decisions to trust or not trust a system or team member. Specific measures that appear to be promising proxy measures of trust include communication-based features of team trust, physiological responses, and even affective cues such as facial expressivity; however, they have yet to be fully validated and more work is certainly needed in this area to develop more standardized metrics for team trust assessment [11]. By leveraging a multimethod, multimodal approach to trust measurement, we can enable a more robust and accurate understanding of trust development, maintenance, and even calibration. In so doing, we attempt to understand the causal structure between the latent, underlying construct of team trust and specific measures to help inform trust-based interventions and even trust calibration. When we measure trust appropriately, we can determine whether levels of team trust are appropriate given the team's performance—if trust is too high or too low, we can then evaluate how interventions affect trust calibration.

For optimal performance, it is understood that human-autonomy teams should exhibit well-calibrated team trust. Team trust emerges and develops based on a continuous process by which the trustor evaluates how well the actions and performance of another human or autonomous system align with their expectations [21]. Trust is a dynamic construct and, once established, it changes as people interact with each other and with autonomous systems. Trust calibration refers to how well a level of trust corresponds to the actual capabilities needed in the situation [28] suggesting that trust can rise or fall to levels that are inappropriate for certain contexts. Poor trust calibration (e.g., trust too high or too low) occurs when there is a mismatch between an operator's trust and the actual trustworthiness of a system [29]. For example, overreliance on autonomous systems to perform certain tasks that are beyond their purpose or capabilities can have serious consequences for team effectiveness [30]. Team members may also become complacent and less aware of the activities and actions of their teammates, potentially leading to costly errors and poor performance. Likewise, inappropriate underreliance on the autonomy creates an unbalanced workload for the team since they feel a need to constantly monitor the system. Undertrust can also influence team members to excessively monitor their teammates' activities, thereby creating unnecessary workload and directing members' attention away from important tasks [14,15]. Therefore trust calibration occurs when the level of trust is appropriate for a particular situation or context [28]. However, the actual process of trust calibration is rather complex due to several influential factors such as propensity, prior history, personality, and expertise [31]. To return team trust to a calibrated state, interventions can be sought and implemented. In the following section, we will address interventions, grounding our discussion in fundamental concepts of teamwork.

4 Interventions and teaming

An intervention is any action that is taken to deliberately influence an outcome. Teamwork interventions are used to address inefficiencies or failures in the processes or interactions inherent to participating in a team. The most familiar example of performing an intervention involves training: specific information can be provided to a team member that is aimed at improving their understanding of how to perform desired teaming functions, or how to address key situations more effectively. In this way, new knowledge will (ideally) improve how the person's attitudes, behaviors, and cognition can mesh with those of the other team members, and subsequently engender improvements in team functioning. However, the process of identifying, implementing, and evaluating an appropriate intervention is far more complex than this example. To discuss interventions, it will be prudent to first review how teams function.

The basic structure for teamwork can be conceptualized by the input-mediator-output–input (IMOI) model [32]. (The IMOI model is a response to the earlier input-process-output (IPO) model [33–35]. The IPO model, while widely used, arguably does not represent team dynamics as well as the IMOI model. The IMOI model is characterized by outputs that feed into subsequent inputs, and the use of mediators, which encompass more team dynamics than just processes. For further discussion, see Ilgen et al. [32].) The model is composed of three parts that then loop back to the first part. Inputs consist of characteristics that exist prior to the team's interactions. For example, these can include a person's level of familiarity with a hand tool, a computer's ability to provide accurate information to the team, one's degree of motivation, and so on. Mediator is a catch-all term that encompasses the many processes, dynamics, and emergent states that the inputs causally act upon to produce variation in teamwork outcomes. For example, if a team is conducting a search-and-rescue task, and they are poorly prepared with no knowledge of the area and few tools, they may exhibit worse decision-making, ineffective coordination of resources, hampered communication, and lower mutual trust. The difficulty present in these processes and emergent states stems from the inputs (low level of preparedness for the task). Outputs encompass all observable team outcomes, such as performance and evaluation scores. Continuing the example of the ill-prepared team, their outputs will likely reflect poorly on the inputs and mediators that came before them: we can expect that the poorly prepared search and rescue team may not find some of their targets, or take too long to respond to certain situations. The last step in the model is input once again, and this represents the extent to which a team's outcomes can inform their subsequent activities. If the search-and-rescue team spent an extensive amount of time searching the area (output), we might expect it to result in a greater familiarity with the search area (input), making their team communications more effective (mediator) and resulting in better search times (output).

A countless number of issues with a team's inputs can lead to problems: for example, the team's membership makeup may not have the appropriate mix of knowledge, skills, or abilities for the task; the team may not have clear goals to work toward; there may not be clear boundaries between each team member's roles, and so on [36]. Likewise, mediators can be associated with issues that result in poor outputs: the team's process for resolving conflicts may be unhealthy and stifle open discussion; teammates may be monitoring or discussing each other's work too much, leaving less time to perform their required tasks; they may not be sharing the right information during key communication exchanges, and so on. It is critical to understand these varied team dynamics, and identify the specific variables associated with the team performance issue, in order to conduct an appropriate teaming intervention and (ideally) improve future team performance.

Let us return to the context of human-autonomy teaming and consider the effects of one specific mediator: trust in an intelligent agent teammate. From prior discussion, we understand the importance of maintaining well-calibrated trust in the agent. However, if it is suspected that the human has too little trust toward the agent, and is underutilizing its capabilities, we can consider an intervention to bring the trust back to a calibrated level. To understand interventions for trust requires an understanding of what causes trust—if we know what variables to increase or decrease in the situation, and are capable of directly manipulating them, then we can move trust in the desired direction (again, assuming we have appropriate measures and metrics). Schaefer and colleagues [37] meta-analytically identified the key antecedents (i.e., inputs) for trust in automation: human-related factors, such as age, gender, propensity to trust, understanding of what the automation can do, and so on; agent-related features, such as its appearance, level of automation, and reliability; and environment-related factors, such as the composition of the team, interdependence among the team's roles, and the degree of risk involved with the task. Thus any of these factors could be contributing to the poorly calibrated trust in the human-agent team.

Of these factors, we might be most capable of manipulating the human's understanding of what the automation can do. This understanding is shaped by past interactions and second-hand knowledge received from others. An after-action review could be conducted between task sessions to identify the specific aspect of the person's knowledge that is mismatched with the agent's capabilities [38]. For example, if the after-action review reveals that our operator is not fully utilizing controls that are used to guide the agent in an autonomous navigation mode, and instead is relying on manual controls, we can then deliver an intervention to improve their knowledge of those functions. Ideally, this intervention will result in improved and better-calibrated trust in the agent, and increased performance of the team.

Our example trust intervention illustrates that the causes and consequences of team trust can be complex and dynamic. Mathematically evaluating the effects of an intervention will require us to move past the IMOI model as a conceptual framework, and account for moderators, mediators, covariates, confounds, and other third-variable problems. To address these challenges, we argue that causal modeling approaches, such as those found in structural equation modeling (SEM) and Bayesian networks (BNs), offer researchers a set of practical and flexible analytical tools for evaluating causal effects. To demonstrate their utility, we developed a simple model of team trust based on robust effects reported in the literature, and then examined how perturbing different causal pathways, as would happen during intervention, would change observed effect patterns.

5 Our model human-autonomy teaming scenario

Here, we present a simplified team scenario between a human and autonomous agent, emphasizing a subset of variables important to the human's trust in the autonomous teammate and the team's performance. Though we understand that both human and autonomy perspectives are important to consider in this trust relationship, our scenario and model will focus more on the human's perspective. Consider a hypothetical scenario involving a human driver traversing an environment with varying terrain (e.g., on- and off-road) and urban development (e.g., city and rural) to reach multiple checkpoints. The human has the aid of an autonomous sensor package that identifies objects of interest in the environment (e.g., obstacles, potential threats). While the human is navigating, the autonomous system sends the human its decisions about possible environmental threats (e.g., insurmountable terrain, possible hostile forces, etc.). The human can accept the autonomy's decisions and circumvent the possible threats or reject the decisions and continue along the shortest path. The human-autonomy team's objective is to note items of interest, avoid hazards, and reach each successive checkpoint as quickly as possible.

Using this scenario, we can then develop a conceptual model that organizes some of the relevant factors contributing to the human's trust in the autonomy. (We note that this conceptual model is not meant to be theoretically complete; rather, it is used to provide an example for discussion.) Prior to the human-autonomy interaction, the human will likely have some prior understanding of autonomous systems, whether through reputation or direct experience, which represents their *familiarity* with the system itself. During the interaction, the human will experience some level of *workload* that is partly due to the human's familiarity with that system. The system's *reliability*, in conjunction with familiarity and workload, will directly affect the human's *trust* in the system, which will in turn affect the extent to which the human accepts or *overrides* (whether explicitly or implicitly) the system's decisions. Adding on to the influence of trust, the human's workload will also have effects on their decisions to override. In addition to these direct effects, there will also be indirect influences; for instance, we can expect workload to partially mediate the effect of familiarity on trust. See Fig. 5.1 for a path diagram of this simplified human-autonomy trust model. Before we consider how to evaluate an intervention designed to influence these trust antecedents, such as a *training intervention* that increases human familiarity with the autonomous system, we will introduce some foundations to modern causal modeling approaches.

6 A brief overview of causal modeling

To have confidence that X causes Y, scholars will rely on specific data collection methods that enable them to compare counterfactual states. For instance, when drug manufacturers are testing their new treatment for an illness, they design randomized-controlled trials (RCTs) wherein some patients are randomly selected to receive the experimental treatment and others are randomly selected to receive treatment-as-usual; this enables them to compare what happens when the experimental treatment is present versus absent. However, not all questions can be answered using perfectly controlled experiments, so researchers use craftier methods to examine these counterfactual states, such as collecting data from several cases that are identical in nearly every way, except for one specific factor. These data collection methods depend on the fundamental assumption that if X causes Y, then changing X will change Y. In many scientific fields, models of just a few factors can rarely account for the totality of our observed variation in outcomes—therefore researchers rely on probabilistic models that describe the processes thought to

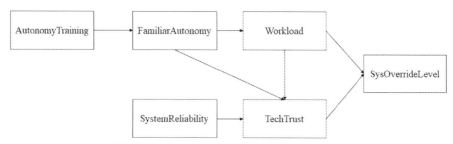

FIG. 5.1 A depiction of our simplified causal model using factors relevant to our human-autonomy trust scenario. Several factors influence the user's trust in technology and the extent to which the user overrides the system's decisions. Recalling the structure of the IMOI model discussed earlier, our model contains inputs (training, familiarity, system reliability), mediators (workload, trust), and an output (system override level).

have generated the observed data. This is what makes causal models fundamentally different from correlational models: causal models specify which factors are responsible for producing variability in outcomes, whereas correlational models only specify that multiple factors have overlapping variability. Importantly, the statistical information derived from a correlational model will be perfectly consistent with the causal models—however, evaluating data against complex causal structures (e.g., mediation, moderation, confounding) will produce statistical results with more nuance than any correlational analysis. Several scholars developed different, but complementary, frameworks and techniques for evaluating models of causal mechanisms that underlie observable phenomena, in a field broadly referred to as causal modeling [39]. Several specific frameworks have been developed with this goal in mind. These include, but are not limited to, Bayesian networks (BNs), structural equation models (SEMs), potential outcomes (e.g., the Rubin causal model [40]), and the unifying structural causal model approach [39]—with BNs and SEMs being arguably the two most common.

Regardless of the approach taken, causal modeling can help researchers consider, and formalize, the breadth of effects between antecedents, mediators, moderators, confounds, and other relevant factors. Take mediation, for instance—mediation occurs when X causes M, which then causes Y, and, in cases of complete mediation, that X only causes Y through M. For example, a vibrating speaker (X) will vibrate air molecules (M) which will then vibrate an eardrum (Y), and the vibrating speaker (X) will only vibrate an eardrum (Y) through air molecules (or some other medium, M). This model could be formally written as X–a $\rightarrow$ M–b $\rightarrow$ Y, where "a" is the magnitude of the effect of X on M and "b" is the magnitude of the effect of M on Y. If this model is correct, then removing air molecules (a = 0) will prevent an eardrum from vibrating (therefore, M and Y = 0), even if the speaker is vibrating and all other factors are the same. Moreover, the magnitude of the effect that the vibrating speaker (X) has on the eardrum (Y) will be a product of the effect that the vibrating speaker has on the air molecules and the effect that the vibrating air molecules have on the eardrum (a * b), thus yielding the indirect effect of X on Y through M. This simple example demonstrates the concordance between path diagrams and theoretical causal models. In addition to modeling direct and indirect effects, path diagrams can incorporate moderators (i.e., constructs that change the magnitude of another effect), covariates (i.e., constructs that share some causal influence on predictors and outcomes), and latent variables (i.e., unobserved constructs that account for the shared variation in several observations).

It is also important to fully appreciate that the models represent the causal processes that generated the observed data. Therefore it is essential that the model incorporates knowledge of the data collection procedure (e.g., random assignment to conditions, the time series of measurement events). This connection between data collection methods and data modeling can help us understand causal effects when randomly assigning participants to conditions is not possible. As a reminder, the purpose of random assignment is to provide methodological control for any extraneous variables that might muddy our interpretation of whether, and how much, X causes Y—that is, by randomly assigning participants to levels of X, there should be no covariation between X and any other variables that might influence Y. Of course, even by random chance, participants in an experimental treatment group could differ meaningfully from participants in a control group (e.g., the experimental group could still be wealthier, taller, more extroverted, etc.), and removing any covariation with random assignment becomes more likely with larger sample sizes. A major benefit of developing causal models is that we can account for extraneous variables in nonexperimental designs by measuring them and balancing their distribution around the focal variable to isolate the causal effect of X on Y. There are also statistical methods of controlling for extraneous variables, but these methods only work when the distributions of the extraneous variables are sufficiently balanced around the focal variable—if they are not balanced, even the most advanced statistical tools would provide meaningless results [41].

In the following sections, we will provide a brief introduction to both BNs and SEMs as a precursor to an example of these models applied to the interaction between humans and autonomous systems that will be presented in a later section. Far more detailed and technical treatments of these models are available in the literature and will be referred to in their respective sections.

6.1 Bayesian networks

A BN (see Refs. [42–44] for more detailed treatments) is a multivariate distribution of discrete variables, commonly depicted as an acyclic directed graph (a.k.a., directed acyclic graph, or DAG) to express the dependence and conditional independence assumptions in the model for the joint distribution. The path diagram in Fig. 5.1 qualifies as a DAG in that it has only directed edges (i.e., no two-headed arrows) and does not contain any directed cycles (i.e., no closed loops). More concretely, a BN models the probability of an event or state, such as a latent proficiency, conditioned on a set of observed states, events, or characteristics. Relationships can be either cross-sectional or longitudinal. In the latter case, the model is referred to as a dynamic Bayesian network [45,46]. A BN consists of a set of variables (often represented as "nodes" in the graph) and a set of "edges" between the variables. These edges are directed (i.e., single-headed arrows) and define the structure of the network.

In typical applications of BNs, each variable included in the model may take on a finite set of mutually exclusive states. The directed arrows from one node to the next represent conditional probability tables (CPTs) of a dimension determined by the number of categories defined for the respective nodes. It is the values in these tables that constitute the parameter set for a given network, which implies that model complexity is largely a function of the number of nodes and edges in the network as well as the number of discrete states specified for each node. The process for determining the values of the CPTs can be driven either by subject matter experts or by data using an algorithm (e.g., maximum likelihood, gradient descent, Markov Chain Monte Carlo) to estimate the values empirically. The initial state for each node is determined via the specification of a prior distribution and our beliefs about the probability distribution are updated via the machinery of Bayes' theorem as we incorporate incoming evidence.

BNs have received increased attention in fields such as training and education in recent years, due in part to the computational efficiency imparted by the ability to streamline the updating of these probability distributions via conditional independence assumptions. That is to say, if a node is conditionally independent from the node for which new evidence is being entered, then that node is essentially skipped in the model updating process.

6.2 Structural equation modeling

Structural equation modeling [47,48] is a statistical framework that allows analysts to specify causal and correlational relationships among both observed and latent variables, and to evaluate these relationships simultaneously in a single model. An SEM typically consists of two parts: a set of equations that defines the causal relationships between the focal variables, which are often latent (i.e., the structural model), and another set of equations that defines, or identifies these variables (i.e., the measurement model [49]). Many common modeling approaches such as regression (including multivariate regression), confirmatory factor analysis, and path analysis can be considered a special case of a structural equation model. In fact, the entire family of models subsumed under the general linear model (GLM) can be considered as structural equation models [50]. SEM is more flexible than the BN framework in terms of the relationships that can be specified in that both bidirectional (indicating a noncausal, or correlational, relationship between two variables) and nonrecursive (e.g., feedback loops) relationships are allowed. Variables can be either continuous or categorical and the form of the relationships between these variables can be either linear or nonlinear. As was the case with BNs, both cross-sectional and longitudinal relationships can be specified. The literature surrounding longitudinal SEM models (e.g., panel models, growth curve models) is very robust and there are several book-length treatments on the topic (e.g., Refs. [51,52]). More broadly, the development and evaluation of new modeling approaches with the SEM framework is a very active area of research. As such, there is an extensive literature on SEM and its applications in the social sciences including several excellent primers (e.g., Refs. [47,53,54]) as well as discussion of its utility as a causal modeling framework (e.g., Ref. [48]). Additionally, Pearl [39] offers a comparison of the BN and SEM frameworks in the context of causal inference.

Now that we have introduced two of the more common methods for causal modeling, we will use information from the trust literature to present an example of a simplified causal model for human-autonomy team trust and use it to illustrate the influence of different causal structures on the implications for designing interventions.

7 Causal modeling in context

The focus of this section is to build a plausible, though hypothetical, larger causal model for a scenario involving trust between humans and an autonomous system, which also incorporates moderators and covariates as a means of grounding some of the relatively abstract concepts discussed in the previous section. Returning to the scenario presented earlier wherein we have a human driver navigating an environment with the help of an autonomous sensor package, we can encode the relationships between the variables used to predict the human user's propensity to override the autonomous system's recommendations using either a BN or a structural equation model (there are other options, of course, but we will stick to these two popular frameworks). As Pearl has noted [39], the two frameworks are closely related and, for most applications, the choice between the two is largely related to what the researcher is hoping to get out of the model. Bayesian networks might be more useful for diagnostic applications exploring the impact that the states of one variable have on other variables throughout the system. The focus in structural equation modeling, on the other hand, is often to better understand population-level relationships between variables. In the context of the current example, the SEM framework might be used to answer the question, "What factors impact (or cause) a human user to accept or reject the system recommendations?" whereas the BN might be used to answer the question, "If we change the value of the intervention variable, what is the resulting impact on the other variables in the model?" The difference between these two questions is subtle and it is possible, perhaps even likely, that a researcher might be interested in both questions during a study. In many cases, a researcher might even use both modeling frameworks to accomplish their goals. For example, the parameter estimates from an SEM model could be used to inform, or perhaps might even be directly translatable to, the values in the conditional probability tables in a BN. The researcher then has the flexibility to leverage estimates from previous literature and/or to capitalize on the strengths of multiple modeling approaches.

Fig. 5.1, provided earlier in the chapter, depicts a path model of the relationships between six variables in our example. Where available, we might also choose to incorporate additional information, such as item-level variables, into the model. Fig. 5.2 shows the same path model modified to include a latent variable representing the human user's level

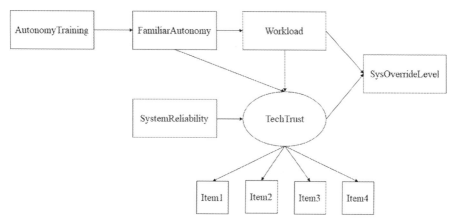

FIG. 5.2 A more detailed path model, with the user's trust in technology characterized as a latent variable.

of trust in technology. This latent variable is indicating individual items from the trust measure. Incorporating this information allows the researcher to account for measurement error which, if left unaccounted for, will attenuate the relationship between the trust variable and other variables in the model. We could perhaps complicate the model further by adding direct effects from the intervention indicator (AutonomyTraining) to the variable representing the propensity for the human user to override the system's recommendations (SysOverrideLevel) in order to test a mediation hypothesis.

Fig. 5.3 shows how the same hypotheses depicted by Fig. 5.1 might be encoded in a Bayesian network using the Netica software package [55]. Note that all the nodes in the BN are categorical. This is typical of BNs and is a contributing factor to their computational efficiency. In contrast, the SEM framework would offer the flexibility to use continuous variables. The latent variable (TechTrust) from Fig. 5.2 would almost certainly be defined by a continuous probability distribution (typically one would assume that the latent variable is normally distributed with a mean of zero and a standard deviation of one). When translating that model for the BN framework (Fig. 5.4), we would need to discretize that node; in this hypothetical example, we use four categories. The models in Figs. 5.3 and 5.4 are represented in a default state where it is assumed that the human user has not participated in the additional training (AutonomyTraining).

In the spirit of the questions posed earlier, a researcher might want to explore the impact that participation in the training intervention would have on the posterior distributions of the other variables in the network. Figs. 5.5 and 5.6 then show how our beliefs about the unobserved variables would change if a human user were to undergo the intervention. In both cases, we see that the posterior probability that the human user would choose to override the recommendations of the autonomous system at an acceptable level would increase. We can also note changes in our expectations related to the values of the other unobserved variables in the model such as the user's workload and their trust in technology.

We hope these examples demonstrate how causal modeling tools can provide an intuitive and flexible approach for specifying causal relationships among variables, evaluating their direct, indirect, and moderated effects, and examining counterfactual states, thus

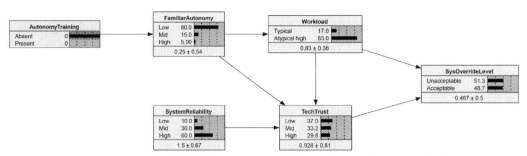

FIG. 5.3 An example Bayesian network of the conceptual model, populated with hypothetical values. This is the "default state," in which no training intervention has yet been implemented.

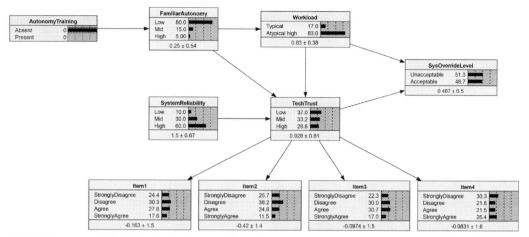

FIG. 5.4 The more detailed latent variable model from Fig. 5.2, represented as a Bayesian network. This also represents the "default state," in which no training intervention has yet been implemented.

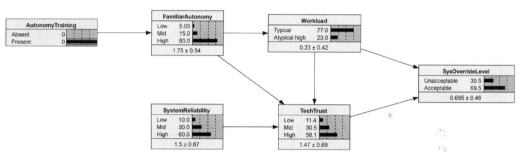

FIG. 5.5 The model from Fig. 5.3., except now a training intervention has been provided to the user. Note the downstream effects on the other factors in the model. Additionally, there has been an increase in the user's acceptable override level, from 48.7 (with no intervention) to 69.5 (with intervention).

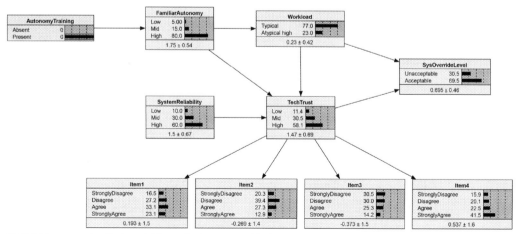

FIG. 5.6 The more detailed model from Fig. 5.4, now with the training intervention applied. Here, observe that the training intervention has also affected the values for the item responses.

producing some clarity for researchers who wish to understand how outcomes might change if predictors took on different values. We also want to emphasize how specifying the causal model can tell us about how to design studies to test the model. For instance, researchers interested in mediation effects, such as the indirect effect from familiarity to system override behavior via system trust in the model here, would benefit from employing research designs that directly test the causal mediation hypothesis (e.g., Ref. [56]). As the complexity of these causal models increases, so too does the number of ways that the model can be tested and refined, ideally leading toward a cohesive literature among trust scholars on how to best define, measure, and influence trust dynamics in human-human and human-autonomy interactions.

8 Conclusions

In this chapter, we presented a causal analysis approach to more comprehensively understanding the effects of interventions in human-autonomy teaming. By considering relationships among antecedents, mediators, moderators, and other relevant factors, such an approach is well-suited to evaluating the effects of a given team intervention on many aspects of human-autonomy interactions, rather than simply comparing outcomes between intervention conditions. Modeling the causal structure of trust interactions within human-autonomy teams will provide a clearer understanding of the team's trust dynamics. This can offer stronger justifications for designing interventions aimed at maintaining appropriate calibration in the human-autonomy team's trust. As we outlined in the previous sections, developing causal models requires that the trust-relevant constructs are defined in causal terms (i.e., what makes it distinct from other constructs, what causes it, and what does it cause), that any measurements used to inform parts of the causal model must be validated, and that causal models must specify extraneous variables, such as confounds and covariates, as well as mechanisms of action, such as moderators and mediators. Ultimately, better tools for understanding team trust interventions will enable researchers and practitioners to identify the interventions most suited for a given context, which can include changes in autonomy behavior, improving communication and transparency elements, providing after-action reviews, and so on. Avenues for ongoing and future work in this area involve the development and validation of better measures of trust that can be used in a multimethod manner for HAT [10,11,27]; this, in conjunction with the advancement of causal modeling as a means for comprehensively evaluating trust interventions, will enable significant inroads into effective maintenance of trust calibration in the complex and dynamic human-autonomy teams of the near future. Other opportunities for expanding on this research will involve implementing causal models in human-autonomy trust intervention experiments to establish their efficacy using real-world data, while utilizing multimethod trust measurement approaches in the collection of the data.

References

[1] M. Demir, A.D. Likens, N.J. Cooke, P.G. Amazeen, N.J. McNeese, Team coordination and effectiveness in human-autonomy teaming, IEEE Trans. Hum. Mach. Syst. 49 (2) (2019) 150–159, https://doi.org/10.1109/THMS.2018.2877482.

[2] L.A. DeChurch, J.R. Mesmer-Magnus, The cognitive underpinnings of effective teamwork: a meta-analysis, J. Appl. Psychol. 95 (1) (2010) 32–53.

[3] J.B. Lyons, K. Sycara, M. Lewis, A. Capioloa, Human-autonomy teaming: definitions, debates, and directions, Front. Psychol. 12 (2021) 589585, https://doi.org/10.3389/fpsyg.2021.589585.

[4] S. Nahavandi, Trusted autonomy between humans and robots: toward human-on-the-loop in robotics and autonomous systems, IEEE Trans. Syst. Man Cybern. Syst. 3 (1) (2017) 10–17, https://doi.org/10.1109/MSMC.2016.2623867.

[5] J. Lee, Park, G.,& Ahn, S., A performance evaluation of the collaborative robot system, in: 2021 21st international conference on control, automation and systems (ICCAS), 2021, pp. 1643–1648, https://doi.org/10.23919/ICCAS52745.2021.9649859.

[6] M. Johnson, A. Vera, No AI is an island: the case for teaming intelligence, AI Mag. 40 (1) (2019) 16–28, https://doi.org/10.1609/aimag.v40i1.2842.

[7] N. McNeese, M. Demir, E. Chiou, N. Cooke, G. Yanikian, Understanding the role of trust in human-autonomy teaming, in: Paper Presented at the Proceedings of the 52nd Hawaii International Conference on System Sciences, Maui, Grand Wailea, 2019.

[8] E. Phillips, S. Ososky, J. Grove, F. Jentsch, From tools to teammates: toward the development of appropriate mental models for intelligent robots, Proceedings of the Human Factors and Ergonomics Society Annual Meeting, vol. 55, SAGE Publications, Los Angeles, CA, 2011, pp. 1491–1495.

[9] E. Salas, M. Rosen, S. Burke, G. Goodwin, The wisdom of collectives in organizations: An update on team competencies, in: E. Salas, G. Goodwin, S. Burke (Eds.), Team Effectiveness in Complex Organizations, Taylor & Francis, New York, NY, 2009, pp. 39–79.

[10] A.L. Baker, S.M. Fitzhugh, L. Huang, D.E. Forster, A. Scharine, C. Neubauer, et al., Approaches for assessing communication in human-autonomy teams, Hum. Intell. Syst. Integr. 3 (2021) 99–128, https://doi.org/10.1007/s42454-021-00026-2.

[11] A. Krausman, C. Neubauer, D. Forster, S. Lakhmani, A.L. Baker, S.M. Fitzhugh, G. Gremillion, J.L. Wright, J.S. Metcalfe, K.E. Schaefer, Trust measurement in human-autonomy teams: development of a conceptual toolkit, J. Hum.-Robot Interact. (2022), https://doi.org/10.1145/3530874.

[12] R. Grossman, J. Feitosa, Team trust over time: modeling reciprocal and contextual influences in action teams, Hum. Resour. Manag. Rev. 28 (2018) 395–410.

[13] A.C. Costa, R.A. Roe, T. Taillieu, Trust within teams: the relation with performance effectiveness, Eur. J. Work Organ. Psychol. 10 (3) (2001) 225–244.

[14] B.A. de Jong, K.T. Dirks, N. Gillespie, Trust and team performance: a meta-analysis of main effects, moderators, and covariates, J. Appl. Psychol. 101 (8) (2016) 1134–1150.

[15] E. de Visser, M.M. Peeters, M. Jung, S. Kohn, T.H. Shaw, R. Pak, M.A. Neerincx, Towards a theory of longitudinal trust calibration in human-robot teams, Int. J. Soc. Robot. 12 (2019) 459–478.

[16] M. Hou, G. Ho, D. Dunwoody, IMPACTS: a trust model for human-autonomy teaming, Hum. Intell. Syst. Integr. (2021) 1–19.

[17] M. Hou, G. Ho, D. Dunwoody, IMPACTS: a trust model for human-autonomy teaming, Hum. Intell. Syst. Integr. 3 (2021), https://doi.org/10.1007/s42454-020-00023-x.

[18] K. Drnec, J. Metcalfe, Paradigm Development for Identifying and Validating Indicators of Trust in Automation in the Operational Environment of Human Automation Integration, HCI, 2016.

[19] S. Jessup, T. Schneider, G. Alarcon, T. Ryan, A. Capiola, The Measurement of the Propensity to Trust Automation, 2019, https://doi.org/10.1007/978-3-030-21565-1_32.

[20] A.L. Baker, E.K. Phillips, D. Ullman, J.R. Keebler, Toward an understanding of trust repair in human-robot interaction: current research and future directions, ACM Trans. Interact. Intell. 8 (4) (2018) 1–30.

[21] K.E. Schaefer, B.S. Perelman, G.M. Gremillion, A.R. Marathe, J.S. Metcalfe, A roadmap for developing team trust for human- autonomy teams, in: J. Lyons, C. Nam (Eds.), Trust in Human-Robot Interaction: Research and Applications, Elsevier, 2021.

[22] D. Borsboom, The attack of the psychometricians, Psychometrika 71 (3) (2006) 425–440.

[23] R.C. Mayer, J.H. Davis, F.D. Schoorman, An integrative model of organizational trust, Acad. Manag. Rev. 20 (3) (1995) 709–734.

[24] T. Mooradian, B. Renzl, K. Matzler, Who trusts? Personality, trust and knowledge sharing, Manag. Learn. 37 (4) (2006) 523–540.

[25] M. Chita-Tegmark, T. Law, N. Rabb, M. Scheutz, Can you trust your trust measure? in: Proceedings of the 2021 ACM/IEEE International Conference on Human-Robot Interaction, 2021, March, pp. 92–100.

[26] R.S. Gutzwiller, E.K. Chiou, S.D. Craig, C.M. Lewis, G.J. Lematta, C.P. Hsiung, Positive bias in the 'Trust in Automated Systems Survey'? An examination of the Jian et al.(2000) scale, in: Proceedings of the Human Factors and Ergonomics Society Annual Meeting, vol. 63, No. 1, SAGE publications, Los Angeles, CA, 2019, pp. 217–221.

[27] K.E. Schaefer, A.L. Baker, R.W. Brewer, D. Patton, J. Canady, J.S. Metcalfe, Assessing multi-agent human-autonomy teams: US Army Robotic Wingman Gunnery Operations, in: Presented at the SPIE Defense + Commercial Sensing, Baltimore, MD, 2019.

[28] B.M. Muir, Trust between humans and machines, and the design of decision aids, Int. J. Man-Mach. Stud. 27 (5–6) (1987) 527–539.

[29] J.D. Lee, K.A. See, Trust in automation: designing for appropriate reliance, Hum. Factors 46 (1) (2004) 50–80.

[30] P. Robinette, W. Li, R. Allen, A. Howard, A.R. Wagner, Overtrust of robots in emergency evacuation scenarios, IEEE Int. Conf. Hum.-Robot Interact. (2016), https://doi.org/10.1109/HRI.2016.7451740.

[31] A.R. Wagner, P. Robinette, An explanation is not an excuse: Trust calibration in an age of transparent robots, in: J. Lyons, C. Nam (Eds.), Trust in Human-Robot Interaction: Research and Applications, Elsevier, 2021.

[32] D.R. Ilgen, J.R. Hollenbeck, M. Johnson, D. Jundt, Teams in organizations: from input-process-output models to IMOI models, Annu. Rev. Polit. Sci. 56 (2005) 517–543.

[33] J.R. Hackman, The design of work teams, in: Handbook of Organizational Behavior, Prentice Hall, 1987, pp. 315–342.

[34] J.E. McGrath, Groups: Interaction and Performance, 1984. Prentice-Hall.

[35] I.D. Steiner, Group Processes and Productivity, Academic Press, New York, NY, 1972.

[36] D. Coutu, M. Beschloss, Why teams don't work, Harv. Bus. Rev. 87 (5) (2009) 98–105.

[37] K.E. Schaefer, J.Y. Chen, J.L. Szalma, P.A. Hancock, A meta-analysis of factors influencing the development of trust in automation: implications for understanding autonomy in future systems, Hum. Factors 58 (3) (2016) 377–400.

[38] R.W. Brewer, A.J. Walker, E.R. Pursel, E.J. Cerame, A.L. Baker, K.E. Schaefer, Assessment of manned-unmanned team performance: comprehensive after-action review technology development, in: Paper Presented at the International Conference on Applied Human Factors and Ergonomics, 2019. Washington, DC.

[39] J. Pearl, Causality, 2009. Cambridge.

[40] D.B. Rubin, Causal inference using potential outcomes: design, modeling, decisions, J. Am. Stat. Assoc. 100 (469) (2005) 322–331.

[41] G.A. Miller, J.P. Chapman, Misunderstanding analysis of covariance, J. Abnorm. Psychol. 110 (1) (2001) 40.

[42] F.V. Jensen, Bayesian Networks and Decision Graphs, 2, Springer, 2009.

[43] R.E. Neapolitan, Learning Bayesian Networks, vol. 38, Pearson, 2004.

[44] J. Pearl, Probabilistic Reasoning in Intelligent Systems: Networks of Plausible Inference, Morgan Kaufmann Publishers, 1988.

[45] R. Reichenberg, Dynamic Bayesian networks in educational measurement: reviewing and advancing the state of the field, Appl. Meas. Educ. 31 (4) (2018) 335–350.

[46] J. Reye, Student modelling based on belief networks, Int. J. Artif. Intell. Educ. 14 (1) (2004) 63–96.

[47] R.B. Kline, Principles and Practice of Structural Equation Modeling, Guilford, 2015.

[48] J. Pearl, The causal foundations of structural equation modeling, in: R.H. Hoyle (Ed.), Handbook of Structural Equation Modeling, Guilford Press, 2012, pp. 68–91.

[49] K.A. Bollen, J. Pearl, Eight myths about causality and structural equation models, in: Handbook of Causal Analysis for Social Research, Springer, 2013, pp. 301–328.

[50] J.M. Graham, The general linear model as structural equation modeling, J. Educ. Behav. Stat. 33 (4) (2008) 485–506.

[51] K.J. Grimm, N. Ram, R. Estabrook, Growth Modeling: Structural Equation and Multilevel Modeling Approaches, Guilford Publications, 2016.

[52] T.D. Little, Longitudinal Structural Equation Modeling, Guilford Press, 2013.

[53] K.A. Bollen, Structural Equations with Latent Variables, vol. 210, John Wiley & Sons, 1989.

[54] G.R. Hancock, R.O. Mueller (Eds.), Structural Equation Modeling: A Second Course, IAP, 2013.

[55] Norsys Software Corp. (1992-2021). Netica 6.09. http://www.norsys.com.

[56] A.G. Pirlott, D.P. MacKinnon, Design approaches to experimental mediation, J. Exp. Soc. Psychol. 66 (2016) 29–38. https://doi.org/10.1016/j.jesp.2015.09.012.

6

Risk management in human-in-the-loop AI-assisted attention aware systems

Max Nicosia and Per Ola Kristensson

DEPARTMENT OF ENGINEERING, UNIVERSITY OF CAMBRIDGE, CAMBRIDGE, CAMBRIDGESHIRE, UNITED KINGDOM

1 Introduction

AI-assisted systems are becoming more pervasive across a variety of fields. The level of assistance ranges from partial assistance to fully autonomous AI control. In the former, the system assists the operator, so they perform at their highest level under a variety of conditions. In the latter, no input from the operator is necessary. Examples of partial assistance include airport baggage threat detection, which uses machine learning classifiers, and attention aware systems, which employ sensors and AI to monitor operator behavior and assist them where possible. Fully automated AI systems can be found in several sectors, such as aircraft, transport, and medicine.

Any complex system, whether it uses AI or not, is susceptible to risks. Risks need to be assessed and managed. As a consequence, many risk assessment and risk management strategies have been developed over the years. These strategies have been developed to identify, capture, and manage risks in various ways. The reason for a plurality of methods is that no single strategy is capable of capturing and managing all risks in all situations.

For instance, failure mode and effects analysis (FMEA) is one of the first developed risk assessment methods. A failure mode is the manner in which something fails to fulfill its function. The method was originally described in US Armed Forces Military Procedures document MIL-P-1629 [1] in 1949. By the 1960s, NASA [2] was using variants of FMEA in their programs (e.g., Apollo [3], Viking, Voyager, Magellan, Galileo [4], and Skylab [5]) and by the 1970s, it had spread to petroleum exploration [6], the automotive industry [7], wastewater treatment plants [8], and the food industry [9].

Another widely used risk assessment method is fault tree analysis (FTA). FTA was developed in 1962 by Bell Labs to evaluate ballistic launch control systems [10]. Boeing incorporated FTA in their civilian aircraft design process in 1966 [11]. In 1970, the US Federal Aviation Administration (FAA) incorporated FTA into the CFR §25.1309 airworthiness regulations for aircraft. The FAA later extended its use to other areas within the US National

Airspace Systems [12]. NASA considered using FTA in their Challenger program but decided against it due to the calculations resulting in unacceptably low reliability values. Instead, they favored the continuation of qualitative risk assessment methods, including the previously mentioned FMEA method. This decision proved to be a major oversight after the Challenger accident. As a consequence, NASA reconsidered the importance of using quantitative risk assessment methods and resumed their use, including the use of FTA [13].

Another popular method is event tree analysis (ETA). ETA was developed as an alternative to FTA. Since ETA makes an assumption on systems units either working or failing, it makes the analysis more manageable [14]. As ETA is used to analyze specific events, it allows the identification of all sequences of events and failures that can occur as a result of an originating event and its subsequent events.

FMEA, FTA, and ETA have evolved from the need to address different challenges presented by various types of systems with varying inherent complexities. History has demonstrated that failing to assess and manage risks can lead to catastrophic failures, such as the Challenger accident. However, risk management techniques can be difficult to transfer across domains as each domain has different requirements. This is particularly difficult for systems involving AI, as they have a level of automation that may not be possible to directly control. To tackle this problem, Falco et al. [15] proposed the creation of a regulatory body that would standardize risk management techniques and enforce compliance as a way to improve responsibility and control of AI-assisted systems.

Attention aware systems are a relatively new class of AI-assisted systems, and their risks are currently not well understood. In this work, we want to draw attention to how risk management strategies can be applied to attention aware systems. Our motivation is to avoid history repeating itself.

We first explain what an attention aware system is and what its capabilities are. We then discuss the importance of managing and mitigating risks for such systems. We later discuss various considerations that should be taken into account and give recommendations on how to develop effective risk management strategies. This work is a first step to understanding how to manage risks in human-in-the-loop AI-assisted attention aware systems.

2 Attention aware systems

The main purpose of an attention aware system is to detect operators' focus of attention and manage it in such a way that the system can ensure optimal task operation under a variety of conditions, such as when operators are experiencing fatigue and/or extraordinarily high workloads.

Fig. 6.1 shows an example of a functional architecture of an attention aware system. The dashed rounded rectangle represents the system boundary. Arrows entering the boundary are input signals from other systems and arrows leaving the boundary are output signals from the system. In this particular functional architecture, the functionality of

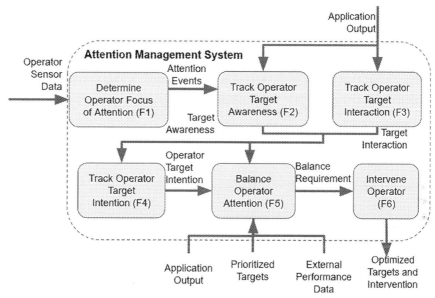

FIG. 6.1 Functional structure of an attention aware system.

the system has been decomposed into six functional units represented by gray rounded rectangles.

The function `Determine Operator Focus of Attention` parses operator sensor data and generates attention events. The function `Track Operator Target Awareness` combines suitable application output and attention events to detect target awareness. The function `Track Operator Target Interaction` collates target interactions from the output provided by the application. The function `Track Operator Target Intention` collates and analyzes target awareness and target interaction information to estimate potential future operator actions. The function `Balance Operator Attention` combines target awareness, target interaction, and target intention information with the application output, target priorities, and task performance data to build and evaluate the current application and operator state in order to output a balance requirement signal. Finally, the `Intervene Operator` function processes a balance requirement signal by turning it into a sequence of instructions containing the targets and interventions that the application needs to deploy. For a more detailed explanation of these functions, see prior work [16].

From the preceding description it is clear that, depending on the particular implementation and application domain, the complexity of each of the previous functions can vary greatly. This, in turn, will affect the difficulty of analyzing and managing the risks of that particular deployment of the attention management system.

The most complex function in the system is `Balance Operator Attention`. It works by processing signals containing operator and application states to construct an overall operator-application state. It then evaluates this state to detect drops in task performance or any other operator aspect that may require additional attention and infers the source causing a drop in performance. There are several failure modes that can arise in such a function. Two of the most obvious ones are arriving at an incorrect state because signals from previous functions failed in some way, and/or incorrect aggregation and/or assessment of the overall state due to parameterization errors.

Another complex function is `Intervene Operator`. This function uses the information in the balance requirement signal to address the sources of poor performance identified by the previous component. To achieve this, it devises an intervention strategy that assists the operator in returning to an acceptable level of operating performance. An intervention strategy consists of coordinated instructions where, for example, the saliency of certain information presented on a display is manipulated in such a way that the operator's focus of attention is directed either toward it or away from it. However, the mechanisms used to influence the operator introduce their own failure modes. For example, saliency changes can appear confusing to the operator or the behavior prompted by the system can be misunderstood by the operator.

Another important aspect consists of the objectives behind interventions. As previously stated, the primary purpose of the system is to manage the operator's focus of attention such that the operator can maintain an appropriate task performance level. However, the system can leverage the operator's focus of attention in a variety of ways depending on the behavior the system wishes to induce in the operator. For example, the system can intervene with the operator to ensure that they balance their actions between all tasks. As a result of stress, fatigue, or misunderstandings, the operator can neglect some of their duties in favor of others. This can lead to failures in monitoring important information or periodically carrying out less frequent tasks.

Another objective of interventions can be to ensure that operators use their time or actions more effectively. For example, the system can help the operator focus on higher-priority targets or tasks. However, for this to be possible, the system needs to understand the tasks to potentially identify more optimal ways of performing them. In other cases, poor task performance may be caused by the operator misunderstanding or misinterpreting information. This can happen due to, for example, the state of the operator, such as being under stress or experiencing fatigue, or due to cognitive overload. In such situations, the system needs to either drive the operator's focus of attention toward relevant information or lower the amount of disruption caused by the information being presented to the operator. Depending on the circumstances, choosing one mechanism over another can lead to additional failure modes.

Finally, the system can manage attention to preempt potential errors by precluding the operator from performing certain actions by estimating the intention of the operator. However, such detection mechanisms need to be reliable within the application domain or they can lead to further unexpected operator behavior and potential failure modes.

An attention management system can be seen as a type of human-machine teaming in which the AI tries to affect operator behavior in order to influence their task performance.

Moreover, as the system is continuously monitoring the operator and application state, it can use this state information to adapt its level of influence accordingly. This adaptation capability means that the system can vary its level of automation during operation. Within the 10-point system automation model of Parasuraman et al. [17], an attention management system alternates between levels 1, 2, and 3 during operation, where level 1 provides no assistance, level 2 makes the system provide all possible alternative actions/decisions to the operator, and level 3 reduces the number of decisions for the operator. However, while level 3 automation reduces the number of decisions for the operator by changing the saliency of certain targets, it still permits the operator to remain in full control as all decision options continue to be visible and accessible to the operator at all times. Ensuring that the system manages the switch between the automation levels is a strength of the system's capabilities, but also a source of risk.

3 Risk management of attention aware systems

Risk management is carried out to ensure risks are understood and under control in order to prevent risks from materializing into incidents. We define risk as the likelihood of undesirable behavior in the joint human-machine system multiplied by the impact of such undesirable behavior happening.

Hubbard [18] defines risk management as: "The identification, assessment, and prioritization of risks followed by coordinated and economical application of resources to minimize, monitor, and control the probability and/or impact of unfortunate events." For reference, risk analysis is the systematic identification of potential sources of harm—hazards—and their risks. Risk management is the entire process of identifying hazards and subsequently analyzing and controlling risks. In other words, risk management encompasses hazard identification, risk estimation, risk evaluation, risk control, and risk monitoring.

Before a risk management strategy can be developed, it is first necessary to perform a risk analysis. The first step in risk analysis is to define the system boundary around the joint human-machine system. We will only consider risks within the boundary. In practice, deciding on a system boundary requires careful reflection, as it is vital any that relevant factor to eventual risk is captured within the system boundary. Having set the system boundary, the system within the boundary can be mapped out so that it is fully understood. This can, for example, involve decomposing the system's overall function into function structures and constructing diagrams indicating the flow of signals and other information.

Having set the system boundary and mapped out the system, it is possible to identify hazards and estimate risks. However, this is not always straightforward, as we discuss later on. Another challenge is quantifying risks in terms of their impact and their frequency of occurrence. Assessing their impact can be difficult, as it may not always be known what other undesirable behavior may manifest as a result of a single risk materializing into an undesirable event. Assessing individual risks is important, as it is critical to be able to understand the level of risk in subsystems and the system as a whole in order to determine an overall acceptable level of risk.

Having identified all hazards and estimated the risks, these risks now need to be managed—monitored and controlled—throughout the operating life of the joint human-machine system. Risk management methods provide established processes for achieving this and they may be qualitative or quantitative. When developing a risk management strategy for an AI-assisted joint human-machine system, both types of methodologies may be required. In general, systematic approaches are required, as risk assessment depends on human judgment and humans have a tendency to underestimate or overlook risks.

Attention aware systems are AI-assisted joint human-machine systems that involve multiple stakeholders. They depend on multiple external systems and require extensive parameterization to function correctly. Since each specific deployment has its own associated specific domain risks, the risk of experiencing undesirable behavior is very high.

In the event of undesirable behavior, the repercussions can affect multiple stakeholders and it may be difficult to attribute responsibility. Let us consider the following simple scenario: the system reduces the saliency of a piece of information that needs to be considered by the operator during a specific task and, as a result of this, the operator fails to perceive the information and does not complete the task in a timely manner. In such a situation, it is not obvious where to attribute responsibility or determine the root cause. The information was always visible to the operator, and while the system increased the saliency of alternative information, or reduced the relevant information's saliency, these actions ultimately amounted to the system incorrectly managing the operator's focus of attention. Nevertheless, throughout this process, the operator still remained in full control, having access to all relevant information.

To ameliorate such scenarios, we will draw attention to some strategies that can be considered when managing the risks of such systems and highlight potential specific risks that may arise in deployment.

4 Risk management considerations

Our purpose is to draw attention to some considerations when developing a risk management strategy for attention management systems. As such, we focus mainly on the potential sources of risks, the stakeholders involved, and the importance of evaluating the effectiveness of the risk management strategy.

As explained in Section 2, an AI-assisted attention aware system is in between a fully manual system (no automation) and a fully autonomous system (full automation). A key difference compared to a fully autonomous system is that operators remain in full control of operation (i.e., there is no action automation). An AI-assisted attention aware system manipulates the presentation of information the operator receives without direct operator control.

Successful operation of an AI-assisted attention aware system will depend primarily on three factors: (1) domain context including task complexity; (2) the joint human-machine system's parameterization; and (3) factors relating to the individual operator, or team of operators, such as skill level, experience, and behavior.

A complication that is unique to such a human-in-the-loop system is that the AI-assisted functionality can possibly generate actions that are unthinkable or otherwise confusing to human operators, such as hiding relevant information from operators because the AI-assistive function disagrees with the operator on its relevance. We briefly mentioned this when we gave an example of a failure scenario in Section 3.

While the risk of any AI-assisted attention aware system is coupled to a particular system configuration and domain, we have identified four general sources of risk that are likely to require substantial attention when risk managing such systems. These are (1) system failures, (2) incorrect parameterization, (3) operator characteristics, and (4) domain context characteristics.

Some risks can arise individually or out of interactions between the stakeholders involved. The stakeholders of such a system include the operators, the people responsible for parameterizing the system, any entity that is directly or indirectly controlled by the operator through the system, the organization providing the system, and the organization managing and employing the operator.

System failures are all failures that involve the system not operating as intended despite being correctly parameterized. This can involve both software failures, such as program bugs, and hardware failures, such as loose cables or sensor errors. As described in Section 2, the system depends on four incoming signals: (1) the external performance data, (2) the output of the application, (3) the prioritized targets, and (4) the operator sensor data. Each of these signals can fail or operate at different reliability levels. Any risk management strategy needs to consider the levels of possible operation and their impact on overall system operability.

Parameterization errors can lead to undesirable behavior as a result of the system being incorrectly configured. For example, depending on the metrics the system is monitoring and the thresholds it uses as the basis for intervention deployment decisions, the system may react in completely different ways under the same circumstances. Similarly, the mechanisms set up for manipulating target saliencies will need to match the context and requirements of the application domain. For example, simple color mismatches can result in catastrophic failure situations with operators misunderstanding the importance of the information they are presented with.

Operator characteristics are another general source of risk. The operator can be out-of-phase with the system, which can result in undesirable behavior as a result of the operator misunderstanding the system's interventions or the system misunderstanding the operator's actions. The former situation can happen if the operator is unaware of the specifics of an intervention, despite the system being correctly parameterized. The latter situation can arise due to a variety of reasons, including parameterization errors, system failures, or unexpected or unknown operator behavior.

Domain context characteristics are another source of risk, as such characteristics can cause the system to react in an unexpected manner to unaccounted inputs and thus result in the system generating undefined behavior. Examples of risks associated with domain context characteristics include targets changing their visual appearance (e.g., if a new type of mine is encountered by a submarine demining vehicle), unexpected task complexities, such as a sudden rapid explosion of targets on a display, or a sudden unexpected situation, such as an unexpected incident situation in an air-traffic-control system.

Further, interactions between stakeholders can cause many undesirable outcomes. These may range from miscommunication to an inability to maintain proper records about what parameterizations are required for deployment. In Section 5, we discuss some diagrammatic methods of representing various aspects of information circulation and communication to help identify any associated risks.

Finally, once risk assessments have been carried out and mitigation strategies have been developed, it is important to monitor and evaluate the efficacy of risk management. Since the behavior of AI-assisted attention aware systems depends on both specific deployment situations and operators, new risks may arise from such system dynamics. It is also possible that emergent behavior caused by interventions may give rise to new unanticipated risks. As such, it is necessary to continuously monitor and evaluate risks throughout the lifetime of the system.

5 Risk management approaches

Risk management consists of carrying out the following tasks: (1) identifying the system boundary, (2) mapping the system, (3) identifying hazards, (4) assessing risks, and (5) devising suitable strategies to manage/mitigate risks.

However, not all of these tasks can always be carried out in this order. For example, sometimes the system boundary is not clear until all relevant components and subsystems have been identified. Therefore, it is necessary to first understand the system and all its components and subsystems. This process is known as system mapping.

Diagrams are very helpful for system mapping. Such diagrams can, for example, capture the people involved and their functions, the information flowing between people, and the system or the organization that the system is embedded within. Organizational diagrams are useful for determining the people involved, their roles, and their relationships with other people. Organizational diagrams can also help with understanding the scope and determining the boundaries of all involved entities. Information diagrams can be used to capture the relationships between documents in the system. Communication diagrams reveal how information flows between stakeholders and other entities in the system. The latter can be of particular importance if system parameterization changes need to be requested or acted upon. These diagrams also provide context for the system operator's role in the entire communication structure of the organization.

Once the system has been mapped out it is possible to assess risks. There are many well-established methods for risk assessment. In this work, we focus on three methods, which were briefly introduced earlier.

FMEA is a method for identifying potential failure modes in products or processes and devising corrective measures to address the resulting associated risks. A failure mode is the manner in which a system, mechanism, or component fails to fulfill its function. Once a failure mode has been identified, its effect, severity, and cause can be determined. FMEA involves identifying the issues related to failure modes, ranking them by their importance, and devising corrective measures for issues with serious concerns. In addition to identifying failure modes, FMEA helps with establishing failure rates and root causes of known failures. The advantage of using FMEA is that it is good at systematically cataloging all possible sources of failure. As such, an FMEA is ideal for collecting information and exchanging it between teams. This helps with the early identification of potential sources of failure. FMEA is not designed to demonstrate how robust a system is to multiple failures, or failures due to external events. Another weakness FMEA shares with many other risk assessment methods is that an FMEA can become too large to be effectively maintained and understood.

FTA is used to trace a failure path to identify all the events that lead to said failure. Fault trees are normally represented graphically, with each event connected through logic gates. Tracing events also allows for identifying the components involved in the failure. Once the components and the events that lead to the failure have been identified, it is possible to develop a strategy to prevent it from reoccurring. FTA is a top-down approach that allows the mapping of the dependencies of each event and the calculation of the probabilities of specific failures, provided that the probabilities of the events involved are known. An advantage of FTA is that it allows analyzing the effects of initiating faults, which is the opposite of FMEA. Additionally, FTA allows analyzing multiple complex failures taking place at the same time. An FTA will consider external events, which is useful for assessing how robust a system is against single and multiple failures. A disadvantage, however, is that FTA is not suitable for finding all possible initiating faults.

ETA is used to determine the probability of a specific event occurring based on the probability of the chronological sequence of events leading up to it. ETA is useful for determining the effect that a particular failure can have on the overall system. The approach is inductive and follows a bottom-up approach. An advantage of ETA is that it enables the assessment of multiple simultaneous functions in both failure and success states. This is useful, as events do not need to be anticipated as they are only the starting point. An ETA is also useful for identifying single sources of failures and system paths that can lead to failure. Additionally, an ETA is suitable for modeling complex systems, as it can visualize cause-and-effect relationships. Finally, an ETA allows tracing faults across boundaries of subsystems. A disadvantage of ETA is that it always starts with one initiating event at a time, which must be identified before commencing the analysis. Another disadvantage is that partial successes and failures are not accounted for.

As with any new technology, AI-assisted attention aware systems present new challenges in terms of configuration and deployment. Organizations deploying such systems need to make sure that they can manage evolving requirements and potential new sources of undesirable outcomes. Most importantly, they need to ensure that appropriate mechanisms are in place to mitigate possible unexpected undesirable outcomes.

To ensure that risk management strategies are effective, they need to be continuously evaluated by systematic monitoring and evaluation of risk at all levels in the system. Exactly how this is to be carried out is dependent on the precise application domain. Nevertheless, ensuring reliable accounting of all registered failures and undesirable events, as well as continuous monitoring of risk rates in the system, is critical to fully understanding the level of risk in the system and the efficacy of any active risk management strategies. Additionally, monitoring the effects of mitigation procedures can also assist system evaluation. Finally, ensuring that the cost and effort of each strategy are yielding a comparable risk reduction benefit is of utmost importance for ensuring organizations optimize their use of available resources.

6 Discussion and conclusions

Applying risk management strategies to AI-assisted attention aware systems provides many benefits. First, it allows early detection of potential human-machine system failures and the creation of appropriate mitigation and monitoring strategies to eliminate, reduce, and track the effects of such failures. Mitigation strategies can range from improving the design of a system to increasing redundancy, which simply incorporates a verification mechanism that prevents miscommunications or errors.

A second benefit is that risk management strategies can provide assurance that some minimum level of safety has been considered. This is of particular benefit for organizations as they want to ensure that their products are reliable and safe and their reputations maintained. Moreover, in some domains, implementing appropriate risk management strategies is part of the obligations necessary for regulatory compliance or adherence to ethical standards.

As explained in this chapter, AI-assisted attention aware systems are human-in-the-loop systems with multiple sources of failure. In addition, they are often meant to be used in safety-critical operations. They present challenges similar to those presented by fully autonomous AI systems, as well as additional risks caused by the reliance on human operators and the resultant human-AI interaction. As with any AI system designed for decision-making, the adaptive nature of the system can make it difficult to identify the root cause of a fault or undesired outcome. For example, the operator can fail to perceive a target due to a system intervening in a way that ends up obfuscating the target. While such a target is always visible to the operator, the system intervention actively diverts the operator's attention away from the target.

We also highlight that identifying risk in AI-assisted attention aware systems can be challenging. This is because it is not always evident what the source of a particular failure or undesirable event is, or how frequently it occurs. We advocate a systematic approach and, as a first step, suggest three widely used risk assessment strategies: FMEA, FTA, and ETA. These methodologies provide both quantitative and qualitative analysis mechanisms. FMEA is suitable for devising corrective measures for all possible faults. ETA is useful for identifying what other failures can occur as a result of specific undesirable events or failures. FTA focuses on identifying the factors that lead to certain faults.

To help the discovery of risks in AI-assisted attention aware systems, Section 4 presents four general sources of risks. These are (1) system failures, (2) incorrect parameterization, (3) operator characteristics, and (4) domain context characteristics. The first source involves failures in software or hardware. The second source encompasses configuration errors that fundamentally affect overall system behavior. The third source accounts for the operator failing to understand the system's interventions or the system not understanding the operator's behavior. The fourth source relates to the joint human-machine system failing to manage unexpected changes or other aspects of the application domain itself.

We predict AI-assisted attention aware systems will be increasingly critical as human-in-the-loop systems become increasingly sophisticated and prominent in application domains ranging from security and air-traffic control to manufacturing and healthcare. However, such systems also introduce additional complexities, which give rise to additional risk. This chapter is a first step toward tailored risk management approaches for such systems.

References

[1] USDD, MIL-P-1629-Procedures for Performing a Failure Mode Effect and Critical analysis, 1949.

[2] R.A. Neal, Modes of failure analysis summary for the Nerva B-2 reactor, Astronuclear Laboratory Westinghouse Electric Corporation, 1962. Technical Report WANL-TNR-042.

[3] Office of Manned Space Flight, Apollo Program, Apollo Reliability and Quality Assurance Office, Procedure for Failure Mode, Effects and Criticality Analysis (FMECA), NASA, Washington, DC, 1966.

[4] NASA, Failure Modes, Effects and Criticality Analysis (FMECA), Practice no. PD-AP-1307 (1999).

[5] NASA, Experimenters' Reference Based Upon Skylab Experiment Management, Technical Memorandum (TM) NASA-TM-X-72397 (1974).

[6] M.K. Dyer, D.G. Little, E.G. Hoard, A.C. Taylor, R. Campbell, Applicability of NASA Contract Quality Management and Failure Mode Effect Analysis Procedures to the USGS Outer Continental Shelf Oil and Gas Lease Management Program, National Aeronautics and Space Administration, 1972.

[7] K. Matsumoto, T. Matsumoto, Y. Goto, Reliability analysis of catalytic converter as an automotive emission control system, SAE Trans. 84 (1975) 728–738.

[8] C.W. Mallory, Application of Selected Industrial Engineering Techniques to Wastewater Treatment Plants, vol. 1, US Government Printing Office, 1973.

[9] W.H. Sperber, R.F. Stier, Happy 50th birthday to HACCP: retrospective and prospective, Food Saf. Mag. (2009) 42–46.

[10] C.A. Ericson, C. Ll, Fault tree analysis, System Safety Conference, Orlando, Florida, vol. 1, 1999, pp. 1–9.

[11] A.F. Hixenbaugh, Fault tree for safety, Boeing Co Seattle WA Support Systems Engineering, 1968. Technical Report.

[12] Federal Aviation Authority, System Safety Handbook, 2000.

[13] M. Stamatelatos, W. Vesely, J. Dugan, J. Fragola, J. Minarick, J. Railsback, Fault Tree Handbook With Aerospace Applications, National Aeronautics and Space Administration, Washington, DC, 2002.

[14] P.L. Clemens, R.J. Simmons, System Safety and Risk Management: NIOSH Instructional Module, US Department of Health and Human Services, 1998.

[15] G. Falco, B. Shneiderman, J. Badger, R. Carrier, A. Dahbura, D. Danks, M. Eling, A. Goodloe, J. Gupta, C. Hart, et al., Governing AI safety through independent audits, Nat. Mach. Intell. 3 (7) (2021) 566–571.

[16] M. Nicosia, P.O. Kristensson, Design principles for AI-assisted attention aware systems in human-in-the-loop safety critical applications, in: Engineering Artificially Intelligent Systems, Springer, 2021, pp. 230–246.

[17] R. Parasuraman, T.B. Sheridan, C.D. Wickens, A model for types and levels of human interaction with automation, IEEE Trans. Syst. Man Cybern. Part A Syst. Hum. 30 (3) (2000) 286–297.

[18] D.W. Hubbard, Healthy skepticism for risk management, in: The Failure of Risk Management, Chapter 1, John Wiley & Sons Ltd, 2020, pp. 1–19, https://doi.org/10.1002/9781119521914.ch1.

7

Enabling trustworthiness in human-swarm systems through a digital twin

Mohammad D. Soorati[a], Mohammad Naiseh[b], William Hunt[a],
Katie Parnell[a], Jediah Clark[a], and Sarvapali D. Ramchurn[a]

[a]SCHOOL OF ELECTRONICS AND COMPUTER SCIENCE, UNIVERSITY OF SOUTHAMPTON,
SOUTHAMPTON, UNITED KINGDOM [b]DEPARTMENT OF COMPUTING AND INFORMATICS,
BOURNEMOUTH UNIVERSITY, POOLE, UNITED KINGDOM

1 Introduction

Manufacturing uncrewed aerial vehicles (UAVs) is becoming more efficient and afford-able, to the extent that even the toy models are proven to have many applications [1–3]. Functionalities as simple as video streaming [4] and as complex as package delivery [5, 6] and uncrewed surveillance [7, 8] show the promising future applications for this technology. Nonetheless, autonomous UAVs are not capable of performing complex tasks that require a strategic decision at critical times. The challenges in using autonomous systems (ASs) are known to us and the same holds for using UAVs. Designing safe and self-aware ASs is very complex, but humans excel at situation management and decision-making in comparison to machines. It is therefore logical to combine the strength of UAV fleets with human operators to utilize the robustness, adaptivity, and scal-ability of UAV swarms as well as the critical decision-making of human operators [9].

Disaster management is an area of application in which UAVs can be quite helpful in providing additional information and situation awareness that assists the human opera-tors in managing the situation [10, 11]. However, humans under stress and uncertainty tend not to trust autonomous agents [12] and the trust may decrease as the task becomes more difficult [13]. As a result, the AS may not be effectively used [14]. While many studies focus on engineering swarms [15–17], there is a lack of human factors studies that inform us about the issues that affect real-world applications. This chapter presents a multidis-ciplinary approach to designing trustworthy human-swarm systems that takes into account the perspectives of experts in swarm robotics, human factors, and software engi-neering, from academia and industry.

We plan to first understand the key issues in establishing trust between human oper-ators and a swarm of UAVs. We start by interviewing experienced UAV operators to gain a better understanding of the barriers to trust and how they can be overcome. We also learn more about the techniques to repair damaged trust in the operation of UAVs. We lay out a list of requirements and use this list as a measuring unit for the trustworthiness of any human-swarm systems.

In a large group of semi-ASs with a high degree of stochasticity, the lack of understanding of the behavior of the swarm is inevitable. It is extremely complex to explain the behavior of a swarm and predict the future states of the system in a given situation. Although explainability has been mentioned in human-swarm systems as a key requirement for trust [18, 19], the literature lacks a shared knowledge and understanding of what questions an explainable swarm is expected to answer or what "explanations" should be generated. Therefore, we follow the approach proposed in recent studies in eXplainable AI (XAI) [20, 21] to outline an early taxonomy of explainability for swarm systems. We use an online questionnaire supported by three elicitation methods to extract a taxonomy for explainable human-swarm systems to serve as the guideline for designing an explainable human-swarm system.

Despite a few existing applications for swarm robotics [22, 23], it is still a challenge to find a viable use case in the industry for human-swarm systems. We organized a workshop with several main users of large ASs (e.g., Thales Group [www.thalesgroup.com], BAE Systems [www.baesystems.com], etc.) to cocreate a real-world use case for a human-swarm system that can be beneficial and applied in the near future. The result of this workshop is presented as a use case with different roles for human and ASs. The cocreated use case is implemented to study different trust aspects of such applications.

Finally, we propose a Java-based online digital twin platform for human-swarm studies designed for multi-UAV operations. Multiple human operators can share a common scenario, operate different tasks, observe the situation, and control a swarm of UAVs in simulation and real time. In order to measure the compliance of our simulator platform with the listed trust and explainability requirements, we use the ISO/IEC 9126 software quality standard [24] to map the requirements into quality characteristics and then evaluate the compliance of our software with these characteristics.

Fig. 7.1 demonstrates the entire process of validating the digital twin for trustworthy human-swarm interaction. The process started by creating a checklist for trust requirements with UAV operation experts (Section 3) and using academic experts' knowledge to understand the key elements for designing explainable swarms (Section 4). We cocreated a use case with industry human-computer interaction experts that is implemented in our simulation platform (Section 5). We transform the collated requirement list for trust and explainability into software quality characteristics that can then be used to evaluate any complying software. Finally, we evaluate the compliance of our digital twin platform against the trust quality characteristics and report the results (Section 6). Before we discuss the design process in detail, the next section defines some of the key terminologies that are used throughout this chapter.

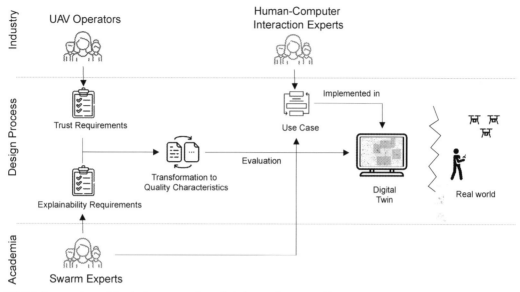

FIG. 7.1 An overview of the design process for a digital twin that is capable of simulating trustworthy human-swarm systems.

2 Trustworthy human-swarm interaction

Trust may refer to various aspects of any entity or process. Therefore, it is essential to define trust before we begin to label any system as trustworthy in our work. In this section, we define trust and what it means to have trust in ASs and multiagent systems specifically. We then move on to our method of obtaining trust requirements.

2.1 What is trust?

There is a general lack of consensus on the meaning and definition of trust, which makes measuring and comparing different studies on trust difficult. Many have tried to identify the components and multiple facets that evoke trust [25, 26]. One difficulty is the multiple applications within which trust is studied and in which it plays a significant role in how a system performs. Research has studied the trusting relationships that we have toward other people [27], and the trust we have in our governments [28] or media channels [29]. Some argue that trust is a belief that is determined by the individual and their characteristics [30]. Others have highlighted the contextual factors that influence how we trust in different situations [31]. With developments in technology and our increased reliance on systems that automate typically human-performed tasks, trust in automation and automated systems is becoming a central concern [32–36].

2.2 Trust in autonomous systems

Trust in automation is a complex issue and the definition of trust in relation to ASs is still unclear. Trust is related to the use, misuse, and disuse of automation [37]. It is highly relevant to the integration, uptake, and success of automated technologies. Sheridan defines trust in automation as "a human's propensity to submit to vulnerability and unpredictability, and nevertheless, to use that automation" (p. 2) [38]. They state that trust can be measured by the propensity to use automation and the open discussions that are had surrounding automated technologies. Sheridan also highlights the reciprocal trust that advanced automation will need to have in individual users, which may be built up through numerous interactions that are stored in computer databases [38]. Others have presented frameworks of trust in automation. Lee et al. [32] provide a qualitative framework for reviewing trust from the operators' perspective. They relate trust to the human effect.

Conversely, another study performed a quantitative analysis to map human-robot interactions to trust and found that trust relates more to robot characteristics, functionality, and performance, rather than factors relating to the individual [39]. They, therefore, recommend that trust can be improved by establishing design and training guidelines for robot development.

Sheridan [38, 40] highlights the distinction between trust and trustworthiness. Trust relates to the subjective human judgment of trust, while trustworthiness is an objective measure of automation. When seeking to study trust in automation, many have sought to review the attributes of trust [38,41–43]. Similarities and overlaps have been observed in these attributes, such as the need for predictability, dependability, competence, responsibility, and understandability (we use the term explainability that refers to the same concept) [38]. Another key attribute that has been linked to the human judgment of trust is the ability of ASs to explain their behavior. Human-AS trust can also be classified into cognition-based and affect-based trust [44]. Cognition-based trust is based on humans' intellectual perceptions and understanding of AI reasoning, whereas affect-based trust is based on humans' emotional responses to ASs. In fact, several studies suggest that explaining the decisions made by ASs can improve both cognition-based and affect-based trust, which would improve the overall performance of the human-AI partnership (Section 4). Many reasons could have contributed to this effect. One is that people are associated with an experiential and rational process and trust cognitively by being exposed to explanations and constituting evidence of trustworthiness. On the other hand, affective trust has a more relational orientation and is closely associated with long-term interaction. If the explanation design, such as its framing or communication method, inspires people's emotional desire to use ASs, then the explanation might impact people's affective trust.

2.3 Trust in multirobot systems

Within multiple-robot systems, it is thought that trust is applied broadly across a system, rather than differentially to different components of that system. This means that an issue with one component can restrict operator reliance and trust across the whole system, termed as the "pull-down effect" [45]. There is evidence for this effect in a realistic multiple UAV operations study then highlighted the complex interdependencies within multiple agent ASs that influence trust [46].

3 Industry-led trust requirements

When designing ASs it is important to understand what is required of a system in order for it to be trustworthy. Yet, as discussed in the previous section, the complexity of trust and its multiple facets mean that developing trust requirements is not straightforward. The user can often be overlooked within the early stages of a design process and their needs and desires may not be fully realized. This can lead to ineffective, unsafe, and unusable systems. It is imperative that the beginning of a design process captures the user population's knowledge, expectations, and the range of tasks that they are involved in. Obtaining this information informs the requirements of a system. Early integration of the users' requirements will lead to improved user experience as well as limit the possibility of incidents and accidents. We present a method that we have developed for generating and understanding trust requirements from a user's perspective.

3.1 Method

We detail a methodological approach that is accompanied by a theoretical model for capturing human-automation interactions that are influenced by, and also influence, trust and trustworthiness. This approach comes from the human factors discipline, which focuses on the interaction between humans and technologies and accounts for the wider systems and operational contexts within which the interactions occur.

3.1.1 Perceptual cycle model

The perceptual cycle model (PCM) [47] accounts for the interaction between the environment (world), human cognition (schema), and system function (action). The model captures these three interacting elements and applies a complex systems perspective to assess system performance in relation to the environment in which it is conducted. A schema is a knowledge cluster that develops from experience and provides a structure to process new events or environments that are similar or have commonalities with previously experienced environments. They provide mental templates to inform our future behaviors. These mental templates are adaptable, they are updated and altered with new experiences, and they modify experiences as new information is received. Within the PCM, the interaction between an individual's schema, the information in the world, and the action that they carry out dynamically interact in a cyclic way (see Fig. 7.2).

Thus the model can account for how the world constrains behavior, as well as how the way in which we think about the world constrains our view of it [48].

The utility of applying this approach to study human-machine interaction is the holistic way in which both the cognitive processing of the individual and the environmental

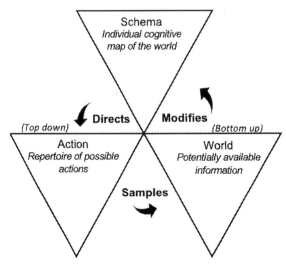

FIG. 7.2 Representation of the PCM. *Modified from K.L. Plant, N.A. Stanton, Why did the pilots shut down the wrong engine? Explaining errors in context using Schema Theory and the Perceptual Cycle Model, Saf. Sci. 50 (2) (2012) 300–315.*

factors that influence behavior can be captured, as well as how each of these can adapt to each other in response to different events [48]. The PCM has been applied across numerous safety-critical domains to understand human-system interactions including aviation, automotive, and automation.

The model requires an understanding of human behavior and the environment within which it occurs in order to map out the key schema, action, and world elements of importance. Often the model is applied to focus on specific events retrospectively within accident analysis to understand what occurred and how to prevent similar events from happening again (e.g., [49, 50]). It can also be used prospectively within the early stages of the design process, with the use of case studies and scenarios to inform requirements [51, 52]. Central to the successful integration of the PCM within the design process is the Schema World Action Research Method (SWARM [53]) which enables data to be collected in a way that can inform PCM development. (In this chapter, the term *swarm* refers to two different concepts. When used in capital letters, SWARM is an interview method and when the term is used in lowercase, it refers to a fleet of ASs.) SWARM is an interview method that provides prompts and cues related to the three components of the PCM, to capture how and why decisions are made. Interviews provide a rich qualitative report on the experiences, actions, and information that occurs during specific situations, either critical events that have happened in the past or key areas of concern for future designs. The schema-action-world (SAW) taxonomy enables a broad range of areas to be covered within the interviews and categorizes them into the schema, world, and action. These prompts can also be applied as qualitative coding metrics to analyze decision-making and review how the process may be refined.

3.1.2 SWARM taxonomy

The SAW taxonomy comprises 6 schema themes, 12 action themes, and 11 world themes relevant to the management of critical events (see Table 7.1). See Plant and Stanton [53] for more information on the taxonomy and its development. Each theme has several interview prompts that allow interviews to be conducted with the human users of a system to build information related to specific scenarios of interest and develop a PCM of the event. There is a total of 95 prompts, but these can be down-selected and tailored to the research questions and scenario under consideration. The taxonomy was originally developed for application to the aviation domain and pilot decision-making within the cockpit. Therefore, the original taxonomy in Table 7.1 has been adapted slightly to make it more applicable to different domains.

3.1.3 Interview questions

The SAW taxonomy focuses on general actions, information, and processing that occurs within events. Yet, as trust is a phenomenon that relates to both the individuals' cognitive judgment as well as the functionality and performance of the automation, its trustworthiness [38], Parnell et al. [51] highlight the ability to model trust with the PCM. They suggest that trust follows the perceptual cycle as it is built from experiences within the cognitive experience of the individual, informed by information that they receive within the world around them and it dictates the actions that they perform. Therefore, Parnell et al. [51] mapped trust onto the SAW taxonomy to incorporate trust into the SWARM method, developing the Trust Schema Action World Research Method (T-SWARM).

There have been numerous attempts within the literature to develop methods to assess trust, with subjective questionnaires developed for multiple different purposes. A popular questionnaire is the "Trust and Liking scale" [54]. This questionnaire has been found to be useful in assessing trust in machines and automation [54–56]. It includes a self-report scale with respondents responding to each item with their agreement on a five-point Likert scale. In assessing trust in automation more specifically, Jian et al. [57] developed

Table 7.1 SAW taxonomy.

Schema themes	Action themes	World themes
Analogical schema	System function	Absent information
Declarative schema	Communicate	System status
Direct past experience	Concurrent diagnosis	Artifacts
Insufficient schema	Decision action	Communicated information
Trained past experience	Navigate	Location
	Nonaction	Natural environment conditions
	Physical actions	Operational context
	Situation assessment	Physical cues
	Standard operating procedure	Severity of problem
	System interaction	Technological conditions
	System monitoring	

the "trust between people and automation scale." This scale provides 12 statements that were developed from a comprehensive review of trust in automation that identified 12 factors that influence trust between people and automated systems. We took the statements within both the trust and liking scale and trust between people and automation scale and reviewed them in relation to the schema, action, world themes, and the PCM. This enabled us to map the different aspects of trust in automation onto the SAW taxonomy. Table 7.2 presents trust questions adapted from the two scales onto the schema, action, and world categories. These questions build on the SAW taxonomy in Table 7.1 and enable us to ask individuals how their experiences, perceptions, and action are informed by, and inform, trust in autonomous technology.

We provide a case study to show the application of these trust questions and SWARM in identifying trust requirements. This case study focuses on multiple UAVs that function together in a swarm (note SWARM refers to the schema-world-action research method, and the swarm is the term used for multiple connected UAVs). "UAV swarm" is the collective term for multiple UAVs working together under a common goal, often to facilitate remote and challenging operations [23]. Trust in UAV technology includes both the

Table 7.2 Trust questions adapted from the two scales onto the schema, action, and world categories.

SAW	Trust questions
Schema	*Past experience*
	1. Can you recall a point in this situation when you did not trust the technology?
	• Please expand on this situation and why you did not trust it
	Current experience
	2. Would you generally tend to trust the technology?
	• Please expand on why this is
	3. Do you have any distrust of the technology?
	• What is the cause of this distrust? How could it be repaired?
	4. Would you be wary or suspicious of the technology at all?
	• At which points in an operation would this be?
	Future expectations
	5. Would you have any reason not to trust the technology in the future?
	6. How reliable/dependable do you view the technology to be?
Action	1. What actions would you be relying on the technology for?
	2. What actions would you **not** be relying on a technology for?
	3. How easy would it be to trust the technology to do its job?
	4. How could your trust in the technology change over the course of the operation?
	5. Could there be any negative outcomes?
	• And how would this affect your trust in the technology for the future?
World	1. Would you ever be uncertain about the reliability or relevance of the information that you had available to you?
	2. What information/knowledge would you need to trust the technology?
	3. Could there be any deceptive information?
	4. What information would you need to repair any lost trust in the technology?

public's trust and the use of airspace [58], as well as the UAV operators' trust in the technology, its performance, the information that it relays, and the interactions that it may have with other systems. The location of the operator and the aircraft are distributed within the system and therefore the performance of the system is challenged by numerous issues such as limited sensory cues, and delays in control and communications [59]. The operator must be able to trust the communications between themselves and the UAV and trust that the information they receive is accurate and valid.

The interaction between the individual UAVs within the swarm, as well as how they communicate with a human operator, is critical to their success. UAV swarms can operate at a higher level of autonomy, interacting with each other in a network without human intervention. Yet, human interaction is still important to ensure the swarm is operating as desired and within the objectives of a set mission [60–62]. The design of the interface between the UAVs and the human operator is therefore highly important to facilitate swarm management as well as trust.

We were interested in identifying how the operator trusts the UAV swarm and what their requirements are for a trustworthy AS. Considering the importance of identifying user requirements early on within the design of the system, we began by identifying how users experience trust within singular UAV operations and then how they feel this may relate to multiple UAVs and UAV swarms. We, therefore, conducted interviews with UAV operators and applied the PCM and trust SWARM methodology presented in the previous section. Full details and results of this study can be found in Parnell et al. [51]. We provide a summary here.

3.1.4 Equipment and procedure

Six male participants were recruited with an average age of 33.83 years (range, 26–52 years). They were qualified UAV drone operators with a mixture of backgrounds; industry (n = 2), academia (n = 3), and military (n = 1). They were interviewed separately. In the interviews, they each spoke about their own experience of trust within a common UAV flight or scenario of their choosing. SWARM prompts from Table 7.1 were selected based on their relevance and the trust questions in Table 7.2 were also posed by the researcher in the interview. This enabled an in-depth discussion of their experiences. Questions relating to their views on trust in UAV swarms were also posed.

3.1.5 Data analysis and results

The interview transcripts were analyzed using thematic coding using the SAW taxonomy classification. This is a qualitative form of analysis that aims to identify, classify, and aggregate meaning from the interview data (see Braun and Clarke [63] for more information on thematic coding and qualitative analysis). The analysis identified the key schema, action, and world factors that the operators discussed when they were discussing trust within UAV and UAV swarms. The top eight SAW factors that were found to influence trust for the UAV operators are listed in Table 7.3. We later map these factors to software quality characteristics (Section 6).

Table 7.3 Key factors that influence UAV operator trust.

SAW factor	Element	Description
Direct experience	Schema	Time and experience in the functioning of the system and how it works across different scenarios
UAV status	World	Update of drone/swarm integrity in real time to diagnose possible issues or confirm expectations
Declarative schema	Schema	Operators must have knowledge of the capabilities and operability of the drone informed by regulation and standardization of drone operating systems
Technological conditions	World	Information on the functionality and capabilities of the system
Analogical schema	Schema	Shared understanding of the intended and unintended consequences as well as possibility for failure
System monitoring	Action	Displays and physical cues from the drone should match and provide a clear picture of the system functioning
System interaction	Action	System interaction should be intuitive and without errors. It should allow the operator to make confirmatory actions as well as inform them of need-to-know information
Standard operating procedure	Action	The standardization and full testing of drone technologies will ensure they are safe for use. Training should also be formalized to ensure operators have standard and well-known procedures to follow across different circumstances
Concurrent diagnosis	Action	Operators must have available capacity to understand any possible failures in the system and manage them

4 Explainability of human-swarm systems

As mentioned in Section 2, an essential requirement for a trustworthy human-swarm team is system monitoring, which requires an approach to communicating the reasons behind swarms' decisions and behavior to humans. This would help human operators to understand the swarm behavior, allow effective intervention when needed, and repair trust in case of failure or unexpected behavior [18]. Explaining the swarm behavior (as individuals and as a collective) is embedded in increasingly complex communication between swarm agents and underlying black-box controller models.

The process of generating explanations from ASs has received a lot of attention from multiple domains, including psychology [64], artificial intelligence [65], social sciences [66], and law [67]. Furthermore, regulatory requirements and increasing customer expectations for responsible artificial intelligence [68] raise new research questions on how to best design meaningful and understandable explanations for a wide range of AI users. To respond to this trend, many eXplainable AI (XAI) models and algorithms have been proposed to explain black-box machine learning models [69], agent-path planning [70], multiagent path planning [71], and explainable scheduling [72]. These explanations can range from local explanations that provide a rationale behind a single decision to global explanations that provide explanations for the overall AI logic [65].

Despite a growing interest in XAI, there are a limited number of studies on how to achieve explainability in swarm robotics. Only a few attempts in the literature have discussed explainability in swarm robotics research. One study has proposed a scalable human-swarm interaction to allow a single human operator to monitor the state of the swarm and create tasks for the swarm [19]. Their study with 100 participants showed that users have different explanation-visualization preferences to observe the state of the swarm depending on the type of the given tasks. However, the explanation in this study was limited to visual representations of the swarm coverage using a heatmap or individual points space that does not capture the wider space of eXplainable Swarm (xSwarm). Another example is the work proposed by Mualla et al. [73]. They proposed a human-agent explainability architecture to filter explanations generated by swarm agents with the aim of reducing human cognitive load. Although these studies provide steps toward the right direction for explainability in swarm systems, the road to implementing explainability in swarm systems is paved with many challenges. According to Roundtree et al. [18], implementing explainability in swarm systems is still one of the unfinished aims of human-swarm interaction literature, as it is an emerging area of research. At the current stage of research, it is still vague what kinds of questions an explainable swarm is expected to answer and what kinds of explanations a swarm system is expected to generate.

In this section, we argue for the need to identify shared understanding across the research community to put things into perspective, for example, to enhance knowledge sharing and pave the way toward more research in this field. Such knowledge is needed given the scarcity of research investigating the explainability of swarm systems. Accordingly, this section presents results from a questionnaire conducted with swarm experts to identify a common understanding of explainable swarms and derive a taxonomy of explanations for swarm systems. We briefly explain our research method, for the scope of this chapter, to derive explanation categories and our results. The full paper can be accessed through [74].

4.1 Research method

In this section, we explain our research method, including our sampling method, equipment, and procedure to create a taxonomy of explanations for swarm systems.

4.1.1 Participants

Twenty-six participants completed our online questionnaire with an average age of 34.52 years (range, 27–59 years). Participants had at least 1 year of experience working in swarm robotics with a mixture of academic (n = 20) and industrial (n = 6) roles. Their experience in swarm robotics ranged between (1–20 years) with a median of 5 years and an average of 8.1 years.

4.1.2 Equipment and procedure

We conducted an online questionnaire with swarm experts to outline an early taxonomy of explanations for swarm systems. First, a draft questionnaire was developed and then the draft was pilot tested with experts (researchers from the same and outside the research group) to assure that question formulation could achieve the research goal and be understandable to study participants. The questionnaire starts by asking participants to read the participant information sheet. The information sheet outlines the motivation of the study, compiles background material about the research goal and problem, and describes the questionnaire structure. The participants' information sheet also contains background material about explainable AI to minimize any availability bias [75]. Then, participants were directed toward the main questionnaire.

The questionnaire was designed to include three elicitation methods to help us obtain representative results from our experts. This approach can also minimize cognitive heuristics bias [75, 76], for example, biased judgments toward recent swarm explanations that our participants encountered. Table 7.4 provides a brief description of the elicitation methods and examples of the questions that were asked of our experts. Since an explanation is an answer to a failure or an unexpected behavior [66], our elicitation methods were motivated by that goal. In the first section of the questionnaire, we asked our participants to give us common examples of swarm failures or unexpected behavior based on their previous experience and knowledge. This was helpful in understanding what types of events would require an explanation and what part of swarm operations may require an explanation. Then, participants were provided with examples and scenarios of swarm failures or unexpected behavior. We asked our participants to list as many questions as they need to debug these failures. We then introduced them again to the same scenarios and asked them to explain why they thought the swarm failed or what the potential reasons were for their unexpected behavior. Fig. 7.3 presents two scenarios provided in the xSwarm questionnaire to debug and explain swarm failure or unexpected behavior.

4.1.3 Data analysis

We used the content analysis method to analyze participants' answers [77]. Our data set included experts' answers to open-ended questions using three elicitation methods outlined in Table 7.4. The analysis was performed by one researcher. After the first coding of the data, we followed an iterative process to increase the trustworthiness of our analysis. This iterative process was put in place to examine and ensure that the taxonomy was representative of participants' answers. Each step of the analysis included several meetings

Table 7.4 Three elicitation methods used in xSwarm questionnaire.

Experience	This method was constructed to elicit participants' prior knowledge and experience about typical failures or unexpected behavior in swarm robotics. We used open-ended questions to ask participants about typical swarm failures and unexpected behavior as the main trigger for explanations, for example, What are the main reasons for swarm failures? and What behavior sparks your concern?
Debug	The Debug method was used to elicit what questions swarm experts use to debug swarm failures or unexpected behavior. We presented two scenarios of swarm failure and asked experts to list as many as questions they could to debug the failure.
Explain	Participants, in this section of the questionnaire, were asked to explain the failures in the same provided scenarios.

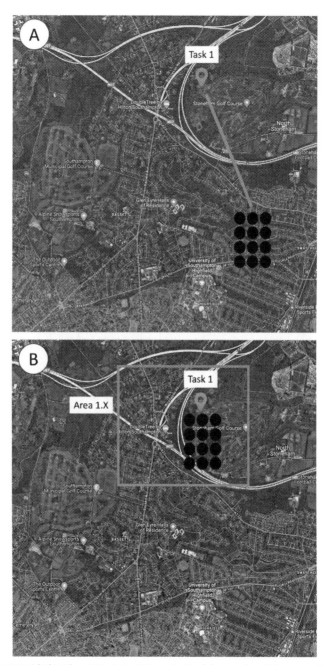

FIG. 7.3 Two scenarios provided in the xSwarm questionnaire to debug and explain swarm failure or unexpected behavior. (A) Picture of Scenario 1 with the following description: The swarm has been tasked to visit the Task 1 location on the map to search for casualties. The swarm is showing the following message: "The swarm is stopped in the middle of the route." What do you think the main reasons for this failure are? [Please list as many reasons as you can.] (B) Picture of Scenario 2 with the following description: The goal is to search for potential causalities; however, the swarm has completed the search and detected five out of six casualties in the selected area. Why do you think that the swarm failed to identify all potential casualties? [Please list as many reasons as you can.]

among the authors to ensure the correct interpretation of each category and the evidence that supports them. These meetings led to splitting, modifying, discarding, or adding categories to ensure that all responses and their contexts were well represented.

4.2 Categories of swarm explanations

In this section, we discuss the general categories of explanations that emerged from participants' answers. Table 7.5 presents a sample of participants' Debug questions and Explanations of failures or unexpected behavior by the swarm.

4.2.1 Consensus-based

The main feature of swarm robotics compared to other multiagent systems is that the swarm can be controlled and managed as a single entity with a shared goal and single task. The agents in the swarm communicate messages to reach a consensus on the swarm's decision and behavior. Our participants frequently debugged or explained the swarm failures or unexpected behavior based on how many or what percentage of agents in the swarm made the decision, for example, P18 mentioned: "How many robots detected the casualties?" and P11 added: "Because of conflict between the swarm agents." (For anonymity, individual participants are referred to as Pn [e.g., P18] in this chapter.) They also went beyond the number of agents and asked questions on the percentage agreement between the agents on the decision features, for example, P18 asked: "Which features of the casualty did the robots agree on?" This theme appeared 24 times in the Debug section and 18 times in the Explain section.

Table 7.5   Explanation categories for swarm robotics.

Category	Explain	Debug
Consensus	Detecting the missing casualty caused a conflict between the agents. Agents were not able to agree on some features of the casualty.	How much of the swarm is consolidating the same view? What is the agreement percentage between agents?
Path planning	Because the swarm initiative plan is to visit Location L, which is a charging location. Because swarm is trying to avoid obstacles.	Is it a collision-free route? Where is the swarm right now? [the probability distribution of possible robots' locations]
Communication	Because the communication is limited between agents. Environmental condition is limiting the connection between agents.	What is the accuracy of information exchange between swarm agents? What is the synchronization status between swarm agents?
Scheduling	Because this casualty should be detected by another cluster. Because there are not enough agents to detect all casualties.	Why there is x number of agents to search this area? Why a cluster c1 is assigned to task t1?
Hardware	Because the casualty is out of sensor coverage distance. Because of low-quality input data from swarm sensors.	What is the communication bandwidth between swarm robots? What is the limitation of the swarm sensors?
Architecture and design	Because robots have limited communication links per robot. Because the swarm has not got enough training in such an environment.	What is the response threshold function? How is the swarm size selected?

4.2.2 Path planning-based

Participants pointed out 22 questions and 14 explanations related to path planning when they were asked to debug or explain the swarm failures or unexpected behavior. Participants' questions were to understand the path that the swarm followed or it was going to follow, for example, P2 mentioned that remote human operators would frequently check what is the next state of the swarm. Participants also asked questions to check properties and features of the path-planning algorithm, for example, P7 asked: "What metrics does the swarm use to follow this route?" Further, participants explained the swarm failure and unexpected behavior based on events related to the path-planning algorithm; P9 stated: "Because the swarm agents are charging their batteries according to the initial plan."

4.2.3 Communication-based

The collective decision-making process is the outcome of an emergent phenomenon that follows localized interactions among agents yielding a global information flow. As a consequence, to investigate the dynamics of collective decision-making, participants went beyond debugging consensus between agents and asked questions related to the communication between swarm agents, for example, P7 pointed out the following question: "What is the synchronization status between swarm agents?" Participants also pointed out that swarm failure in our examples could be critically related to a failure in ensuring the appropriate flow of behavioral information among agents, both in terms of information quantity and accuracy, for example, P21 commented on the failure in Fig. 7.3: "Because the environmental condition is limiting the connection between agents" and P18 added: "Perhaps one of the casualties moved on/no longer is, therefore state was not updated."

4.2.4 Scheduling-based

Task scheduling in the swarm is a collective behavior in which robots distribute themselves over different tasks. The goal is to maximize swarm performance by allowing the robots to spread across different tasks dynamically. Yet, in many applications, human operators are still required to intervene in the task scheduling and adapt accordingly. Participants' understanding of particular swarm failure was associated with initial task scheduling; 14 questions and 8 explanations related to task scheduling appeared in our data. For instance, P2 explained casualty detection failure (Fig. 7.3) with the following explanation, "Because this casualty should be detected by another cluster." Participants often suggested questions implying why this schedule plan is optimal and they were also interested to understand why another scheduling plan is not feasible. For instance, P19 asked the following question: "Why is a cluster c1 better than cluster c2 in performing task t1?"

4.2.5 Hardware-based

Even though the quality and robustness of the hardware are increasing, hardware failure is still quite common. Participants explained the unexpected behavior of the swarm with a link to a hardware problem. For instance, when the swarm was not able to detect all the casualties in the area, P17 commented: "Because of faulty in the sensors." Our participants also discussed that explanations of hardware are necessary to give human operators actionable explanations to either better utilize the swarm or to improve the swarm performance, for example, send more swarm agents to the location to increase the performance of the task. There was also a typical pattern among participants' feedback to explain swarm behavior or failure based on its hardware performance boundaries and limitation as well as environmental and training data limitations. For instance, P11 explained unexpected behavior in Fig. 7.3: "Because some agents have limited speed capabilities, so they are waiting for each other," and similarly, P14 explained Fig. 7.3 failure: "The swarm sensors have low image input."

4.2.6 Architecture and design-based

Participants also found many reasons for swarm failures based on the swarm architecture or design decisions of the system. This category did not appear frequently in the data, and participants mentioned five explanations and five questions. Participants recognized that potential failure or unexpected behavior can be related to initial design decisions made by the systems engineers, for example, P12 answered: "Because the swarm has not got enough training in such an environment" and similarly, P22 commented: "Because the transition probability between states is fixed." Participants also took a broader view of swarm explainability when they discussed questions that can be answered through descriptive information about the swarm system design. For instance, P1 and P4 asked: "How is the swarm size selected?," "What is the design method? Behavior-based or Automatic design?"

5 Use-case development

Kolling et al. [61] presented several use cases of human-swarm interaction from the perspective of remote human supervisors by reviewing existing academic literature. While these generalized frameworks are steps in the right direction, two key gaps remain undefined in most existing work on human-swarm interaction:

- Studies that are often from a "general-purpose" perspective with a broadly defined goal of interaction, not to address specific needs of real-world use cases.
- Studies that are not evaluated adequately reflect how effective their approaches are in real-world settings. Barring a few exceptions [78, 79], much of the existing work is designed and developed for benchmark and validated with user studies limited to users in research settings such as Amazon Mechanical Turk [80].

The current stage of the literature is a body of methodological work, for example, control and interaction techniques, without clear use cases and established real-world utility. In this chapter, we argue that real-world case studies from current industrial organizations are needed to inform the efficacy of human-swarm interaction research. This study aims to explore real-world use cases from industrial organizations of human-swarm partnerships where humans are responsible for controlling and monitoring the swarm. We seek to define operational boundaries, operators' roles, and their implications for human-swarm trust. We argue that such knowledge is essential to understand how global organizations intend to make use of robotic swarms, and to what extent they foresee barriers in trust and operation. To that end, this research study seeks to:

- gain insight into the needs and requirements of global organizations working in swarm robotics;
- provide the community with industry-developed scenarios that are suitable for design and evaluation tasks;
- identify how a user may interact and develop trust with the swarm, what communication links will be required, and how a robotic-swarm system may operate within real-world situations;
- discuss research gaps by comparing the existing body of work to the need of the use cases; and
- propose research direction to develop effective human-swarm interfaces that would lead to improved human-swarm performance and consequently improved trust scores.

The use case presented in this chapter aims to provide researchers, practitioners, and manufacturers with a basis on which autonomous robotic swarms can be designed and regulated within future operational scenarios. The specified tasks and activities identified within each use case are collated to provide an accessible reference tool for the deployment of robotic swarms. Our use cases can also be used as a probe in human-swarm studies to explore new areas of research that remain undefined in human-swarm settings. For instance, explainability has been thoroughly discussed in the previous section. However, it remains unclear if the explanations studied in human-robot settings would be applicable to human-swarm teaming. More fundamentally, what kind of questions an explainable swarm should be expected to answer, and what those explanations would look like, are currently unknown. To that end, our generated use cases may be used to feed into scenario-based elicitation methods [81] to explore how a swarm of robotics can explain their behavior to human operators in dynamic environments.

5.1 Cocreation process

Ethical approval was acquired via the University of Nottingham ethics panel (Ethics No. CS-2020-R54). Participants were recruited by advertising to the project partners of the Trustworthy Autonomous Systems Hub representing various expertise within the industry. Eleven participants took part in the workshop (10M, 1F), comprised of experts in the research, development, operation, and management of unmanned vehicle systems. Experts were members of either defense or maritime organizations. Participants collaborated on "Miro boards" (Miro Board, https://miro.com/, accessed June 13, 2022) an online visual interface for collaborating on shared projects. The workshop consisted of three break-out rooms, each tasked with generating its own use case that could include homogeneous or heterogeneous vehicle systems consisting of aerial, land, and maritime vehicles. Swarms were delineated from multi-UAV systems by outlining that swarm robotics include autonomous behaviors of multiple agents (e.g., flocking, autonomous search patterns, auto-allocation of tasks), rather than individually being controlled. The construction of scenarios was unrestricted, as long as a task involving at least one human operator and a swarm that were not able to be individually controlled to ensure task success was defined. Participants were encouraged to use their professional experience to guide the construction of use cases. Break-out groups were tasked with addressing three core themes: the scenario, operational boundaries, and implications for trust defined in Table 7.6.

5.1.1 Procedure

The workshop lasted 3 h in total. Participants were given a 10-min introduction to inspiration material consisting of links to news articles related to swarms and a table outlining an introduction to the various domains in which swarm robotic behaviors are currently in place [23]. Once introduced to the workshop and the agenda, participants were split into three separate groups consisting of three to four members. Each group took part in three activities for use-case cocreation involving: scenario formation, operational boundaries, and identifying trust factors when interacting with the swarm in their identified scenario. After each section, each group presented back to the workshop.

5.1.2 Analysis

Micro-Boards were compiled into task diagrams split across three time periods: planning, exploring, and response. All references to actions, communication, and requests were

Table 7.6   Group defined scenarios and tasks.

Scenario generation	"Outline the scenario or problem your swarm will be tasked with. Specify the swarm-based solution, the people involved, and an overview of the system (i.e., how it works, what information is being processed, how it addresses the task)"
Operational boundaries	"Identify the limits, boundaries and capabilities of the system including the physical and cognitive arrangements of the system (e.g., roles, objects), how communication will occur, and what limitations and boundaries are present (e.g., range, missing information, dropouts, workload)"
Implications for trust	"Define the main factors that may affect trust during the interaction with the swarm in the for Trust use-case. Please refer to your previous"

included to populate the diagrams, separated by agents (e.g., swarm, supervisor, analyst, first responders). Use cases were then compiled to define a single scenario that can be applied to simulation platforms. We followed an iterative process across several research meetings to formulate, combine, and conceptualize emerging concepts. This iterative process was put in place to examine and ensure that the final use case was representative of a possible real-life scenario.

5.2 Collated use case

Each break-out group outlined mission descriptions, timelines, team roles, and tasks for their scenario. Each use case focused on casualty response tasks as a result of a disaster, two of which defined a search-and-rescue task, whereas one focused on the delivery of medical equipment to appropriate locations. Table 7.7 outlines each group's defined scenario and task definition.

Each break-out group provided in-depth information about their scenario including a task timeline, interface requirements, potential trust issues, and how mission tasks contribute toward mission success. The key recurring features of each use case were then compiled into a single use case. The following use-case scenario represents the collation of the recurring tasks and key components of human-swarm interaction. A task description and outlined activities for mission success are provided here to give researchers and UV swarm developers an informed example in which to test simulations and consider the multiagent network of human-swarm operations.

5.2.1 Compiled use-case background

A hurricane has affected a settlement. Many were evacuated, but some did not manage to escape and have been caught in debris and fallen structures. A UAV swarm is being deployed to identify locations of survivors, to evaluate their condition, and to advise on where unmanned ground rescue vehicles should be deployed to evacuate casualties.

Table 7.7 Group defined scenarios and tasks.

Group	Scenario description	Operator's tasks
1	A multihuman team consisting of an operator (tasked with controlling the swarm) and an analyst (tasked with processing processed data) seek to identify civilians and threats in a reconnaissance mission.	Allocate swarm, receive object classifications, query object classifications, and send mission information.
2	Operators work with coordinators and the swarm to evaluate target areas and suitable landing zones for medical supplies, while ensuring that airspace and other parties are managed.	Allocate swarm, search for target zone, identify suitable drop-off zone, call-in delivery drone, and deliver the payload.
3	Locating and evacuating casualties. Identify, classify, and prioritize casualties based on criticality. Deploy ground vehicles to appropriate locations.	Allocate swarm, relay conditions, classify casualties, plan rescue route, deploy rescue vehicle.

5.2.2 Compiled use-case agents and tasks

Fig. 7.4 outlines the sequence of tasks required to ensure task success, drawn from each group's use-case definitions. The figure presents four agents responsible for mission success, represented by four vertical orange bars within the figure:

- The operator responsible for the operational performance of the swarm including the allocation of UAVs to airspace, allocation of tasks, and monitoring of the connectivity and battery status of the swarm.

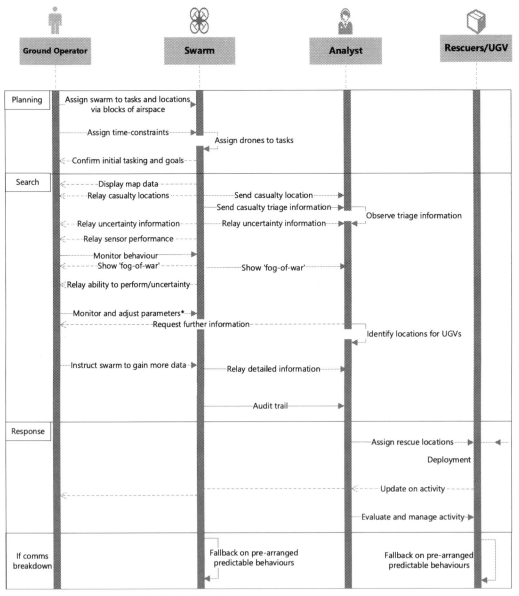

FIG. 7.4 Task-flow diagram for the final collated use case.

- The swarm—responsible for locating casualties and relaying data about the condition of the casualty and the environment—will communicate information related to uncertainty, object classifications, and images of the scene.
- The analyst tasked with sifting through incoming data from the swarm and screening for accuracy. Hazardous objects, route blockages, and the location and criticality of the casualty (calculated by movement, heat signatures, etc.) are presented to the analyst during the mission. The analyst will then identify suitable locations for first responders or unmanned ground vehicles (UGVs) to rescue high-priority casualties.
- The responders/UGVs will take on the commands or recommendations (if a human responder) of the analyst to arrive at the target location and evacuate casualties.

There are three stages in this use case, which are populated with tasks that relay to another agent within the diagram, representing the flow of information and activities.

- The planning stage: The period when the operator and analyst familiarize themselves with the scene and any information made available to them prior to the search. The operator allocates the swarm to their search locations and gives groups of UAVs tasks to complete. The swarm relays that they have received their tasks. This can be done via a map to decide which areas may be more likely to have people (schools, housing blocks, leisure sites).
- The search stage: The main component of the use case. The swarm enacts the search for casualties while relaying casualty positions and images of the scene. Casualty information is relayed to the analyst for triaging casualties into rescue priority. Activity information such as the task being conducted, the location of UAVs, and future intentions are relayed to the operator. Additional information such as uncertainty (of object classification or location of objects) and operational state (such as the battery, connectivity, and structural integrity) are relayed so that the operator can monitor and adjust parameters accordingly. These parameters might include battery usage and swarm density to increase connectivity, or the operator might simply recall UAVs that are low in battery back to the base station to reduce the frequency of dropouts. Both the operator and analyst can request additional information regarding the objects in the scene, or request images to be taken for additional clarity. The operator may adapt the actions of the swarm in line with the analyst's needs so that detailed information can reach the analyst when making decisions on where to send rescue vehicles. During this process, an audit trail is logged so that the analyst can keep track of swarm activities.
- The response stage: Once the analyst is satisfied with the incoming data, they can allocate rescue locations for either first responders or ground vehicles to evacuate as many critical casualties as possible. This decision is to be made by selecting locations that are accessible and that can optimize the distance between itself and casualties.

Should communication deteriorate within the swarm, behaviors will fall back on predefined instructions set by the operator (e.g., follow the last instruction and patrol the area in which it no longer has an instruction until connectivity is restored).

5.3 Discussion

Despite the existence of human-swarm interaction studies investigating human-swarm trust, their efficacy in improving real-world use cases is yet to be sufficiently explored. The use-case scenarios and the compiled use cases in this research study are the first steps toward that goal by providing researchers, practitioners, and manufacturers with a basis upon which autonomous robotic swarms can be designed and regulated within future operational scenarios. This use case is driven by professional UAV operators and designers to reflect the current trends in unmanned-vehicle technological development. Communication requirements and distributed role allocation are provided to map out the potential for multiple humans to work with a swarm and outline what a future human-swarm mission might entail. It is critical that the human-swarm interaction domain develops past that of human-robot interaction, as much of the research in human-robot explainability focuses on individual robot agents, which may not be directly applicable to human-swarm interaction. For example, swarm display methods have an increased requirement for the concatenation of information and succinct display methods to reduce workload and optimize trust and usability [19, 82].

Explainability in swarms can be used to better calibrate trust by giving the operator and analyst a better understanding of the actions and decisions of the swarm. Our integrated use case can be used as a baseline scenario in human-swarm studies to explore new areas of research that remain undefined in human-swarm settings. The compiled use case is presented to provide a representation of a real-world scenario with a defined storyline and phases of the mission. This can be used to test multiple research questions that are facing the human-swarm interaction community as autonomy increases. Many users take part in the same simulated mission. Additionally, each phase of the mission can be addressed separately when managing human-swarm interactions. For example, the planning phase can be explored to develop efficient communication methods and task allocation, and how to support information (e.g., map displays). Individual components of the use case can also be drawn to test hypotheses, for example, focusing on what information will need to be shared or exclusive to either an operator or an analyst (e.g., should an analyst be aware of how the swarm is performing, or should an operator know what casualties are being investigated by the analyst?). Additionally, communication modalities should be considered; for example, how the analyst and the operator could communicate to manage the task and swarm activities, and how this could be facilitated through interfacing (e.g., text, voice, or in-interface communication). Additionally, how objects are classified and relayed to the human operators should be considered (e.g., how many swarm members agree or disagree on the object classification or incoming information?).

The use case serves as a scenario to test these pressing questions and allows multiple communication links and performance metrics to be assessed. In this scenario, the speed at which the team locates casualties' identifies criticality, and selects an optimal pick-up location can be factored in when analyzing performance. Additionally, researchers can analyze and monitor communication between each agent to understand what and when information is communicated. The use case can be applicable to civil applications (e.g., environment monitoring, search and rescue, surveillance) as well as defense applications. To that end, the use case can be flexibly applied to many working domains.

6 Human-swarm teaming simulation platform

There are a number of simulators for multiagent systems such as Gazebo [83], Webots [84], and ARGoS [85], each of which has advantages and disadvantages for a variety of applications. Many new developments in this area focus on scalability [85] and simulation quality, with live 3D renderings of the agents being of great importance to the multijointed robots often modeled in standard robotics simulators. The appetite for real-time human interaction is increasing. For instance, ARGoS supports "loop functions," which facilitate basic waypoint controls or manual operation of individual robots. There are more recent movements to integrate human interaction into Gazebo [86]. This being said, none of these simulators has human interaction as a focus, nor do they have deliberate interface design for continuous command and control from the user during runtime.

Our simulation platform, *Hu*man-swarm *t*eaming *Sim*ulator (HutSim), was first used to study the performance and workload impacts of different allocation strategies for UAV swarms [87]. HutSim was designed with human interaction as the core focus, and the back-end of the platform is built around this idea. A variety of task and allocation approaches were included from the outset. The user interface was also designed for one or more users to operate several windows with varying degrees of control over the swarm and the mission. It also supports features such as random agent dropouts and provides the efficient MaxSum algorithm [88] for easily allocating agents to tasks without causing the user additional cognitive load. It also supports the integration of real drones that can be teleoperated alongside simulated ones, although this is beyond the scope of this work.

Since the original publication [87] it has been expanded significantly, with more allocation strategies including dynamic and ad hoc approaches, and options for simulated image scans that can be reviewed by humans in order to test human-robot trust as well as human-human trust in the case of a multiuser scenario. Additionally, the interface has been expanded, with toggleable interface lenses to view additional contexts such as positional uncertainty for agents, overlayed heatmaps of explored areas, and path prediction arrows to communicate agent intentions. It is run on a server, with user access via a web browser. This makes it easy to access without special hardware or software. The GUI is based on the Google Maps API (Google Maps Javascript API, https://developers.google.com/maps/documentation/javascript, accessed June 13, 2022) and models a real-world location, and natively provides road and satellite maps if desired.

Fig. 7.5 shows a typical scenario in HutSim.[a] Users can use various views to monitor the swarm, as well as creating and assigning tasks to agents.

---

[a]New version of HutSim, HARIS, is available online: https://uos-hutsim.cloud.

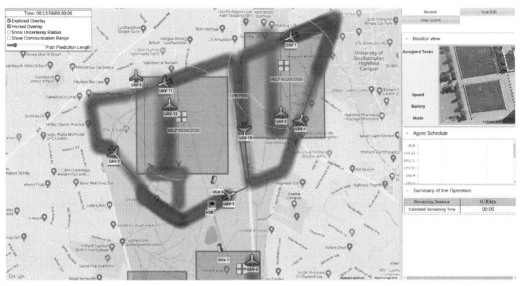

FIG. 7.5 HutSim allows multiple operators to access a shared scenario and perform different tasks simultaneously. Operators can switch between different overlays displayed on the top-left corner, assign tasks to the UAVs manually or automatically, and monitor the task and the UAVs.

6.1 Simulated use case

We provide separate views for the operator and analyst roles, such that there is an appropriate separation of concerns for the two users. The operator can create two types of tasks:

- Region task: Create a rectangular region that agents will search the interior area of by sweeping back and forth in a "lawnmower" (Boustrophedon) pattern [89] (Fig. 7.6).
- Patrol task: Create an enclosed polygon in which agents will trace the perimeter continuously.

The operator can then allocate each agent to a task that they will then carry out. To reduce cognitive load, the operator can use an auto-allocate button that runs the MaxSum algorithm to suggest an allocation. They can also monitor information such as wind details. A chat window is provided for communication between users. Fig. 7.7 shows a typical view for the operator.

The analyst can observe the swarm and use a separate panel to triage potential casualties based on the observed simulated scan. The image panel supports pan and zoom controls but is deliberately low resolution such that the analyst has to use their best judgment to classify the image with a moderately high margin of error. Fig. 7.8 shows a typical view for the analyst.

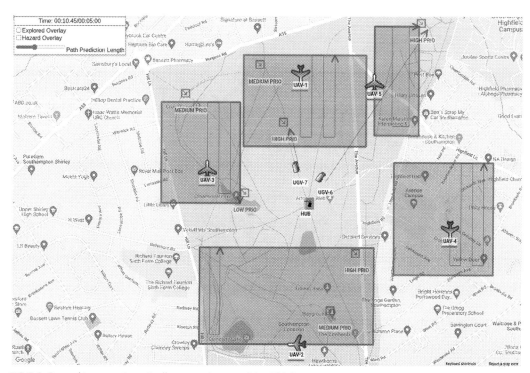

FIG. 7.6 Ground teams automatically respond to casualties. This is a separate autonomous swarm that moves from the hub to identified casualties in order of triage priority; they then collect the casualty and return to the hub.

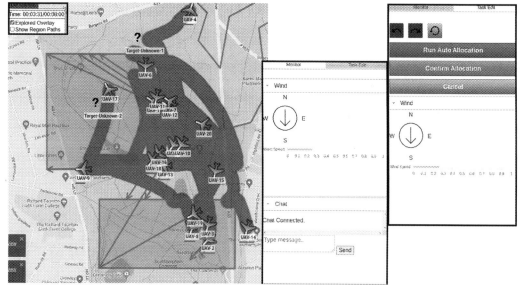

FIG. 7.7 Map and panels for the Operator view. The Operator can monitor the swarm and factor in wind information (shown *right*) and use a chat window to communicate with the analyst. They can also create and assign tasks.

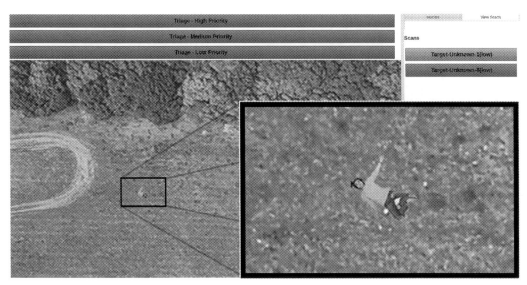

FIG. 7.8 Scan review panel for the Analyst view. The Analyst is presented with a low-resolution image taken by a UAV. They can use pan and zoom controls to search for a possible casualty, and use their certainty to inform a triage decision with the buttons at the top. Note the *black boxes* have been added for clarity.

7 Compliance with requirements

In order to evaluate the compliance of HutSim or any other simulation platform against the listed trust requirements, we need to transform the requirements into software quality characteristics that can then be easily evaluated in any software.

7.1 Requirements for software quality characteristics transformation

ISO/IEC 9126 [90] is an international standard for software quality that characterizes the quality of different software-based metrics including Functionality, Reliability, Usability, Efficiency, Maintainability, and Portability. Table 7.8 maps each requirement generated from the user interviews in Section 3 to the corresponding software quality. For instance, direct experience can be facilitated by reliability and the usability of the software.

7.2 Simulation software quality characteristics

We go through all six quality characteristics and discuss to which extent HutSim meets these qualities.

7.2.1 Functionality

HutSim models a real-world environment by using the Google Maps API, along with physics-derived movement based on SI units for accurate specification of agents; as such the Simulation-to-Reality gap is reasonably small. As a self-contained platform, there is little requirement for interoperability, but the flexible Java at the core permits easy

Table 7.8 Mapping trust requirements and explanation categories to simulation quality characteristics.

	Functional	Reliable	Usable	Efficient	Maintainable	Portable
Direct experience		●	●			
UAV status	●	●	●			
Declarative schema	●		●			
Technological conditions	●		●			
Analogical schema	●		●			
System monitoring		●	●	●		
System interaction	●		●	●		
Standard operating procedure	●	●	●	●		
Concurrent diagnosis		●	●		●	
Explainable		●	●		●	
Consensus		●	●		●	
Path-planning		●	●		●	
Communication		●	●		●	
Scheduling		●	●		●	
Hardware		●	●		●	●
Architecture and design		●	●		●	●

integration of external libraries and by exporting the state representation other platforms can interface with it. There is currently no direct focus on security, but as it is based on standard technologies like the REST API with no requirement for a formal database, this is unlikely to be a concern.

7.2.2 Reliability

The core code of the software is stable and generally bug-free, but as a continuously developing platform, it is subject to the standard drawbacks of bleeding-edge software with respect to new changes impacting the platform. Most errors and exceptions are handled during runtime, so the fault tolerance is generally quite good. As the UI is browser-side and separated from the Java backend, recoverability can usually be achieved by refreshing the browser window. Logging also offers the ability to rebuild failed scenarios, although no automated saving and backup are currently implemented.

7.2.3 Usability

Since the software is targeted at a very specific set of use cases and exists in a small landscape of platforms, it is easy to understand whether the software is suitable for a set of requirements, but it may require adaptation to adjacent use cases. A straightforward graphical user interface makes it easy to learn how to operate it and accommodates different methods and degrees of control if desired. We are yet to collect extensive data on the user experience and interaction interface design of the platform, but the features and layout were designed with the user's requirements in mind through several trials with users from industry and academia.

7.2.4 Efficiency

The backend of the platform is written in Java, which is lightweight, and the GUI is run by the client. This makes it very easy to run, and it has been tested on both low-power AWS hosting and a Raspberry Pi, both handling multiple concurrent instances with several users, and at multiple times the real speed with low memory usage.

7.2.5 Maintainability

The extensive logging allows for easy analysis of failures, and the modular nature of the software makes it easy to adapt the software for new requirements, and if implemented correctly these changes will not negatively impact stability. While there is no formal test framework, a collection of standard scenarios makes it easy to check that functionality remains after such changes.

7.2.6 Portability

As the platform is built on Java, and all libraries and dependencies are included, it can be adapted to any environment that supports Java, even in the form of a precompiled jar file. Multiple instances can be run in separate Java Virtual Machines and the sharing of resources is handled by the OS. It covers most functionality of other simulation platforms (with the exception of high-level physics simulation for each individual drone) and so could be used in place of these if required, although few other simulators are likely to include the human-focused components of HutSim.

As the simulator is targeted at a specific set of use cases and is built with standard tools, the portability is high as well. Currently, our HutSim platform can offer path-planning explanations that emerged from our expert questionnaire. The explanation shows the path that the swarm is going to take to ensure that the route is collision-free. Our future improvements to the simulation will include offering HutSim users a wide range of explanations.

8 Discussion and conclusion

With the rising demand for more efficient human-swarm collaboration, several issues such as trust need to be addressed for technology adoption. This chapter presented the key trust requirements from the perspective of experienced UAV operators. We outlined the main factors for establishing and repairing trust in working with single or multiple ASs, in this case, UAVs. We also performed a user study with academic experts to provide a better understanding of the issue of explainability. Our online simulation platform was presented as the first human-swarm operation platform. We studied the quality characteristics of our simulation platform and showed to which extent our platform meets the outlined requirements of trustworthiness and explainability. Despite the reported progress, the platform is still in its early stages of development and in future work, we will continue to work on it to fully comply with the trust and explainability requirements.

Acknowledgments

This work was conducted as part of the Trustworthy Autonomous Systems Hub, supported by the UK Engineering and Physical Sciences Research Council (EP/V00784X/1). The authors thank Joel Fischer, Marisé Galvez Trigo, Mario Brito, Adrian Bodenmann, and Miquel Massot-Campos for their contribution to the project. This chapter is derived, in part, from an article published in *International Journal of Human–Computer Interaction* on January 2022, available online: http://www.tandfonline.com/10.1080/10447318.2022.2108961 and a paper titled "Outlining the design space of eXplainable swarm (xSwarm): experts' perspective" presented during the *International Symposium on Distributed Autonomous Robotic Systems* in November, 2022.

References

[1] S.I. Jiménez-Jiménez, W. Ojeda-Bustamante, M.d.J. Marcial-Pablo, J. Enciso, Digital terrain models generated with low-cost UAV photogrammetry: methodology and accuracy, ISPRS Int. J. Geo Inf. 10 (5) (2021) 285.

[2] T. Hu, X. Sun, Y. Su, H. Guan, Q. Sun, M. Kelly, Q. Guo, Development and performance evaluation of a very low-cost UAV-LiDAR system for forestry applications, Remote Sens. 13 (1) (2020) 77.

[3] S. Granados-Bolaños, A. Quesada-Román, G.E. Alvarado, Low-cost UAV applications in dynamic tropical volcanic landforms, J. Volcanol. Geotherm. Res. 410 (2021) 107143.

[4] L.A. Binti Burhanuddin, X. Liu, Y. Deng, U. Challita, A. Zahemszky, QoE optimization for live video streaming in UAV-to-UAV communications via deep reinforcement learning, IEEE Trans. Veh. Technol. 71 (5) (2022) 5358–5370.

[5] M. Liu, H. Ma, J. Li, S. Koenig, Task and path planning for multi-agent pickup and delivery, in: Proceedings of the International Joint Conference on Autonomous Agents and Multiagent Systems (AAMAS), 2019.

[6] O. Salzman, R. Stern, Research challenges and opportunities in multi-agent path finding and multi-agent pickup and delivery problems, in: Proceedings of the 19th International Conference on Autonomous Agents and MultiAgent Systems, 2020, pp. 1711–1715.

[7] A. Puri, A Survey of Unmanned Aerial Vehicles (UAV) for Traffic Surveillance, Citeseer: Department of Computer Science and Engineering, University of South Florida, 2005, pp. 1–29.

[8] H. Shakhatreh, A.H. Sawalmeh, A. Al-Fuqaha, Z. Dou, E. Almaita, I. Khalil, N.S. Othman, A. Khreishah, M. Guizani, Unmanned aerial vehicles (UAVs): a survey on civil applications and key research challenges, IEEE Access 7 (2019) 48572–48634.

[9] J.W. Crandall, N. Anderson, C. Ashcraft, J. Grosh, J. Henderson, J. McClellan, A. Neupane, M.A. Goodrich, Human-swarm interaction as shared control: achieving flexible fault-tolerant systems, in: International Conference on Engineering Psychology and Cognitive Ergonomics, Springer, 2017, pp. 266–284.

[10] M. Erdelj, E. Natalizio, UAV-assisted disaster management: applications and open issues, in: 2016 International Conference on Computing, Networking and Communications (ICNC), IEEE, 2016, pp. 1–5.

[11] C. Luo, W. Miao, H. Ullah, S. McClean, G. Parr, G. Min, Unmanned aerial vehicles for disaster management, in: Geological Disaster Monitoring Based on Sensor Networks, Springer, 2019, pp. 83–107.

[12] M.A. Daly, Task load and automation use in an uncertain environment, Retrieved from ProQuest Dissertations and Theses, 2002.

[13] M. Franklin, E. Awad, D. Lagnado, Blaming automated vehicles in difficult situations, iScience 24 (4) (2021) 102252.

[14] M.K. Heinrich, M.D. Soorati, T.K. Kaiser, M. Wahby, H. Hamann, Swarm robotics: robustness, scalability, and self-X features in industrial applications, Inf. Technol. 61 (4) (2019) 159–167.

[15] M. Dorigo, G. Theraulaz, V. Trianni, Reflections on the future of swarm robotics, Sci. Rob. 5 (49) (2020) eabe4385.

[16] H. Hamann, Swarm Robotics: A Formal Approach, vol. 221, Springer, 2018.

[17] M. Dorigo, G. Theraulaz, V. Trianni, Swarm robotics: past, present, and future [point of view], Proc. IEEE 109 (7) (2021) 1152–1165.

[18] K.A. Roundtree, M.A. Goodrich, J.A. Adams, Transparency: transitioning from human-machine systems to human-swarm systems, J. Cognit. Eng. Decis. Making 13 (3) (2019) 171–195.

[19] M.D. Soorati, J. Clark, J. Ghofrani, D. Tarapore, S.D. Ramchurn, Designing a user-centered interaction interface for human-swarm teaming, Drones 5 (4) (2021) 131.

[20] Q.V. Liao, D. Gruen, S. Miller, Questioning the AI: informing design practices for explainable AI user experiences, in: Proceedings of the 2020 CHI Conference on Human Factors in Computing Systems, 2020, pp. 1–15.

[21] M. Brandao, G. Canal, S. Krivić, P. Luff, A. Coles, How experts explain motion planner output: a preliminary user-study to inform the design of explainable planners, in: 2021 30th IEEE International Conference on Robot & Human Interactive Communication (RO-MAN), IEEE, 2021, pp. 299–306.

[22] B. Septfons, A. Chehri, H. Chaibi, R. Saadane, S. Tigani, Swarm robotics: moving from concept to application, Hum. Centred Intell. Syst. 310 (2022) 179–189.

[23] M. Schranz, M. Umlauft, M. Sende, W. Elmenreich, Swarm robotic behaviors and current applications, Front. Rob. AI 7 (2020) 36.

[24] H.-W. Jung, S.-G. Kim, C.-S. Chung, Measuring software product quality: a survey of ISO/IEC 9126, IEEE Softw. 21 (5) (2004) 88–92.

[25] E.L. Glaeser, D.I. Laibson, J.A. Scheinkman, C.L. Soutter, Measuring trust, Q. J. Econ. 115 (3) (2000) 811–846.

[26] P. Sapienza, A. Toldra-Simats, L. Zingales, Understanding trust, Econ. J. 123 (573) (2013) 1313–1332.

[27] J. Delhey, K. Newton, C. Welzel, How general is trust in "most people"? Solving the radius of trust problem, Am. Sociol. Rev. 76 (5) (2011) 786–807.

[28] J.S. Nye, P.D. Zelikow, D.C. King, Why People Don't Trust Government, Harvard University Press, 1997.

[29] T.A. Zimmer, The impact of Watergate on the public's trust in people and confidence in the mass media, Soc. Sci. Q. 59 (4) (1979) 743–751.

[30] D. Gambetta, et al., Can we trust trust, in: Trust: Making and Breaking Cooperative Relations, vol. 13, Citeseer, 2000, pp. 213–237.

[31] M.S. Cohen, R. Parasuraman, J.T. Freeman, Trust in decision aids: a model and its training implications, in: Proc. Command and Control Research and Technology Symposium, Citeseer, 1998.

[32] J.D. Lee, K.A. See, Trust in automation: designing for appropriate reliance, Hum. Factors 46 (1) (2004) 50–80.

[33] R. Parasuraman, T.B. Sheridan, C.D. Wickens, Situation awareness, mental workload, and trust in automation: viable, empirically supported cognitive engineering constructs, J. Cognit. Eng. Decis. Making 2 (2) (2008) 140–160.

[34] K.E. Schaefer, J.Y.C. Chen, J.L. Szalma, P.A. Hancock, A meta-analysis of factors influencing the development of trust in automation: implications for understanding autonomy in future systems, Hum. Factors 58 (3) (2016) 377–400.

[35] K.A. Hoff, M. Bashir, Trust in automation: integrating empirical evidence on factors that influence trust, Hum. Factors 57 (3) (2015) 407–434.

[36] A.D. Kaplan, T.T. Kessler, J.C. Brill, P.A. Hancock, Trust in artificial intelligence: meta-analytic findings, Hum. Factors 65 (2) (2021) 337–359.

[37] R. Parasuraman, V. Riley, Humans and automation: use, misuse, disuse, abuse, Hum. Factors 39 (2) (1997) 230–253.

[38] T.B. Sheridan, Individual differences in attributes of trust in automation: measurement and application to system design, Front. Psychol. 10 (2019) 1117.

[39] P.A. Hancock, D.R. Billings, K.E. Schaefer, J.Y.C. Chen, E.J. De Visser, R. Parasuraman, A meta-analysis of factors affecting trust in human-robot interaction, Hum. Factors 53 (5) (2011) 517–527.

[40] T.B. Sheridan, Extending three existing models to analysis of trust in automation: signal detection, statistical parameter estimation, and model-based control, Hum. Factors 61 (7) (2019) 1162–1170.

[41] T.B. Sheridan, Trustworthiness of command and control systems, in: Analysis, Design and Evaluation of Man-Machine Systems 1988, Elsevier, 1989, pp. 427–431.

[42] J. Haidt, The Righteous Mind, Vintage, New York, 2012.

[43] B.M. Muir, N. Moray, Trust in automation. Part II. Experimental studies of trust and human intervention in a process control simulation, Ergonomics 39 (3) (1996) 429–460, https://doi.org/10.1080/00140139608964474.

[44] M. Madsen, S. Gregor, Measuring human-computer trust, in: 11th Australasian Conference on Information Systems, 53, Citeseer, 2000, pp. 6–8. vol.

[45] D. Keller, S. Rice, System-wide versus component-specific trust using multiple aids, J. Gen. Psychol. Exp. Psychol. Comp. Psychol. 137 (1) (2009) 114–128.

[46] J.C. Walliser, E.J. de Visser, T.H. Shaw, Application of a system-wide trust strategy when supervising multiple autonomous agents, in: Proceedings of the Human Factors and Ergonomics Society Annual Meeting, vol. 60, SAGE Publications, Los Angeles, CA, 2016, pp. 133–137.

[47] U. Neisser, Cognition and Reality, WH Freeman, San Francisco, CA, 1976.

[48] N.A. Stanton, P.M. Salmon, G.H. Walker, D. Jenkins, Genotype and phenotype schemata and their role in distributed situation awareness in collaborative systems, Theor. Issues Ergon. Sci. 10 (1) (2009) 43–68.

[49] K.L. Plant, N.A. Stanton, Why did the pilots shut down the wrong engine? Explaining errors in context using Schema Theory and the Perceptual Cycle Model, Saf. Sci. 50 (2) (2012) 300–315.

[50] V.A. Banks, K.L. Plant, N.A. Stanton, Driver error or designer error: using the Perceptual Cycle Model to explore the circumstances surrounding the fatal Tesla crash on 7th May 2016, Saf. Sci. 108 (2018) 278–285.

[51] K.J. Parnell, J. Fischer, J. Clarck, A. Bodenman, M.J. Galvez Trigo, M.P. Brito, M.D. Soorati, K. Plant, S. Ramchurn, et al., Trustworthy UAV relationships: applying the Schema Action World taxonomy to UAVs and UAV swarm operations, Int. J. Hum.-Comput. Interact. (2022) 1–17.

[52] V.A. Banks, C.K. Allison, K.L. Plant, K.J. Parnell, N.A. Stanton, Using the Perceptual Cycle Model and Schema World Action Research Method to generate design requirements for new avionic systems, Hum. Factors Ergon. Manuf. Serv. Ind. 31 (1) (2021) 66–75.

[53] K.L. Plant, N.A. Stanton, The development of the Schema World Action Research Method (SWARM) for the elicitation of perceptual cycle data, Theor. Issues Ergon. Sci. 17 (4) (2016) 376–401.

[54] S.M. Merritt, Affective processes in human-automation interactions, Hum. Factors 53 (4) (2011) 356–370.

[55] S.M. Merritt, H. Heimbaugh, J. LaChapell, D. Lee, I trust it, but I don't know why: effects of implicit attitudes toward automation on trust in an automated system, Hum. Factors 55 (3) (2013) 520–534.

[56] C.J. Pearson, C.B. Mayhorn, The effects of pedigree and source type on trust in a dual adviser context, in: Proceedings of the Human Factors and Ergonomics Society Annual Meeting, 61, SAGE Publications, Los Angeles, CA, 2017, pp. 319–323. vol.

[57] J.-Y. Jian, A.M. Bisantz, C.G. Drury, Foundations for an empirically determined scale of trust in automated systems, Int. J. Cogn. Ergon. 4 (1) (2000) 53–71.

[58] J. Nelson, T. Gorichanaz, Trust as an ethical value in emerging technology governance: the case of drone regulation, Technol. Soc. 59 (2019) 101131.

[59] J.S. McCarley, C.D. Wickens, Human factors implications of UAVs in the national airspace, US Federal Aviation Administration, 2005. https://www.tc.faa.gov/logistics/Grants/pdf/2004/04-G-032.pdf.

[60] A. Hussein, H. Abbass, Mixed initiative systems for human-swarm interaction: opportunities and challenges, in: 2018 2nd Annual Systems Modelling Conference (SMC), IEEE, 2018, pp. 1–8.

[61] A. Kolling, P. Walker, N. Chakraborty, K. Sycara, M. Lewis, Human interaction with robot swarms: a survey, IEEE Trans. Hum.-Mach. Syst. 46 (1) (2015) 9–26.

[62] D.S. Brown, M.A. Goodrich, S.-Y. Jung, S. Kerman, Two invariants of human-swarm interaction, Air Force Research Lab, Rome, NY, United States, 2016. Technical Report.

[63] V. Braun, V. Clarke, Using thematic analysis in psychology, Qual. Res. Psychol. 3 (2) (2006) 77–101.

[64] S.T. Mueller, R.R. Hoffman, W. Clancey, A. Emrey, G. Klein, Explanation in human-AI systems: a literature meta-review, synopsis of key ideas and publications, and bibliography for explainable AI, arXiv (2019), https://doi.org/10.48550/arXiv.1902.01876.

[65] S.M. Lundberg, G. Erion, H. Chen, A. DeGrave, J.M. Prutkin, B. Nair, R. Katz, J. Himmelfarb, N. Bansal, S.-I. Lee, From local explanations to global understanding with explainable AI for trees, Nat. Mach. Intell. 2 (1) (2020) 56–67.

[66] T. Miller, Explanation in artificial intelligence: insights from the social sciences, Artif. Intell. 267 (2019) 1–38.

[67] K. Atkinson, T. Bench-Capon, D. Bollegala, Explanation in AI and law: past, present and future, Artif. Intell. 289 (2020) 103387.

[68] A.B. Arrieta, N. Díaz-Rodríguez, J. Del Ser, A. Bennetot, S. Tabik, A. Barbado, S. García, S. Gil-López, D. Molina, R. Benjamins, et al., Explainable artificial intelligence (XAI): concepts, taxonomies, opportunities and challenges toward responsible AI, Inf. Fusion 58 (2020) 82–115.

[69] M. Naiseh, N. Jiang, J. Ma, R. Ali, Personalising explainable recommendations: literature and conceptualisation, in: World Conference on Information Systems and Technologies, Springer, 2020, pp. 518–533.

[70] S. Anjomshoae, A. Najjar, D. Calvaresi, K. Främling, Explainable agents and robots: results from a systematic literature review, in: 18th International Conference on Autonomous Agents and Multiagent Systems (AAMAS 2019), Montreal, Canada, May 13–17, 2019, International Foundation for Autonomous Agents and Multiagent Systems, 2019, pp. 1078–1088.

[71] S. Almagor, M. Lahijanian, Explainable multi agent path finding, in: AAMAS, 2020.

[72] S. Kraus, A. Azaria, J. Fiosina, M. Greve, N. Hazon, L. Kolbe, T.-B. Lembcke, J.P. Muller, S. Schleibaum, M. Vollrath, AI for explaining decisions in multi-agent environments, in: Proceedings of the AAAI Conference on Artificial Intelligence, vol. 34, 2020, pp. 13534–13538.

[73] Y. Mualla, I. Tchappi, T. Kampik, A. Najjar, D. Calvaresi, A. Abbas-Turki, S. Galland, C. Nicolle, The quest of parsimonious XAI: a human-agent architecture for explanation formulation, Artif. Intell. 302 (2022) 103573.

[74] M. Naiseh, M.D. Soorati, S. Ramchurn, Outlining the design space of eXplainable swarm (xSwarm): experts perspective, arXiv preprint arXiv:2309.01269 (2023).

[75] M.G. Morgan, Use (and abuse) of expert elicitation in support of decision making for public policy, Proc. Natl. Acad. Sci. 111 (20) (2014) 7176–7184.

[76] G.A. Klein, R. Calderwood, D. Macgregor, Critical decision method for eliciting knowledge, IEEE Trans. Syst. Man Cybern. 19 (3) (1989) 462–472.

[77] J.W. Drisko, T. Maschi, Content analysis, Pocket Guide to Social Work Research Methods, 2016.

[78] J. Patel, Y. Xu, C. Pinciroli, Mixed-granularity human-swarm interaction, in: 2019 International Conference on Robotics and Automation (ICRA), IEEE, 2019, pp. 1059–1065.

[79] O. Bjurling, R. Granlund, J. Alfredson, M. Arvola, T. Ziemke, Drone swarms in forest firefighting: a local development case study of multi-level human-swarm interaction, in: Proceedings of the 11th Nordic Conference on Human-Computer Interaction: Shaping Experiences, Shaping Society, 2020, pp. 1–7.

[80] R. Liu, Z. Cai, M. Lewis, J. Lyons, K. Sycara, Trust repair in human-swarm teams+, in: 2019 28th IEEE International Conference on Robot and Human Interactive Communication (RO-MAN), IEEE, 2019, pp. 1–6.

[81] H. Holbrook III, A scenario-based methodology for conducting requirements elicitation, ACM SIGSOFT Softw. Eng. Notes 15 (1) (1990) 95–104.

[82] F. Saffre, H. Hildmann, H. Karvonen, The design challenges of drone swarm control, in: International Conference on Human-Computer Interaction, Springer, 2021, pp. 408–426.

[83] N. Koenig, A. Howard, Design and use paradigms for gazebo, an open-source multi-robot simulator, in: 2004 IEEE/RSJ International Conference on Intelligent Robots and Systems (IROS)(IEEE Cat. No. 04CH37566), vol. 3, IEEE, 2004, pp. 2149–2154.

[84] O. Michel, Cyberbotics Ltd. Webots^{TM}: professional mobile robot simulation, Int. J. Adv. Robot. Syst. 1 (1) (2004) 5.

[85] C. Pinciroli, V. Trianni, R. O'Grady, G. Pini, A. Brutschy, M. Brambilla, N. Mathews, E. Ferrante, G. Di Caro, F. Ducatelle, et al., ARGoS: a modular, parallel, multi-engine simulator for multi-robot systems, Swarm Intell. 6 (4) (2012) 271–295.

[86] L. He, P. Glogowski, K. Lemmerz, B. Kuhlenkötter, W. Zhang, Method to integrate human simulation into gazebo for human-robot collaboration, IOP Conf. Ser. Mater. Sci. Eng. 825 (1) (2020) 012006.

[87] S.D. Ramchurn, J.E. Fischer, Y. Ikuno, F. Wu, J. Flann, A. Waldock, A study of human-agent collaboration for multi-UAV task allocation in dynamic environments, in: Twenty-Fourth International Joint Conference on Artificial Intelligence, 2015, pp. 1184–1192.

[88] A. Rogers, A. Farinelli, R. Stranders, N.R. Jennings, Bounded approximate decentralised coordination via the max-sum algorithm, Artif. Intell. 175 (2) (2011) 730–759.

[89] H. Choset, P. Pignon, Coverage path planning: the boustrophedon cellular decomposition, in: Field and Service Robotics, Springer, 1998, pp. 203–209.

[90] ISO/IEC, ISO/IEC 9126. Software Engineering—Product Quality, ISO/IEC, 2001.

8

Building trust with the ethical affordances of education technologies: A sociotechnical systems perspective

Jordan Richard Schoenherr[a,b], Erin Chiou[c], and Maria Goldshtein[c]

[a]CONCORDIA UNIVERSITY, MONTREAL, QC, CANADA [b]CARLETON UNIVERSITY, OTTAWA, ON, CANADA [c]ARIZONA STATE UNIVERSITY, TEMPE, AZ, UNITED STATES

1 Introduction

Technologies used to support learning objectives (collectively referred to as learning technologies) offer many promises, including greater accessibility (geographically and socioeconomically), scalability, consistency, and flexibility of instruction (e.g., Refs. [1,2]). In particular, artificial intelligence (AI) as a supplement or replacement to traditional instruction and assessment has received considerable attention [3–7]. However, by extending certain human capabilities and capacities through technology, trade-offs can be created. These trade-offs have the potential to affect people's trust and subsequent use of those learning technologies at multiple levels of the education sociotechnical system (STS), e.g., students, teachers, administrators, organizations, and communities. Stakeholders involved in the design, development, and implementation of these learning technologies must therefore seek to understand the factors that affect technology adoption [8–10].

To ensure integrity in educational activities, learning engineering (the process of creating instructional and assessment technologies) must consider the ethical affordances that mediate instructor-learner and institution-learner relationships. Despite the availability of ethical norms proposed by professional organizations (e.g., IEEE, ACM, APA [11]), increasing ubiquity and complexity of learning technologies present unique social and ethical challenges. These challenges stem from the novel functional affordances of learning technologies and, therefore, are not always addressed by existing ethical norms and conventions. Consequently, relevant ethical norms are not always incorporated in the development of learning technologies. Furthermore, ethical and technical norms and conventions will coevolve over time as societies and technologies change [12], requiring updates to their assessment, codification, and implementation. Concurrently, the

Putting AI in the Critical Loop. https://doi.org/10.1016/B978-0-443-15988-6.00003-0

availability of informal learning activities and the need for data portability between organizations pose specific challenges for higher education (HE).

Coupled with these comparatively novel concerns are traditional issues that extend
into the virtual domain, including records management, data integrity, educator competencies, and learner academic misconduct (e.g., Refs. [13,14]). The existence of concerns
at multiple levels of a networked system highlights the need for an STS analysis for determining technology trustworthiness in HE.

Engaging the many stakeholders of the education STS and promoting their trust in
technology is not necessarily a straightforward task. Complex, AI-enabled technologies
are often described as "black boxes" due to the knowledge that is required to understand
how they operate, and the lack of access end-users have to that information [15–17]. These
issues have received increasing attention due to potential biases in consequential
decision-making, such as university candidate selection [18] or the assignment of grades
to learners [19]. While there has been a drive for more "explainable" and "transparent" systems, these criteria are not solutions themselves. Operationalizing explainability and
transparency across levels of a system requires correctly identifying the goals, knowledge,
and relationships of the respective stakeholders (e.g., Refs. [20–25]).

Higher education as a system. HE can be represented as a complex STS defined by a
plurality of technologies (e.g., learning management systems, AI-based tutors) and social
organizations (e.g., universities, regional and national government, professional organizations; [26–29]). Educators, learners, and administrators might accept the presence of these
technologies without formal consideration of their reliability, accuracy, and other performance metrics. This passive acceptance can lead to a failure to consider the social and
ethical issues that are inextricably bound to instruction and assessment (e.g., Refs.
[7,30]). For instance, while the contents of a course reflect a degree program's curriculum,
the knowledge required to use technologies (e.g., software, learning management systems,
etc.) is often part of an informal curriculum defined by implicit expectations that learners
and instructors will acquire this knowledge independently (for discussions of the differences between explicit and implicit curriculum, see Refs. [31–34]).

If learners do not trust institutional learning technologies, or if learners have greater
trust in alternative systems, they might go outside the formal institutional structure, making use of informal social networks and technologies to facilitate learning goals and bypass
institutional bureaucracy. A straightforward means to promote trust is to ensure that technologies are designed, evaluated, and refined using considerations of ethics [11,12,35]. Yet,
ethical analyses of educational technologies remain limited. For instance, in a survey of
learning analytics, Viberg et al. [36] found that only 18% of studies considered ethical
issues, with the majority focusing on privacy issues and data ownership (e.g., Refs.
[37,38]).

A survey of social and ethical issues must also go outside the formal structures of HE by
examining the products offered by third-party service providers. Educators have begun to
realize the importance of back channels that exploit existing pathways of interpersonal
communication (e.g., CourseHero, Discord, or Facebook). While adaptation of

technologies that function as back channels in HE has many benefits, including familiarity, limited or no training (i.e., if these are technologies already being used for other purposes), and rapid dissemination of information, they can also be used to facilitate academic misconduct and could lead to the spread of misinformation and disinformation in populations of learners, adversely impacting grades.

The number of potential social and ethical issues are as vast as the technologies used in HE. While ethical issues of learning technologies can include recruitment and candidate selection [39], this chapter focuses on the STS of learning engineering that affects learners, educators, and administrators in HE (Fig. 8.1). In Section 2, we review the value frameworks that can be applied to learning technologies with special attention given to psychometric frameworks that highlight the value-laden nature of assessment (i.e., Ref. [30]). We argue that this framework presents a clear means to inform value-based design in learning technologies, especially AI-based technologies used for assessment. Then, in Section 3, we survey AI-based learning technologies that facilitate instruction and learner assessment, i.e., adaptive instruction systems (AIS), computerized adaptive testing (CAT), and automated writing evaluation (AWE). In Section 4, we consider the ethical requirements of Integrated/Comprehensive Learner Records (ILRs/CLRs) in terms of privacy and security issues. In Section 5, we review the general features of learning management systems (LMS) and informal back channels. Here, we highlight the importance of understanding academic misconduct strategies adopted by learners in online learning environments (especially MOOCs) and student preferences for third-party back channels. These latter resources create clear vulnerabilities to academic integrity even if they are

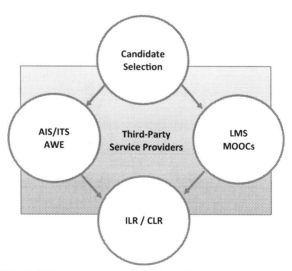

FIG. 8.1 Ethical vulnerabilities in higher education associated with learning and support technologies. Social and ethical issues can arise by introducing technologies into the candidate selection and recruitment process, the use of instruction and assessment methods and learning management systems, and collection, retention, and use of learner information in databases.

monitored by HE institutions. Finally, in Section 6 we propose a framework for learning engineering that incorporates principles of responsive and resilient design and ethical sensemaking, to address the social and ethical issues discussed.

2 Operationalizing ethics in learning engineering: From values to assessment

Ethical issues in learning engineering are exceptionally broad. They include general scientific norms such as authorship/contributorship [40–42], publication practices [43–46], research ethics [47], and data integrity issues [48,49]. Ideally, these issues would be directly addressed in the formal curriculum, which consists of the required courses, educational activities, and milestones that a student is expected to complete. However, ethics is often not addressed in methods courses or textbooks (e.g., for an analysis in the behavioral and social sciences, see Ref. [50]), leaving the development of these competencies and regulation of their practice to supervisors or mentors. This is arguably problematic as some ethical affordances (e.g., privacy engineering methodologies; [51–53]) are not necessarily seen as relevant or feasible by engineers [54,55].

Learning engineering faces many unique challenges, given the need to incorporate norms developed in science, engineering, and education (see Table 8.1; [11,12,56]). These norms, when codified as frameworks, can identify unrelated or overlapping value dimensions. For instance, the IEEE has created both the *Code of Ethics* for engineers as well as the guidance document, *Ethically Aligned Design*. In these documents, developers are directed toward creating systems that empower individuals to access and control their

Table 8.1 Value dimensions identified by professional ethics frameworks.

Principle	General description
Accountability/responsibility	Ensure that responsibility is effectively allocated between users, developers, and distributors of AIs taking into consideration any power asymmetries and historical inequalities that might exist
Explainability/transparency	Ensure that the datasets and operations of AIs are intelligible/traceable to stakeholders by identifying the level of detail required to address their goals
Fairness/nondiscrimination	Ensure that the operations, outcomes, and consequences of AI use benefit all stakeholders equally regardless of their power and status
Integrity	Ensure that the datasets and design are without the types of common biases that can limit access or disproportionately harm marginalized groups, including biases that may benefit the developer or distributor at the expense of the user
Privacy/confidentiality	Ensure that personal or identifiable information remains private and under their source's control, while additionally taking into account the use context and cultural norms
Safety/security	Ensure that the accumulation of data or the use of AI enables the safety of users and does not threaten the physical or psychological safety of the user or their community

Adapted from J.R. Schoenherr, Learning engineering is ethical, in: J. Goodell (Ed.), Learning Engineering Toolkit, Routledge Publishing, 2022b; J.R. Schoenherr, Ethical Artificial Intelligence from Popular to Cognitive Science: Trust in the age of Entanglement, Taylor & Francis, 2022a; J.R. Schoenherr, Whose privacy, what surveillance? Dimensions of the mental models for privacy and security, IEEE Technol. Soc. Mag. 41 (2022c) 54–65.

data [57]. Focusing on software development, the ACM [58] also acknowledges the need to address the inevitability of coding errors, how software updates will affect users, and the need to consider the implementation context.

Learning engineers must also consider codes of ethics that apply to assessment. The American National Council on Measurement in Education developed the *Code of Fair Testing Practices in Education* [59], which considers ethics from the perspective of those who construct tests (test developers) as well as those who commission, select, administer, and make decisions on the basis of those tests (test users). For instance, test developers must be able to describe how assessment instruments were developed and the evidence that supports their use, and provide support for their competent administration as well as modified forms of the test for people with disabilities [60]. In contrast, test users must take into consideration implementation issues (e.g., testing conditions and modifications), establishment of performance standards, avoidance of using a single test and corresponding score for assessment, and they must ensure that they take all these factors into consideration when interpreting and using test results. These requirements highlight the need to develop explainable instruction and assessment systems, criteria that are made more problematic when examining sophisticated learning technologies, e.g., AI-enabled systems that use algorithms and large, dynamic datasets.

2.1 Ethical sensemaking in design

In addition to the technical and functional affordances (Here, following Schoenherr [11,12,56], we assume that technical affordances reflect the technical features (e.g., hardware and software) whereas functional affordances reflect the goal-directed features of design. Analogous to the distinction made by Marr between algorithmic and implementational levels.) of a system, we must also be capable of identifying its *ethical* affordances. Ethical affordances reflect those features of a system which relate to social and ethical issues that are relevant within a community of users and practitioners. For instance, users might be encouraged to accept the terms and conditions that might not be in their best interests by making the default option "agree to all." Alternatively, if the creation of a user account requires providing demographic information, users might be forced to select ethnic or gender categories to classify themselves that do not accurately align with their identity. While these choices might be pragmatic—and users might even perceive them as such—their long-term use can greatly constrain the perceived choices of users, discouraging and disempowering them.

To identify ethical affordances, values and ethics must first be *operationalizable* for them to be included in the process or products of design [12,35]. Further complicating this requirement, learning engineers must identify and address multiple values, i.e., ethical pluralism [35]. For instance, software developers are frequently expected to resolve issues of security, privacy, and transparency concurrently in their design solutions. Many would argue that values are qualitatively different, i.e., they are incommensurable and nonsupervenient. However, if we consider values as a multidimensional space, it is also likely that these values need not be orthogonal and can be complementary. For instance, network surveillance can pre-serve privacy by ensuring that user data is not exfiltrated and used by malicious actors.

Learning engineers must acknowledge that the design and development of learning technologies are necessarily informed by value judgments that are made through the design process. Value-centered design frameworks have been proposed as an effective means to *explicitly* address these concerns, e.g., value-sensitive design, [61], reflective design [62], and design justice [63] (for reviews, see Refs. [11,12]), especially within or for specific populations and their needs. Yet, scaling these learning technologies can also mean that products are developed in one country but adopted in many. Thus cross-cultural differences in values should be acknowledged and relevant adaptations should be explored, tested, and implemented (see Chapter 4, [11]). For instance, cultural norms can affect the nature and extent of codification of ethical standards [64], leading to situ-ations in which many norms and conventions are implicit or might not be addressed. This is likely to be a complicated task, as cross-cultural surveys have found different rates and motivations associated with cheating [65–68] cf. [69]. Special attention should be directed toward values that apply to engineering and those associated with software development [70–72] and use [73].

Supporting these design considerations, given increasing investment and develop-ment of AI-enabled systems, the IEEE [74] suggests that system designers should identify the relevant values and norms of the groups affected by the autonomous and intelligent systems, how these norms are implemented within the AI, and evaluate the compatibility and alignment of the norms and values between humans and AI. Complementing ethical frameworks in education and engineering, governments (e.g., African Union, European Union, South Korea, Taiwan, United Kingdom; for reviews, see Refs. [56,75]) and nongo-vernmental groups (e.g., UNESCO; [76]) have also developed guidelines that address infor-mation ethics and AI.

At the same time, emphasis on cross-cultural differences ignores the many parallels that exist between value frameworks and underlying professional values shared by engi-neers [77,78]. For instance, the fact that individualistic and collectivistic societies might emphasize the benefits of individual learners or of society as a whole reflects separate rather than conflicting value dimensions [79]. Moreover, both Eastern and Western phil-osophical traditions have emphasized the value-laden nature of educational practices [80,81]. Similarly, moral education traditions exist in Western and Eastern societies that can be used to inform educational practices in engineering [82]. Concurrently, ethical tra-ditions incorporate aspects of virtue theory, which places an emphasis on ethical sense-making rather than solely focusing on principles or consequence (Chapter 4, [12]).

2.2 Psychometrics and validity

Validity represents a core concept and value in psychometrics. The challenges associated with demonstrating validity serve as a useful analogy for addressing the social and ethical concerns and design and implementation challenges in learning engineering. Historically, psychometricians argued for the existence of multiple *kinds* of validity (e.g., convergent, concurrent, predictive). Integrating these accounts, Messick [30] suggested that each kind of validity was in fact a source of *evidence for* the existence of a construct, e.g., competency of a test-taker in a particular area that is being assessed by a test. With a sufficient quantity of evidence from diverse sources, researchers can support a validity argument ([83]; see Table 8.2).

Learning engineers must not only address whether an instructional method or assessment instrument *could* be used to improve or assess performance. Rather, they must also consider *how* and *why* a test is being adopted (e.g., why an institution is conducting an assessment, and what the benefits are relative to the costs; [84]) as well as the social consequences of test-taking on the learner (Fig. 8.2). By considering the motivation for the adoption of an instructional method or assessment instrument, alternative solutions can be evaluated that might perform a task equally well.

By introducing a learning technology, many value-laden assessment decisions will be rendered opaque, embedding many implicit value judgments within the test content, data structures (i.e., "black data"; [85]), as well as learning and assessment criteria (i.e., "black box" algorithms; for a review of these general issues, see Chapter 2, [12,24]). In the context of HE, provisions must be made to ensure that the validity of constituent testing and selection criteria are theoretically and empirically supported, and that the selection process is transparent and explainable.

Beyond these social and ethical issues, the presence of antidiscrimination legislation introduces regulatory and legal issues in terms of compliance with existing standards (e.g., General Data Protection Regulation in the European Union; Title VII of the Civil Rights Act of 1964 in the United States; IEEE P7010 Well-being Metric for AIS). For instance, the use of learning analytics in candidate selection assumes that the errors (e.g., misses) and biases (i.e., discrimination) can be overcome, leading to more effective decision-making. Yet, biases are often imported with the datasets, meaning that AI-based selection mechanisms might obscure, rather than solve, these problems [39]. This issue is equally concerning for

Table 8.2 Source of validity evidence for a learning or assessment instrument.

Validity evidence	Description
Content	The extent to which the content effectively samples the full range of tasks or items that are required to assess a construct
Response process	Whether the learning or assessment instrument uses the mental processes that are believed to be associated with a competency
Internal structure	How the features or dimensions of a learning or assessment instrument relate to one another
Relation to other variables	How a learner's responses on a learning or assessment instrument compare to their performance on existing instruments or tasks that should be related
Generalizability	The extent to which the assessment instrument's efficacy is comparable in multiple usage contexts
Social consequences	How the outcomes of test-taking affect the learner within a larger social framework

Modified from Messick's Construct Validity Framework.

	Test Interpretation	Test Use
Evidential Basis	*Construct Validity (CV)*	*CV + Relevance/Utility*
Consequential Basis	*CV + Value Implications (VI)*	*CV + R/U + VI + Social Consequences*

FIG. 8.2 Facets of construct validity. *Adapted from S. Messick, Validity of psychological assessment: validation of inferences from persons' responses and performances as scientific inquiry into score meaning, Am. Psychol. 50 (1995) 741–749.*

AI-based instruction and assessment technologies. Learning engineers must ensure that these standards are accounted for during design and development. The use of learning technologies that can be adopted globally, and that are not transparent, requires that learning engineers directly address these questions.

2.3 Operationalizing ethics in psychometrics: Differential item function

At the core of psychometrics is the notion that items assess the attitudes, relevant traits, or competency of learners. Item response theory (IRT; [86,87]) provides one means to determine the effectiveness of an assessment technique and individual items by examining the response of an individual to a given item and their overall performance on an assessment instrument. Differential item function (DIF; [88–90]) further extends this idea by considering the statistical properties of a specific test item that demonstrates differential assessment of performance for different groups (e.g., "race," "gender," socioeconomic status) within a population. It requires identifying individuals' overall competency, and then examining whether specific items differ from one group to another. For instance, if individuals from Group A and Group B obtain the same average score on a test, we might find that a specific question produces high scores for one group and low scores for another. However, simply because DIF is observed does not necessarily mean that items are biased. For instance, students' prior knowledge and experience might *justify* their performance in a course.

Schoenherr [35] has argued that psychometric techniques like DIF can be used to operationalize ethical concerns such as fairness, accuracy, transparency, and ethics, or FATE. Consequently, adaptive instruction systems (AIS) or intelligent tutoring systems (ITS) could be granted a *degree* of ethical autonomy if these psychometric techniques were embedded within their operations. However, human designers and developers must still act as the primary adjudicators of social and ethical issues, in that they must select relevant social categories to assess, select the degree of reliability that is acceptable, and decide what trade-offs are acceptable to adjust scores. (For instance, Kearns and Roth [91] argue that reducing biases for one group might increase bias for another group. While not all decisions will reflect such zero-sum problems, these concerns must be considered.) Learning engineers must also go beyond the current variables that are used to assess the construct and implementation validity (Implementation validity reflects the extent to which an instructional method or assessment instrument can be reliably integrated into a programmatic context [60].) and consider the potential unintended consequences of adopting an assessment instrument, e.g., increasing economic inequality, digital divides, etc.

In the remainder of the chapter, we use this basic ethical framework to consider specific learning technologies that are currently in use.

3 AI-based technologies for instruction and assessment

AI-based learning technologies have increased in prevalence within recent decades. Whereas computer adaptive testing (CAT) represents one of the earliest learning technologies [92–96], learning analytics [36,97], adaptive instructional systems (AIS) / intelligent tutoring systems (ITS; [98–102]), and automated writing evaluation (AWE) are more recent additions to the STSs of education.

AI-based systems have the potential to provide individualized and efficient learning experiences beyond the capacities of an overburdened educational system [103,104]. Techniques such as learning analytics can be used to identify students whose performance suggests that they might require additional educational interventions. For instance, the University of South Australia used learning analytics to identify a cohort of students who were at risk of failing or attrition. Of those contacted as a result of this identification process, 66% passed their course, compared to 52% who were not contacted [105].

In the long run, AI-based instructional and assessment technologies might be more reliable than human instructors, being capable of providing learners with personalized feedback in large courses. However, when using AI-based systems such as LA, CAT, AIS/ITS, and AWE, the curriculum for educators will need to change to ensure that the strengths and limitations of these systems are well understood and that these tools are evaluated and made responsive to the various stakeholders' needs [35]. The relevant stakeholders in this case include the students, instructors, and administrators [106]. Moreover, as with massive online open courses (MOOCs), the use of these systems also raises concerns for the future experiences that might otherwise be given to students (as teaching assistants) and postgraduates (as instructors), and whether these systems can be misused to spread disinformation, or to cheat.

3.1 AI-enabled learning and feedback

Learners produce copious amounts of data that are often neglected. Cognitive strategies and behavioral traces associated with studying, writing, and test taking are often left unexamined. For instance, a key feature of expertise that is often not considered in professional education is the speed of the learner's responses, which might suggest a response strategy, e.g., speed-accuracy trade-off [35,107]. Another area that remains comparatively under-used by instructors is the nature of the errors made by learners. Incorrect responses selected by the learner are often as informative, as they suggest what the learner currently knows. For instance, techniques for the assessment of partial knowledge assumes that response alternatives can be differentially weighted [108–110].

The provision of "partial marks" provides more than just assessing learning performance. Partial marks also represent an important source of feedback to learners [111–113]. In contrast to conceptualizing feedback as unidirectional information transmission to the learner, feedback can also reflect an interactive cycle of receiving information, responding to information, and altering response selection strategies [114,115]. This iterative cycle is at the core of AIS/ITS. At the level of social cognition, this reflects an understanding of a learner's mental model of the subject matter that can then be used to improve the efficacy of feedback and learning strategies, and personalize those things to improve the learner's experience and gains [35]. This technological responsivity has been suggested to improve trusting relationships with technology [116].

In the context of AIS/ITS the identification of learning strategies [117] could be facilitated by accounting for systematic patterns (e.g., errors, speed-accuracy trade-offs) in a way that exceeds the capacity of educators with limited time [118]. For instance, some ITSs have used gaze patterns to identify what information and responses learners are attending to in a display [99,119]. By examining the features of a problem or task that a learner is considering prior to response selection as well as the amount of vacillation between response alternatives, a more comprehensive understanding of the learner's mental models of the test material can be developed. In a similar vein, affective computing considers the importance of users' affective responses [120,121]. The ability to detect anxiety or stress [122] in learners can provide valuable information in terms of when learners perform poorly for reasons other than their content knowledge. However, learning engineers must be mindful that by recording biometric data (e.g., gaze patterns) and affective responses, they are increasing the sensitive nature of the dataset. Finally, like all AIs, we must consider the training sets used to develop systems like AIS/ITS: if the dataset used to train the system is not representative of the learner population that AIS/ITS will (or could) be used within, then there is a distinct possibility that learning strategies might differ, producing suboptimal learning outcomes.

3.2 AI-enabled assessment

AI can also be used to evaluate learner performance in terms of CAT and AWE. Computerized testing remains at the core of learning technologies such as CAT and LMS [123]. Some ethical issues remain outside the test-taking context. For instance, a perennial issue in test administration is authentication of a test-taker's identity. If we cannot rapidly and accurately authenticate a test-taker's identity, it is irrelevant how effective instructional or assessment strategies are, as performance cannot be accurately assigned to a learner. Concurrently, we might also have concerns regarding the security of the information obtained from the learner. Finally, following general concerns of validity, most ethical issues concern the structure and content of the assessment instrument.

3.2.1 Computerized adaptive testing

Whereas test items can be provided to learners from a random subset of questions from a database in LMSs, CAT provides a more systematic and optimal administration of items for each learner. Optimality is judged in terms of a theoretical trait or competency. Test items are selected from a pool with the assumption that the learner will be equally likely to respond correctly for a set of items. Over time, item difficulty is increased or decreased until the learners' level of competency is identified. To achieve this, CAT developers must address validity issues in terms of estimating a trait or competency, creating appropriate test item banks, and developing item selection procedures.

Ethical issues related to CAT are most clearly illustrated in terms of test integrity and item security [124]. A clear benefit of the computerized testing is that each learner is presented with a unique testing experience. However, the test items within a pool reflect a finite set, meaning that the probability of any two learners receiving the same item is nonzero. If a sufficiently large number of learners take the same test wherein items are not retired or replaced, a test-taker with malicious intent could conceivably reconstruct the content and items that were used in the CAT. This very issue emerged in the 1990s with the Graduate Record Examination (GRE), such that the Educational Testing Service halted the use of the CAT until this vulnerability was addressed [125], a decision that affected 1000 test-takers [126]. However, many strategies can be adopted to address these concerns, such as identifying subsets of test items and varying the frequency of item exposure or temporarily removing items from a test bank (for a review, see Ref. [124]). Similarly, small changes in wording that affect which response alternative is correct can be used to increase the confusability of question variants for test-takers who would share or use CAMEO strategies (see Section 5.2, Massive Open Online Courses).

3.2.2 Automated writing evaluation (AWE)

As the number of learners in a course increases, an educator's ability to provide written assignments decreases if additional resources are not provided, e.g., teaching assistants. AWE tools are meant to streamline the work of human evaluators and provide consistent feedback across samples. AWE systems provide a straightforward means to address this concern. AWE typically uses natural language processing (NLP) to extract critical features from a text (e.g., semantic, syntactic) and uses algorithms to correlate those features with human raters' evaluation of essays and make generalizations about the properties of good and bad writing. Evidence suggests that AWE can provide a valid means to assess a learner's performance. However, learner motivation and multiple feedback modalities (specifically, a combination of AWE and teacher feedback) appear to be the most effective determinants of improvements (e.g., Refs. [104,127,128]).

Biases in writing assessment. Ethical concerns related to the development and use of AWE tools are not unique to those tools. Potentials for bias are inherent to any system that uses language data or other types of data that allow for identity categorization (e.g., photos of individuals, their audio data, demographic data) for automation. In addition, ethical concerns pertaining to source data and internal rubrics are relevant to all automated tools that perform evaluation that affects people's lives. Negative outcomes of algorithmic bias have been shown to have serious effects when unregulated and not sufficiently tested [16,129–131].

Writing tasks are commonly used to evaluate students' verbal abilities and reasoning abilities. However, writers are sometimes unfairly evaluated due to linguistic variables that do not pertain to the quality of their writing. Compared to written language, spoken language has more properties that allow a listener to speculate about a speaker's identity, e.g., their accent, pitch, intonation. Written language still has properties that can be tied to different identities by a reader, e.g., vocabulary, spelling, syntax, life examples used to support arguments. There is extensive evidence of linguistic biases that may affect writers' evaluation, either positively or negatively [132]. Readers and listeners have language attitudes toward different varieties of English, often prioritizing Standard American English (SAE) over varieties associated with less hegemonic ethnic, racial, and regional identities, e.g., African American, southern, Hispanic, and other varieties of English. By inferring writers' identities from their use of language, human evaluators might judge different authors' writing in a qualitatively different manner. Given that bias can be both explicit and implicit [133,134], the mere awareness of linguistic bias does not ensure that it can be eliminated.

As human raters are likely to have linguistic biases [135], these biases can be significant enough to affect the machine learning process, depending on their prevalence. This means that AWE algorithms can be importing biases into the system [136], a problem these tools are purported to alleviate. In addition, the makeup of the training dataset determines the AWE tool's standards of evaluation. This means that for equitable evaluation, the dataset must be representative of the populations that the AWE tool will evaluate. Given the existence of various dialects and varieties of English, judging one group and their language using linguistic data from another group would be infelicitous and unfair. To guarantee equity in AWE tool performance, the training datasets should be comparable to the data the AWE tool is expected to evaluate.

The implementation and use of AWE present additional ethical concerns. The properties an AWE algorithm associates with good and bad writing are used as an internal "rubric," a standard to which new writing samples are compared as they get evaluated. One ethical issue with this practice is that the internal rubric of the AWE tool is usually not communicated to the writer. The writer, then, has to produce what they perceive to be a good writing sample and hope that the assessment tool has conceptions of writing quality similar to their own. This lack of transparency and explainability puts writers at a disadvantage, not knowing what *exactly* is expected of them. Moreover, within the US educational system, where there are significant gaps in the quality of education different regions receive, some students are likely to receive poorer quality training, and be less pre-pared for standardized tests [137]. Not receiving detailed instruction on what is expected from them or specific exemplars, writers from minority populations of disadvantaged backgrounds might not have the same explicit knowledge of standard assessment criteria nor any implicit norms that might result from socialization.

In addition, we cannot assume that educators or institutions will receive sufficient information to be capable of understanding the technical affordances of AWE. Conse-quently, they might lack the ability to assess the efficacy of these systems. If the AWE tool's evaluation outputs are in line with their own practices and conceptions of what makes good writing good, it is unlikely that they will detect an existing problem; this would reflect confirmation bias. However, if there are differences, the lack of transparency and explain-ability of internal rubrics would make it hard for teachers to supplement the tool's eval-uation, and for institutions to locate the source of the misalignment. Automated tools often go through internal quality assurance procedures. However, that quality assurance does not normally involve future users. Thus detectable issues may go unnoticed in the process of developing the AWE, and remain unaddressed after the tool's deployment. Con-forming to conceptions of validity [30,83], the validity of AWE methods must be assessed throughout their development and implementation [60].

Without access to clear information about the AWE tool's operation and communica-tion channels for providing feedback, stakeholders might find themselves unmotivated to use the tool and not trusting its outputs. Another set of trust-related issues stems from language attitudes [138,139]. Human stakeholders in AWE systems might have concerns about standard language biases even if linguistic bias is not significant or not present in the AWE tool's assessment practices. The lived experiences of minoritized individuals and speakers of nonstandard dialects can cultivate the expectation of being judged unfairly [140]. Having experiences of bias based on one's use of language can lead them to expect that bias from people, institutions, and tools they interact with, even if bias does not end up occurring, i.e., overgeneralization. Thus an AWE system would need to be designed to identify or reduce biases, and to communicate those efforts to stakeholders interacting with said system, to increase trustworthiness [116].

Mapping out potentials for bias and distrust/mistrust [141,142] is crucial in the devel-opment and implementation of AI-based instructional and assessment systems. Interac-tion with potential stakeholders (students, teachers, and institutions) must inform bias and trust assessments, beyond internal assessment of bias and trust issues that companies perform in their development of tools. We return to the requirements of responsibility and accountability in the final part of this chapter.

3.3 Plagiarism and automated plagiarism detection

Plagiarism is a perennial problem within higher education, and relevant for academic integrity and research misconduct. Rather than a strict definition, plagiarism is likely best understood in terms of a family resemblance structure – a set of interrelated forms that share some, but not all, features [143,144]. For instance, verbatim plagiarism takes the form of copying of text in whole or in part, whereas structural plagiarism occurs when a writer presents a line argumentation in the same manner. In some cases, plagiarists make a marginal attempt at mirroring appropriate authorship practices by copying material from one source and attributing it to another source. (Appropriation of another program's code without attribution would fall within this definition. In both cases, material might reflect a violation of intellectual property.) Whereas plagiarism typically reflects representing the product of another's work as one's own, self-plagiarism occurs when an individual uses significant portions of text from their own work rather than attempting to reformulate or restructure content to align with the purpose of a specific document [145].

The internet and its associated search engines were developed to facilitate identification of relevant information. Consequently, these same systems have been exploited to identify resources that can be plagiarized in whole or part in a manner that students were not capable of doing previously [146,147]. A prominent example is the availability of student term-paper mills. Term-paper mills are not a new phenomenon [148,149]. Early estimates of this method of cheating suggest that between 2% and 14% of students purchased or otherwise obtained a paper and submitted it as their own work, suggesting that the behavior is reasonably common [150–152]. Third-party service providers have enabled more systematic and widespread availability of these practices. In contrast to note-sharing websites that provide general course information and test banks (e.g., NoteBro), paper mills are services that facilitate the interaction between a would-be paper writer and a learner who requires a paper on a specific topic [153]. Paralleling human-based paper mills, concerns have also been raised that the capabilities of AI can be used to write convincing responses to exam questions and papers, creating significant threats to learner's competency development and academic integrity (see discussions concerning generative AI such as ChatGPT; [154]). Thus, AWE might also need to be adapted to identify the possibility that a paper was written by an AI, creating an academic arms race.

3.3.1 Plagiarism detection

Like AWE, automated plagiarism detection provides another example of AI-enabled pattern recognition in assessment [155,156]. Elementary forms of automated plagiarism detection scan a document and examine its similarity to other target documents. Contemporary methods of plagiarism can also include changes in text style (stylometry), the co-occurrence of words (latent semantic analysis), comparing multiple documents within a set or a learner's submission to a database [155,157–159]. Hybridized methods can also be used to combine these approaches to provide a more robust means to identify anomalous materials that might reflect plagiarism [160].

Plagiarism detection mechanisms can be integrated in LMSs for automatic scanning. For instance, the popular open source LMS, Moodle, contains a plagiarism plug-in that detects paraphrased content and can provide a detailed report on the percentage of plagiarism present within a submitted document. This process occurs automatically after a learner has uploaded their submission. The percentage of similar text is then annotated next to the student's submission. By adopting such a process, educators can more strategically choose the order and nature of paper assessment. For instance, if the add-on indicates that it is probably a paper that contains plagiarism, educators can allocate their grading time accordingly. However, by relying on AI-based screening, educators might inadvertently engage in confirmation bias concerning the existence of plagiarism when assessing papers, perceiving plagiarism when none has occurred. Moreover, by using a percentage rating of content, educators might mistake this for a degree of confidence or reliability that the educator should place in the system.

When plagiarism is confirmed, we must acknowledge the value-laden nature of the decision-making process that went into developing these detection mechanisms as well as the social consequences. If penalties are to be assigned to students as a result of these detection mechanisms, they must be commensurate with the degree of certainty that learning engineers have in their system. However, automated plagiarism detection can reduce the subjectivity of plagiarism detection, identify students who might have incidentally copied information, and those who have intentionally committed acts of misconduct. Moreover, if students are aware of the system's abilities to successfully detect plagiarism, the knowledge of being monitored might act as a deterrent.

4 Knowledge management, learner records, and data lakes

Once learners have been assessed, learner records remain. Like assessment, learner records reflect one of the earliest forms of educational technologies. Historically, learner records are covered by records management policies describing standards for the retention, storage, access, and disposition of records [161]. Learner records can include information pertaining to a learner's performance in a course, their academic achievements, academic misconduct, student loans and internal payments, grant money and research funding, addresses, and date of birth (see also "data backpacks"; [162]).

Learner records can also be situated with larger concerns of organizational knowledge management (KM). KM reflects an often-neglected area of HE. KM is defined by a variety of methods directed toward the collection, storage, sharing, and processing of knowledge and information within an organizational context [163,164]. Effective and ethical KM strategies are critical to the mandate and operations of HE: once assessments have been conducted (e.g., quiz or written assignment), they are used to determine whether learners have successfully completed a milestone (e.g., passed a course, comprehensive), and the learner's achievement is retained in a learner record. Educators and institutions are then responsible for these data lakes and must ensure the integrity of these records in terms of accuracy, accessibility, and confidentiality given the social consequences of these records, e.g., access to graduate school, employment opportunities.

4.1 Integrated and comprehensive learner records

In contemporary HE, record retention must also consider new facets of issues associated with data integrity. Innovations such as Comprehensive Learner Records (CLRs) or Integrated Learner Records (ILRs) constitute novel approaches to ensure that standardized learner records can be developed [165,166]. These approaches are in response to a perceived need to ensure methods of credentialing can be consistently applied inside and outside HE. These records can include grades, credits, credentials, experiential learning, internships, externships, or other evidence of achievement.

Organizations such as IMS Learning Consortium and the American Workforce Policy Advisory Board have identified eight interrelated features that are required of CLRs. These features emphasize both the individual and collective value of this information (Table 8.3). For instance, while shareability allows learners to decide who has access to their records, verifiability allows an organization to confirm the authenticity of the record of achievement. Crucially, these standards also reinforce Messick's [30] observation that assessment instruments must add value to the learner and the organization and have important social consequences (e.g., continuing professional development credits).

Table 8.3   Interrelated features that defined CLR/ILR.

Feature	Description
Privacy	Accessibility of learner records is restricted by the learner to specific individuals/organizations, for specific purposes, and for a specific duration. Relevant privacy engineering methodologies and standards are incorporated into the design, product, and implementation of these systems
Relevance	Learner records should be meaningful and valuable to specific activities such as employment, advancement, and continued professional development. Recognized experts or credentialing organizations can endorse these records. Time limitations on records can lead to continued updating of credentials
Transparency	Learner records allow for direct comparison based on common frameworks, ontologies, and standards. Contextual information (e.g., data started/completed, credentialing context) should also be available
Security	Learner records conform to relevant (e.g., institutional, state/province, federal, international) security standards for the protection of data from unauthorized access, editing, or erasure
Equitability	Learner records afford users with educational, economic, and social mobility. Records should enable learners to advance or maintain their status within relevant organizations
Portability	Learner records can be used in multiple learning environments (i.e., provincial/state, private/public institutions) and for multiple purposes (e.g., education, employment). Users have control over how records are organized, where they are stored, and how they are combined
Interoperability	Common frameworks, ontologies, and standards are used to ensure that records are readable, exchangeable, and actionable across multiple technological systems. Data from multiple sources can be combined. Users from various sectors should be capable of understanding learner records
Shareability/verifiability	Learners can share CLRs/ILRs with prospective or current academic institutions and employers. Learners control accessibility of records. Organizations can digitally confirm the authenticity and integrity of the learner record

ILR/CLR standards present a useful case study to illustrate how multiple values or design features could come into conflict. For instance, while there is a clear need for transparency, portability, shareability, and interoperability, these values must also be balanced with those of security and privacy. This balancing requires practices such as privacy by design [167]. Beyond data anonymization techniques (e.g., *k*-anonymity, *l*-diversity, *t*-closeness [168]), privacy engineering methodologies (PEMs; [51–53]) reflect strategies that focus on reducing the amount of user data that is collected, conduct a privacy impact assessment of design features, and consider how the design process, policies, or product could adversely affect a user. However, principles alone are likely to be unsatisfactory if learning engineers do not see the value or practicality in including these design features [54,55].

In addition to issues of accuracy, retention standards, and privacy, records like CLR/ILR create novel issues associated with cybersecurity. University databases are tempting targets for cybercriminals, with the FBI reporting widespread credential theft [169]. Ransomware can be used to render learner records (and other data) inaccessible or lead to their destruction if demands are not met. These attacks are not uncommon, with one survey suggesting that 44% of institutions have experienced an attack and that the average payment was $112,435 USD for those institutions that paid it [170].

Cyberattack can also lead to data exfiltration. For instance, in 2019 the Australian National University discovered that a data breach had occurred in 2018 leading to a loss of 19 years of data [171]. In that learner records contain the preferences (e.g., courses attended) and behavior (e.g., grade achieved) of learners, the information can be used to blackmail the learner or to target them in social influence campaigns. As learner records could soon include biometric data and behavioral traces, this process of datafication must be considered by HE institutions.

Institutions that have faced these attacks have suggested the adoption of several strategies to address these concerns [172]. These strategies include the use of multifactor authentication, using a tier network model to compartmentalize information, establishing collaborative networks with national institutions and agencies, and the use of endpoint detection and response (EDR) or, ideally, extended detection and response (XDR) strategies. (These approaches focus on network surveillance and anomaly detection of the endpoints (EDR) or the network as a whole (XDR).) The prevalence, evolution, and consequentiality of cybercrimes highlight the importance of ILR/CLR and KM as well as the need to perceive HE as an STS of heterogeneous systems.

5 Learning systems inside and outside higher education

Social network analysis typically differentiates between formal and informal social networks within an organization. HE is no different in this regard. Learning management systems (LMSs) present a basic infrastructure for the STS of HE. However, any consideration of these systems quickly reveals that back channels are frequently used by students and third parties, which can both complement and conflict with the learning objectives of HE and the private sector.

5.1 Learning management systems

LMSs remain the most common technology allowing learners, instructors, and other members of the HE community to interact. These systems can now be complemented with other services like cloud storage for media and the plagiarism detection and AWE systems noted earlier. LMSs have become an essential mediating technology in HE in that they are used to upload assignments, post grades, instant message, and contribute to asynchronous discussion boards. To be usable, LMSs need to be reliable and trustworthy, reflected through file integrity during upload, grading completeness, and system stability, especially during critical milestones or timeframes, like the delivery of a quiz. By using LMSs, academic integrity issues are also introduced in terms of authenticating learners' performance when submitting assignments by using identity tracking or plagiarism detection software. This is manifested most recently in the phenomenon of ghost-students, individuals that are paid to complete the course for a particular learner [173].

LMSs can also have indirect effects on learners in terms of the experience of anonymity and reductions in engagement, and can alter the kind of learning strategies that are adopted. For instance, while promoting flexibility, asynchronous online courses (especially massive open online courses, or MOOCs) might decrease the experience of being monitored and promote anonymity, leading to reduction in performance and increased antisocial behavior. Similarly, the asynchronous modality can affect long-term outcomes by reducing the need for educators, thereby limiting the availability of jobs for future academics. Nevertheless, these approaches to delivering instruction can increase accessibility of HE, introducing a need to consider how these alternative educational activities can be credentialed and whether course delivery will become a free market.

An important feature of LMSs is their ability to provide formative and summative online quizzes and examinations. These platforms typically allow educators to present questions in a randomized order, select questions for a pool, and determine whether questions can be viewed simultaneously or in isolation, and whether learners can return to examine their responses. Asynchronous exams can be used to empower learners, in that they can select the time and place of test administration, and better incorporate learning with other activities (e.g., travel, jobs, parenting). Moreover, such examinations can demonstrate the same level of performance as traditional examinations [174].

Despite their utility, asynchronous tests provided on LMSs introduce the possibility of collaborative cheating where students can share information via back channels. Educators can examine the data collected by LMSs for patterns indicative of cheating. LMSs typically record the duration of the test as well as completion time and date. *Sequential collaborative cheating* is evidenced by two (or more) learners consistently taking the test sequentially (with small delays), with one learner taking the test first and another taking the test later. Learners might alternate whose test is complete first. In contrast, *concurrent collaborative cheating* is evidenced by learners consistently completing the test at the same time and taking the same duration. Here, back channels or direct communication are used to coordinate this activity. When using these patterns to determine potential academic misconduct, educators need to be mindful that the kinds of correlation described here do not reflect a causal relationship. These patterns can nevertheless be used in conjunction with other evidence.

Cheaters can be quite prosperous. For instance, Chen et al. [175] observed that cheaters typically have 13% advantage over noncheating students. However, they also observed that this advantage was considerably reduced (2%–3%) when multiple question variants were used. Similarly, whereas randomized question presentation can produce biases due to random variation on question difficulty, historical information on student performance can be used to create equitable pools of questions (e.g., Ref. [176]). Following Messick, educators must continuously question difficulty, retiring and replacing questions when necessary.

5.2 Massive open online courses

MOOCs represent the best illustration of the issues faced by using learning technologies on a large scale. While these courses present a low-cost, highly accessible approach to education that can engage historically disadvantaged communities, MOOCs can vary in the quality and quantity of their content, whether learners' performance can be adequately monitored and authenticated, as well as whether the resulting credentials provide added value to learners and institutions [177–180]. For instance, many MOOCs rely on peer assessment that can vary in terms of its quality due to differences in expertise, time, and motivation [181].

Concerns over academic misconduct in MOOCs have loomed large. One of the more common forms of academic misconduct is CAMEO: Copying Answers using Multiple Existences Online [182–185]. Rather than sequential collaborative cheating, CAMEO consists of users creating multiple accounts on an online learning platform and obtaining answers from assessments from one MOOC account (harvest) and using them to complete assessments on another MOOC account (master). Learners' CAMEO strategies work because MOOCs typically use a finite number of questions and provide learners with feedback following an assessment. With comparatively low thresholds for passes (e.g., 60%–80%) and the provision of multiple attempts, random responding to assessments across multiple harvest accounts can provide feedback that can be pooled together to pass the assessments in a master account. In the absence of special techniques, CAMEO presents an especially important concern as all learners can easily adopt it. The prevalence of CAMEO is unclear. For instance, studies have suggested that the prevalence of CAMEO is comparatively small [183] whereas other have found that one-quarter of the learners who obtained numerous certificates (20 or more) employed this strategy [184].

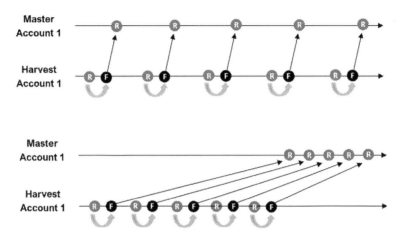

FIG. 8.3 Two prototypical examples of CAMEO strategies. MOOC users can select a response (R) and immediately use feedback (F) from their harvest account to select a response in their master account (distributed collection, short-harvest lag; upper figure) or can select multiple response and complete feedback collection in the harvest account and then use all responses concurrently to respond in the master account (mass collection, long-lag; lower figure). *Adapted from C.G. Northcutt, A.D. Ho, I.L. Chuang, Detecting and preventing "multiple-account" cheating in massive open online courses, Comput. Educ. 100 (2016) 71–80.*

Detection methods for strategies like CAMEO are reasonably straightforward. The underlying assumption is that users' responses on their master account should be preceded by responses on one or more harvest accounts (Fig. 8.3). For instance, a distributed collection, short-harvest lag strategy consists of a short-lag between a response-feedback event on a harvester account and the master account whereas a mass collection, long-lag strategy consists of a long-lag between a response-feedback event on a harvester account and the master account such that the user collects all feedback prior to responding on their master account.

5.3 Communication back channels

Students have become increasingly reliant on back channels: platforms that facilitate communication between learners and educators outside the structure of the course [186–193]. While educators can organize and participate in back channels using institutional LMSs, back channels can also be self-organized by students using social media websites. However, even when educators attempt to participate or monitor back channels, the sheer volume of communication can be overwhelming [194]. Some evidence suggests that the use of back channels can increase learner engagement [195,196] and a sense of community [197] that might otherwise be lacking for distant and absent students [186]. Back channels can be used to facilitate or undermine their learning experience: in addition to discussion of course content and deadlines, back channels are used as heuristics to find short-cuts for the completion of assignments or used to commit or coordinate academic misconduct.

Educators do not necessarily perceive the harm in the use of back channels [192]. Yet, challenges are created by creating a stable social network that parallels the formal institutional network. First is the spread of misinformation and disinformation concerning course content. Misinformation is observed when one learner fails to understand the requirements of a task or course and communicates their beliefs to others. While students can self-correct, contradictory information can create confusion. In contrast, disinformation is evidenced when a student intentionally provides erroneous information to undermine other students' performance. For instance, by providing inaccurate submission dates, formatting instructions, or assessment requirements, the performance of other learners can be reduced. At present, it is not clear to what extent these phenomena occur. (Over the course of many years, the first author (JRS) has encountered an increase in reliance on back channels. This social proof is sufficiently convincing that the learner will refer to information in back channels as equally credible to that of the instructor or teaching assistants. Some learners also seemingly prefer back channels to institutional discussion boards, often noting the benefits of using existing social media platforms. In some cases, learners have expressed disbelief that information communicated in the back channels could be erroneous.)

Second, students use these channels to coordinate, in real-time, the completion of online tests over the course of a semester. For instance, Crump [198] described the use of back channels in his course that led to widespread academic misconduct during the pandemic. (Crump's extensive multiday account has subsequently been deleted and replaced with references.) In his case, students had inadvertently added him to the back channel, allowing him to monitor the nature of their exchanges. He observed extensive coordinated cheating during online quizzes. To address this concern, Crump subsequently provided quiz questions in a random sequence and disabled the students' ability to return to previous answers; students were thus less capable of using simple cheating strategies.

It is also important to highlight that such discussions do not necessarily represent intentional academic misconduct. Students often have anxieties concerning assignments and assessments, leading them to discuss them once completed. They might inadvertently provide information to other students. For instance, when an asynchronous quiz remains open for a class over a lengthy period, students who have completed it might discuss difficult questions with others. In an open back channel, other students might knowingly or unknowingly encounter or solicit this information and use it to improve their performance. More generally, a disaffordance of social media is that it can often lead users to experience anonymity and reduce their situational awareness due to absence of normal social cues.

Educators must not only seek to engage students, but must also develop countermeasures to accommodate this reality that takes advantage of the affordances of learning technologies. For instance, by using slight changes in the wording of multiple choice questions or true and false questions (e.g., "Which is the correct (or incorrect) answer?") paired with changes to response alternatives (e.g., "The United Nations was officially created on October 24, 1945" and "… created October 12, 1945"), can create many question variants that are confusable if shared by learners. (One of the authors (JRS) is frequently approached by two or more students who have "the same" question and wonder why one answered correctly whereas the other answered incorrectly.) By varying orthographic and semantic similarity, a wide variety of variants can be created. Similarly, these changes can be implemented such that the correct answer on one variant is an incorrect answer on another. In addition to increasing the attention that students will give to a question, it likely discourages some forms of collaboration.

5.4 The business ethics of virtual learning

The accessibility of online education options, from MOOCs to entire degree programs conducted online, is largely a positive development that can further the democratization of education through accessibility to populations who are historically excluded from receiving higher education opportunities. However, there are unique ethical concerns related to online programs, mainly related, but not limited, to the quality of education that students receive for the fees and time they invest in this opportunity. Importantly, remote or virtual education and learning opportunities might disproportionately harm vulnerable groups and individuals through targeting marginalized students through specific advertisements and aggressive recruitment strategies, targeting students who did not qualify for in-person admission, and obfuscation of important information about online programs and their ties to universities (for a review, see Ref. [199]). The problematic practices of online program managers (OPMs) can at least partially be attributed to their business models that focus on revenues and profits. An association with established universities benefits and legitimizes OPMs, even if they are not directly tied to universities and do not offer the same level of education.

Students receiving online education have been shown (During, but not limited to the COVID-19 lockdowns.) to pay less attention to the content provided, and to be less engaged in the learning process than in-person students [200]. One can rightfully argue that learning online, though less optimal, is still preferred to not being able to receive any education at all due to life limitations, e.g., having to work inside or outside the home to support a family, or health considerations such as compromised immunity or during the global pandemic. Online classes and programs provide a legitimate means to extend educational opportunities to those who might otherwise be unable to receive them. However, the educational content (e.g., slides, lectures, assignments, quizzes) is not likely to be equivalent to the quality of in-person programs. If learners are led to believe that their attainments are comparable to those attending the in-person, or could be used as equivalencies for program entry, it represents a clear instance of misrepresentation.

Educators and distributors should clearly communicate the equivalencies and limitations of these online programs (e.g., MOOCs). Several different cases of online universities and programs defrauding students (e.g., Trump University, Corinthian Universities) have garnered widespread attention. Some incidents of students signing up for online programs that fit their academic aspirations, financial abilities, and their ability to commit time have ended up leaving students poorer and without the desired education [201,202]. These cases show the dire need for regulation within virtual education options, if not for the humanist perspective benefiting students and the education system, then at least from the capitalist perspective of protecting customers and investors' funds and penalizing predatory businesses.

A relevant recommendation to address some of the preceding ethical concerns is the formulation of regulatory standards to be enforced for the accreditation of online classes, programs, and universities. By lending their name to a course, universities must be accountable for the academic integrity of the course and the content. Regulations can serve to represent the interest of students and universities collaborating with OPMs and protect their financial investment.

6 Responsive and resilient design in learning engineering

In this section we review these themes of this chapter and propose a design framework that incorporates the principles of ethical sensemaking that promotes responsive and resilient design in learning engineering. We have highlighted the importance of viewing learning technologies as a sociotechnical system (STS), defined by those adopted by academic institutions and learners, and those offered by third parties. We now demonstrate how this proposed framework can address the social and ethical issues created by introducing learning technologies into existing systems.

6.1 Operationalizing social and ethical norms

Operationalizing social and ethical norms will require the efforts of interdisciplinary teams that understand the concerns of stakeholders (e.g., learners, parents, educators, administrators), legal requirements within specific regions, and technical requirements of design. Meanwhile, Messick's [30] validity framework provides a comprehensive means to assess construct validity, particularly for assessment instruments. To ensure that these technologies effectively assess the desired constructs, we can begin by evaluating AI-based technologies in terms of multiple forms of evidence, e.g., content, internal structure, generalizability. In acknowledging that there are real, social consequences of assessment for students and their communities, this additionally emphasizes the importance of continuously evaluating learning technologies throughout their lifecycle from development and implementation to deimplementation [60].

Conceptual validity frameworks directed toward AI more generally can also be adopted to address social and ethical requirements of learning technologies. In addition to considering ethics during the design process, ethical features can be directly incorporated into AI-based instruction and assessment. Methods such as DIF and PEM are established means to operationalize specific value dimensions (i.e., fairness and privacy) within psychometrics and design [35]. Designers and developers must make more efforts to operationalize and explain the validity of their systems.

6.2 Trust and trustworthiness

Though many have proposed frameworks or instruments for assessing trust in technology [203–210] and associated outcomes of trust like technology acceptance [211,212], these frameworks focus predominantly on capturing individual assessments of the technology, i.e., their perceptions and attitudes toward the technology. For example, one instrument asks respondents to what extent the respondent's specific goals could be met by that technology [203]. Similarly, frameworks for evaluating the *trustworthiness* of technology tend to focus on capturing individual assessments of the technology's features and to what extent it addresses a customer's needs (e.g., Ref. [213]). Through the scientific method, these assessments can be administered across a population sample and statistical analyses can be applied to find convergence around some central tendency if there is disagreement or variance.

While such frameworks have been useful for evaluating trust in specific use contexts, as well as evaluating general societal trust in technology, our discussion of ethical affordances demonstrates that future assessments of technology (specifically of trust and trustworthiness) would benefit from broadening the unit of analysis. Rather than simplifying the system to focus on the perceptions of a single entity (a person, organization, or population) a more wholistic and comprehensive view that considers the trade-offs of various stakeholders with potentially competing goals, across various situations, should result in a more robust assessment of trust and trustworthiness. More recent work describing how to assess trust transfer across networked stakeholders [214] and how trusting relationships can be supported by responsive technology designs [116] remain limited but hold promise in this regard.

Furthermore, from an STS and human systems integration perspective [215], it is critical to consider the entire technology lifecycle for trust and trustworthiness, such as the support systems and continuous improvement channels that are available for troubleshooting, auditing, improving, and adapting. Reliable and robust question-and-answer communities such as Stack Overflow, user-generated reviews, or a carefully curated FAQ page, as well as consumer protection or customer service departments that take meaningful action following feedback or complaints, are examples in which the surrounding system can affect technology responsivity and trust [116,216].

6.3 Reconceptualizing the role of learners and educators

The introduction of learning technologies as mediators necessarily changes the relationship between learners, educators, and academic institutions. By migrating educational activities to online environments, the boundary between the learning context and other environments (e.g., social media) becomes progressively blurred. This could also lead to the further gamification of test-taking, which might alter learning strategies or promote academic misconduct by reducing the social cues that the context of education might normally be associated with. Concurrently, educators and administrators must develop new competencies to be able to create, modify, and manage learning technologies. The demands of learning engineering include an understanding of both assessment validity [30] and learning technologies, which has unwittingly created a skills gap for educators.

Forms of credentialing outside of HE also create new challenges. Nonacademic credentialing presents a parallel pool of complementary knowledge and skills, ones that are likely much more specific and attuned to the demands of organizations. By obtaining these external credentials, learners can have a more robust (and personalized) set of competencies. However, alternative forms of credentialing can reflect an incommensurable and unregulated (or underregulated) free market of certificates. Learners might seek out these credentials due to the ease of completion or perceptions of applicability or equivalency. To address this, accreditation organizations might emerge to monitor and regulate these services; however, it is unclear whether this would improve upon HE or lead to a convergent organizational structure that might lose some of the advantage of these alternative forms of learning.

6.4 Explainability and knowledge translation

Education is ultimately about empowerment. Learners should acquire new knowledge and skills that allow them to engage with the world, advance their status within society, and contribute to the larger community. A key aspect of this empowerment is ensuring that they understand how they are being assessed. In addition to traditional facets of instructions (e.g., providing syllabi and rubrics in advance), the introduction of advanced learning technologies requires that we address the explainability of these systems. Explainability must ultimately be judged in terms of the goals of the learners and educator relative to their competencies [24,25]. For instance, a learner might simply want to know that a test was graded appropriately, whereas an educator might wish to ensure that a student's score reflects their performance and not that of another. While supplementary training and information can be provided by institutions, learning engineers must also consider incorporating these elements into their design to allow stakeholders to understand the systems for themselves.

Yet, explainability introduces its own challenges. For instance, there are inherent barriers in speaking to learners (or administrators) concerning psychometric techniques due to their lack of competency in this domain. Moreover, by making the assessment process more transparent, it increases the likelihood that unscrupulous test-takers or third parties will identify means to circumvent the integrity of the assessment process. Thus criteria such as explainability and transparency must be qualified to strike a balance between ensuring that learners and educators are aware of the assessment methods and their reliability, and a sufficient level of opacity to ensure that proprietary methods and question content remain inaccessible.

6.5 Implementation validity of learning technologies

The recognition that learning technologies in HE should not be black boxes highlights the need to evaluate the implementation validity of a technology within a given context [60]. Among other factors, implementation validity of an instructional method or assessment instrument requires that we consider the available human, material, and financial resources within a programmatic context and assess the use of new learning technologies in these terms. For instance, even if a learning technology has extensive evidence supporting its validity in particular institutional contexts, if other institutions do not have personnel with the appropriate skills to monitor, evaluate, and regulate these systems, the technology might be used inappropriately.

Like construct validity, this necessitates the collection of evidence, in an ongoing effort of program evaluation within an organization. Thus even if an institution adopts a learning technology that has the necessary technical or functional affordances, an institution must continue assessing the technology's fit-for-purpose. For instance, the LMS Blackboard has been the focus of considerable criticism concerning the system's reliability [217,218]. In other cases, a system's inability to accommodate increased size of academic courses or failures of vendors to support software reinforce the need to consider implementation validity on an ongoing basis [219]. Problems associated with implementation and de-implementation demonstrate the importance of perceiving learning technologies as part of a larger STS.

6.6 Knowledge management and common standards

The adoption of a plurality of instructional and assessment technologies in conjunction with the vast quantities of data that can be accumulated on the learners and education create significant demands on knowledge management. ILR/CLR and related innovation such as credentialing wallets highlight the changes to the STS of education and the importance of interoperability. By developing services beyond traditional HE, learners can have educational experiences that might otherwise be prohibitively costly, inaccessible due to geographic remoteness, or unrealistic given the limited resources available to educators. This potential must also be weighed against the need to ensure that the credentials have integrity. Flooding markets with low-cost, low-quality credentials will do little to alter education ecosystems and can lead to exploitation if learners falsely believe that investments of time and money will yield equitable outcomes. These systems will no doubt be abandoned as the demand decreases.

International organizations can assume leading roles in developing these technologies. In general, organizations such as the WHO have developed guidelines for educational technologies. Standards organizations can also play a leading role. For instance, IEEE has created, or is in the process of creating, a variety of standards for incorporating ethics into design. These standards are nevertheless very general. Concurrently, many value-based design frameworks have also been proposed (Value-Sensitive Design, [61]; Reflective Design, [62]; Design Justice, [63]; Trusting Automation, [116]; Ethical Sensemaking Design, [12]), but none of these processes are requirements.

6.7 Ethical pluralism and cross-cultural issues

A tacit assumption of adopting common standards is that it is both possible and desirable to establish a single set of criteria. The corresponding issue is that many of the values and ethics frameworks that are provided by professional societies, universities, and accreditation organizations are arguably underspecified. For instance, while there might be a consensus that data should remain private, privacy must be qualified. The adoption of technologies and corresponding privacy standards are likely to be influenced by sociocultural norms [11,35,56,220]. Concurrently, while high-level parallels can be drawn between educational traditions, differences in learning styles, emphases, and content (e.g., regional history) are likely to create a variety of opportunities and challenges in adoption of technology across borders. For instance, studies of analogical reasoning have demonstrated that culture-specific folktales can increase performance in specific problem-solving tasks [221].

A focus on cross-cultural issues should not lessen our focus on *intra*cultural differences in values [222,223]. Reconciling stakeholders' values and expectations will not necessarily be an easy task. In many cases, learning technologies are likely to be adopted due to their cost-effectiveness. While more advanced forms of AI-based instruction and assessment might offer learners a personalized educational experience, it is more likely the case that approaches like MOOCs will promote more homogeneity in the provision of instruction. Similarly, by deskilling the provision of instruction and assessment (e.g., promoting peer-based assessment), learners might not receive the same level of quality as traditional instructions. Concurrently, while learners might prefer AI- or online-based instruction due to convenience, they might devalue the benefits accrued from developing personal relationships with students and instructors.

7 Conclusion

Education is about the empowerment of learners. The critical thinking skills, knowledge, and virtues acquired by learners are essential features for effective democracies [224–227]; cf. [228]; education also promotes social mobility [229,230]. Learning technologies have the potential to increase learner access and accessibility to high-quality instruction (e.g., MOOCs) and personalized educational experiences (e.g., adaptive instruction systems and intelligent tutoring systems). While digital divides remain, creating barriers for learners who do not have access to these technologies [176,231,232], their availability creates promises of redressing past inequities.

Education and learning engineering are not value-free. To ensure that learning technologies add value to learners, educators, and society, these technologies must be designed with consideration of the *ethical affordances* of these systems. Like all affordances, ethical affordances need not be apparent. Developers can create technologies without being aware of the impact they will have on users [12], resulting in a need for learning engineers to critically reflect on the ethical principles that should inform design and how best to operationalize them [11,35]. Concurrently, the basis for users' trust must also be considered [116], as trust is the responsibility of designers, developers, and distributors. Trustworthiness can be cultivated through behaviors like transparency and explainability of learning technologies. However, these competencies must be explicitly acknowledged and developed, requiring integration of training concerning social and ethical issues related to system design and implementation in the curriculum [82].

In this chapter, we have identified the ethical affordances of several prominent learning technologies. Like other cybersecurity issues, we have suggested that the comprehensive and integrated learning records (CLRs/ILRs) and learning management systems (LMSs) present clear and present threats to learners' privacy. These threats make the adoption of appropriate cybersecurity measures and policies essential. In an analogous manner to network security, we also assume that it is necessary to identify and understand the vulnerabilities of these systems to cheating to ensure fairness assessment and integrity of these systems. Specifically, learners can hack grades in MOOCs by using alternative accounts and collaborate across back channels. We reviewed some ways to address these security issues. For example, learning engineers can design question variants that increase the probability of confusability between collaborators. Similarly, artificial intelligence can itself be used to reduce the prevalence of cheating. Specifically, plagiarism tools (e.g., AWE, LMS plagiarism detection plug-ins) can provide high-level assessments of whether a paper contains repeated materials, with some of these systems embedded within LMSs for the convenience of educators.

References

[1] T. Farrell, N. Rushby, Assessment and learning technologies: an overview, Br. J. Educ. Technol. 47 (2016) 106–120.

[2] D. Laurillard, Rethinking University Teaching: A Conversational Framework for the Effective Use of Learning Technologies, Routledge, 2002.

[3] M. Jones, Applications of artificial intelligence within education, Comput. Math. Appl. 11 (1985) 517–526.

[4] P. Langley, An integrative framework for artificial intelligence education, in: Proceedings of the AAAI Conference on Artificial Intelligence, vol. 33, No. 01, 2019, pp. 9670–9677.

[5] O. Zawacki-Richter, V.I. Marín, M. Bond, F. Gouverneur, Systematic review of research on artificial intelligence applications in higher education–where are the educators? Int. J. Educ. Technol. High. Educ. 16 (2019) 1–27.

[6] T. Bates, C. Cobo, O. Mariño, S. Wheeler, Can artificial intelligence transform higher education? Int. J. Educ. Technol. High. Educ. 17 (1) (2020) 1–12.

[7] X. Zhai, X. Chu, C.S. Chai, M.S.Y. Jong, A. Istenic, M. Spector, Y. Li, A review of artificial intelligence (AI) in education from 2010 to 2020, Complexity 2021 (2021) 1–18.

[8] T. Buchanan, P. Sainter, G. Saunders, Factors affecting faculty use of learning technologies: implications for models of technology adoption, J. Comput. High. Educ. 25 (2013) 1–11.

[9] Q. Liu, S. Geertshuis, R. Grainger, Understanding academics' adoption of learning technologies: a systematic review, Comput. Educ. 151 (2020) 103857.

[10] B. Wilson, L. Sherry, J. Dobrovolny, M. Batty, M. Ryder, Adoption of learning technologies in schools and universities, in: Handbook on Information Technologies for Education & Training, Springer-Verlag, New York, 2000.

[11] J.R. Schoenherr, Learning engineering is ethical, in: J. Goodell (Ed.), Learning Engineering Toolkit, Routledge Publishing, 2022.

[12] J.R. Schoenherr, Ethical Artificial Intelligence from Popular to Cognitive Science: Trust in the age of Entanglement, Taylor & Francis, 2022.

[13] L.Z. Bain, How students use technology to cheat and what faculty can do about it, Inf. Syst. Educ. J. 13 (2015) 92–99.

[14] T.S. Foulger, K.J. Graziano, D. Schmidt-Crawford, D.A. Slykhuis, Teacher educator technology competencies, J. Technol. Teach. Educ. 25 (4) (2017) 413–448.

[15] T. Mühlbacher, H. Piringer, S. Gratzl, M. Sedlmair, M. Streit, Opening the black box: strategies for increased user involvement in existing algorithm implementations, IEEE Trans. Vis. Comput. Graph. 20 (2014) 1643–1652.

[16] C. O'Neil, Weapons of Math Destruction: How Big Data Increases Inequality and Threatens Democracy, Crown, 2016.

[17] F. Pasquale, The Black Box Society: The Secret Algorithms that Control Money and Information, Harvard University Press, 2015.

[18] T. Chamorro-Premuzic, R. Akhtar, D. Winsborough, R.A. Sherman, The datafication of talent: how technology is advancing the science of human potential at work, Curr. Opin. Behav. Sci. 18 (2017) 13–16.

[19] T. Evgeniou, D.R. Hardoon, A. Ovchinnikov, What happens when AI is used to set grades? Harv. Bus. Rev. (2020). https://hbr.org/2020/08/what-happens-when-ai-is-used-to-set-grades.

[20] A. Adadi, M. Berrada, Peeking inside the black-box: a survey on explainable artificial intelligence (XAI), IEEE Access 6 (2018) 52138–52160.

[21] M. Ananny, K. Crawford, Seeing without knowing: limitations of the transparency ideal and its application to algorithmic accountability, New Media Soc. (2016) 1–17. https://doi.org/10.1177/1461444816676645.

[22] F.K. Došilović, M. Brčić, N. Hlupić, Explainable artificial intelligence: a survey, in: 2018 41st International Convention on Information and Communication Technology, Electronics and Microelectronics (MIPRO), IEEE, 2018, pp. 0210–0215.

[23] D. Gunning, M. Stefik, J. Choi, T. Miller, S. Stumpf, G.Z. Yang, XAI—Explainable artificial intelligence, Sci. Robot. 4 (2019), https://doi.org/10.1126/scirobotics.aay712.

[24] J.R. Schoenherr, Black boxes of the mind: from psychophysics to explainable artificial intelligence, in: Proceedings of Fechner Day, 2020, International Society for Psychophysics, 2020, pp. 46–51.

[25] R. Thomson, J.R. Schoenherr, Knowledge-to-information translation training (KITT): An adaptive approach to explainable artificial intelligence, in: International Conference on Human-Computer Interaction, Springer, Cham, 2020, pp. 187–204.

[26] R.H. Adams, I.I. Ivanov, Using socio-technical system methodology to analyze emerging information technology implementation in the higher education settings, IJEEEE 5 (2015) 31–39.

[27] E. Dahlstrom, D.C. Brooks, J. Bichsel, The Current Ecosystem of Learning Management Systems in Higher Education: Student, Faculty, and IT Perspectives, 2014. Research Report.

[28] M. Legemaate, R. Grol, J. Huisman, H. Oolbekkink-Marchand, L. Nieuwenhuis, Enhancing a quality culture in higher education from a socio-technical systems design perspective, in: Quality in Higher Education, 2021, pp. 1–14.

[29] E. Navarro-Bringas, G. Bowles, G.H. Walker, Embracing complexity: a sociotechnical systems approach for the design and evaluation of higher education learning environments, Theor. Issues Ergon. Sci. 21 (2020) 595–613.

[30] S. Messick, Validity of psychological assessment: validation of inferences from persons' responses and performances as scientific inquiry into score meaning, Am. Psychol. 50 (1995) 741–749.

[31] M.A. Alsubaie, Hidden curriculum as one of current issue of curriculum, J. Educ. Pract. 6 (33) (2015) 125–128.

[32] C. Cornbleth, Beyond hidden curriculum? J. Curric. Stud. 16 (1984) 29–36.

[33] J.P. Portelli, Exposing the hidden curriculum, J. Curric. Stud. 25 (4) (1993) 343–358.

[34] D.J. Wren, School culture: exploring the hidden curriculum, Adolescence 34 (135) (1999).

[35] J.R. Schoenherr, Designing ethical agency for adaptive instructional systems: The FATE of learning and assessment, in: International Conference on Human-Computer Interaction, Springer, Cham, 2021, pp. 265–283.

[36] O. Viberg, M. Hatakka, O. Bälter, A. Mavroudi, The current landscape of learning analytics in higher education, Comput. Hum. Behav. 89 (2018) 98–110.

[37] A. Rubel, K. Jones, Student privacy in learning analytics: an information ethics perspective, Inf. Soc. 32 (2016) 143–159.

[38] S. Slade, P. Prinsloo, Learning analytics: ethical issues and dilemmas, Am. Behav. Sci. 57 (2013) 1510–1529.

[39] M. Ekowo, I. Palmer, The Promise and Peril of Predictive Analytics in Higher Education: A Landscape Analysis, 2016. New America.

[40] T. Bates, A. Anić, M. Marušić, A. Marušić, Authorship criteria and disclosure of contributions: comparison of 3 general medical journals with different author contribution forms, J. Am. Med. Assoc. 292 (2004) 86–88.

[41] R. Bhopal, J. Rankin, E. McColl, L. Thomas, E. Kaner, R. Stacy, H. Rodgers, The vexed question of authorship: views of researchers in a British medical faculty, Br. Med. J. 314 (1997) 1009–1012.

[42] J.R. Schoenherr, Social-cognitive barriers to ethical authorship, Front. Psychol. 6 (2015) 877.

[43] K.A. Amos, The ethics of scholarly publishing: exploring differences in plagiarism and duplicate publication across nations, J. Med. Libr. Assoc. 102 (2) (2014) 87.

[44] D. Ding, B. Nguyen, K. Gebel, A. Bauman, L. Bero, Duplicate and salami publication: a prevalence study of journal policies, Int. J. Epidemiol. 49 (2020) 281–288.

[45] G. Norman, Data dredging, salami-slicing, and other successful strategies to ensure rejection: twelve tips on how to not get your paper published, Adv. Health Sci. Educ. 19 (2014) 1–5.

[46] M.G. Tolsgaard, R. Ellaway, N. Woods, G. Norman, Salami-slicing and plagiarism: how should we respond? Adv. Health Sci. Educ. 24 (1) (2019) 3–14.

[47] N.W. Sochacka, J. Walther, A.L. Pawley, Ethical validation: reframing research ethics in engineering education research to improve research quality, J. Eng. Educ. 107 (2018) 362–379.

[48] E. Cohen, L. Cornwell, A question of ethics: developing information system ethics, J. Bus. Ethics 8 (6) (1989) 431–437.

[49] G. Sivathanu, C.P. Wright, E. Zadok, Ensuring data integrity in storage: techniques and applications, in: Proceedings of the 2005 ACM workshop on Storage security and Survivability, 2005, pp. 26–36.

[50] J.R. Schoenherr, Scientific integrity in research methods, Front. Psychol. 6 (2015) 1562.

[51] Y. Al-Slais, Privacy engineering methodologies: A survey, in: 2020 International Conference on Innovation and Intelligence for Informatics, Computing and Technologies (3ICT), IEEE, 2020, December, pp. 1–6.

[52] S. Gürses, J.M. Del Alamo, Privacy engineering: shaping an emerging field of research and practice, IEEE Secur. Priv. 14 (2016) 40–46.

[53] N. Notario, A. Crespo, Y.S. Martín, J.M. Del Alamo, D. Le Métayer, T. Antignac, D. Wright, PRIPARE: Integrating privacy best practices into a privacy engineering methodology, in: 2015 IEEE Security and Privacy Workshops, IEEE, 2015, pp. 151–158.

[54] A. Senarath, M. Grobler, N.A.G. Arachchilage, Will they use it or not? Investigating software developers' intention to follow privacy engineering methodologies, ACM Trans. Priv. Secur. 22 (2019) 1–30.

[55] A. Senarath, N.A. Arachchilage, Why developers cannot embed privacy into software systems? An empirical investigation, in: Proceedings of the 22nd International Conference on Evaluation and Assessment in Software Engineering 2018, 2018, June, pp. 211–216.

[56] J.R. Schoenherr, Whose privacy, what surveillance? Dimensions of the mental models for privacy and security, IEEE Technol. Soc. Mag. 41 (2022) 54–65.

[57] IEEE, in: Ethically Aligned Design: A Vision for Prioritizing Wellbeing With Artificial Intelligence and Autonomous Systems [online], 2016. Available from: http://www.standards.ieee.org/develop/indconn/ec/autonomous_systems.html.

[58] D.W. Gotterbarn, B. Brinkman, C. Flick, M.S. Kirkpatrick, K. Miller, K. Vazansky, M.J. Wolf, ACM code of ethics and professional conduct, 2018. Retrieved October 2, 2022, from https://dora.dmu.ac.uk/server/api/core/bitstreams/1e5b3cb8-2d77-4ab4-885b-d5996b74605f/content.

[59] The American National Council on Measurement in Education, in: Code of Fair Testing Practices in Education, Joint Committee on Testing Practices, 2018.

[60] J.R. Schoenherr, S.J. Hamstra, Validity in health professions education: from assessment instruments to program evaluation, in: Fundamentals and Frontiers of Medical Education and Decision-Making, Taylor and Francis, 2023.

[61] B. Friedman, P.H. Kahn, A. Borning, Value sensitive design and information systems, in: The Handbook of Information and Computer Ethics, 2008, pp. 69–101.

[62] P. Sengers, K. Boehner, S. David, J.J. Kaye, Reflective design, in: Proceedings of the 4th Decennial Conference on Critical Computing: Between Sense and Sensibility, 2005, pp. 49–58.

[63] S. Costanza-Chock, Design justice: towards an intersectional feminist framework for design theory and practice, in: Proceedings of the Design Research Society 2018, University of Limerick, 2018, pp. 1–14.

[64] D.R. Forsyth, E.H. O'Boyle Jr., Rules, standards, and ethics: relativism predicts cross-national differences in the codification of moral standards, Int. Bus. Rev. 20 (2011) 353–361.

[65] A. Chudzicka-Czupała, A. Lupina-Wegener, S. Borter, N. Hapon, Students' attitude toward cheating in Switzerland, Ukraine and Poland, New Educ. Rev. 2 (2013) 66–76.

[66] D. Pascual-Ezama, T.R. Fosgaard, J.C. Cardenas, P. Kujal, R. Veszteg, B.G.G. de Liaño, P. Branas-Garza, Context-dependent cheating: experimental evidence from 16 countries, J. Econ. Behav. Organ. 116 (2015) 379–386.

[67] S.B. Salter, D.M. Guffey, J.J. McMillan, Truth, consequences and culture: a comparative examination of cheating and attitudes about cheating among US and UK students, J. Bus. Ethics 31 (2001) 37–50.

[68] G.M. Diekhoff, E.E. LaBeff, K. Shinohara, H. Yasukawa, College cheating in Japan and the United States, Res. High. Educ. 40 (1999) 343–353.

[69] A. Chudzicka-Czupała, D. Grabowski, A.L. Mello, J. Kuntz, D.V. Zaharia, N. Hapon, D. Börü, Application of the theory of planned behavior in academic cheating research–cross-cultural comparison, Ethics Behav. 26 (2016) 638–659.

[70] F.B. Aydemir, F. Dalpiaz, A roadmap for ethics-aware software engineering, in: 2018 IEEE/ACM International Workshop on Software Fairness (FairWare), IEEE, 2018, May, pp. 15–21.

[71] V.J. Calluzzo, C.J. Cante, Ethics in information technology and software use, J. Bus. Ethics 51 (2004) 301–312.

[72] J.H. Moor, What is computer ethics? Metaphilosophy 16 (1985) 266–275.

[73] M.E. Whitman, A.M. Townsend, A.R. Hendrickson, Cross-national differences in computer-use ethics: a nine-country study, J. Int. Bus. Stud. 30 (1999) 673–687.

[74] IEEE, The IEEE Global Initiative on Ethics of Autonomous and Intelligent Systems. Ethically Aligned Design: A Vision for Prioritizing Human Well-being with Autonomous and Intelligent Systems, first ed., 2019. (IEEE) Retrieved November 17, 2019, from https://standards.ieee.org/content/ieeestandards/en/industry-connections/ec/autonomous-systems.html Accessed 28 May 2019.

[75] G. Greenleaf, Asian Data Privacy Laws: Trade & Human Rights Perspectives, Oxford University Press, 2014.

[76] F. Miao, W. Holmes, R. Huang, H. Zhang, AI and Education: A Guidance for Policymakers, UNESCO Publishing, 2021.

[77] P.H. Wong, Global engineering ethics, in: The Routledge Handbook of the Philosophy of Engineering, Routledge, 2020, pp. 620–629.

[78] Q. Zhu, B.K. Jesiek, Engineering ethics in global context: Four fundamental approaches, in: 2017 ASEE Annual Conference & Exposition, 2017.

[79] T.M. Singelis, H.C. Triandis, D.P. Bhawuk, M.J. Gelfand, Horizontal and vertical dimensions of individualism and collectivism: a theoretical and measurement refinement, Cross-Cult. Res. 29 (1995) 240–275.

[80] N.T. Feather, Values in Education and Society, Free Press, 1975.

[81] J.M. Halstead, M.J. Taylor, Values in Education and Education in Values, Psychology Press, 1996.

[82] J.R. Schoenherr, J. DeFalco, Moral education and A/IS standardization: responsible and ethical design through education, in: 2021 IEEE International Symposium on Technology and Society (ISTAS), IEEE, 2021, pp. 1–8.

[83] M.T. Kane, An argument-based approach to validity, Psychol. Bull. 112 (1992) 527–535.

[84] L.J. Cronbach, G.C. Gleser, Psychological Tests and Personnel Decisions, second ed., University of Illinois Press, Urbana, 1965.

[85] A.G. Ferguson, Illuminating black data policing, Ohio State J. Crim. Law 15 (2017) 503–525.

[86] J.C. Nunnally, I.H. Bernstein, Psychometric Theory, McGraw-Hill, New York, 1994.

[87] G. Rasch, Studies in Mathematical Psychology: I. Probabilistic Models for Some Intelligence and Attainment Tests, Nielsen & Lydiche, 1960.

[88] A.S. Cohen, D.M. Bolt, A mixture model analysis of differential item functioning, J. Educ. Meas. 42 (2005) 133–148.

[89] P.W. Holland, D.T. Thayer, Differential Item Functioning and the Mantel-Haenszel Procedure, ETS Research Report Series, 1986. i-24.

[90] S.J. Osterlind, H.T. Everson, Differential Item Functioning, vol. 161, Sage Publications, 2009.

[91] M. Kearns, A. Roth, The Ethical Algorithm: The Science of Socially Aware Algorithm Design, Oxford University Press, 2019.

[92] G.G. Kingsbury, D.J. Weiss, A comparison of IRT-based adaptive mastery testing and a sequential mastery testing procedure, in: D.J. Weiss (Ed.), New Horizons in Testing: Latent Trait Theory and Computerized Adaptive Testing, Academic Press, New York, 1983, pp. 237–254.

[93] F.M. Lord, A theoretical study of two-stage testing, Psychometrika 36 (1971) 227–242.

[94] F.M. Lord, A broad-range tailored test of verbal ability, Appl. Psychol. Meas. 1 (1977) 95–100.

[95] M.D. Reckase, A procedure for decision making using tailored testing, in: D.J. Weiss (Ed.), New Horizons in Testing: Latent Trait Theory and Computerized Adaptive Testing, Academic Press, New York, 1983, pp. 237–254.

[96] D.J. Weiss, G.G. Kingsbury, Application of computerized adaptive testing to educational problems, J. Educ. Meas. 21 (1984) 361–375.

[97] N. Sclater, A. Peasgood, J. Mullan, Learning Analytics in Higher Education, vol. 8(2017), Jisc, London, 2016, p. 176. Accessed February.

[98] J.R. Anderson, Cognitive psychology and intelligent tutoring, in: Proceedings of the Cognitive Science Society Conference, Lawrence Erlbaum Associates, Inc., Boulder, Colorado, 1984, pp. 37–43.

[99] J.R. Anderson, S. Betts, J.L. Ferris, J.M. Fincham, Neural imaging to track mental states while using an intelligent tutoring system, Proc. Natl. Acad. Sci. 107 (2010) 7018–7023.

[100] J. Lee, O.C. Park, Adaptive instructional systems, in: Handbook of Research on Educational Communications and Technology, Routledge, 2008, pp. 469–484.

[101] O.C. Park, J. Lee, Adaptive instructional systems, in: Handbook of Research on Educational Communications and Technology, Routledge, 2013, pp. 647–680.

[102] M.C. Polson, J.J. Richardson, Foundations of Intelligent Tutoring Systems, Psychology Press, 2013.

[103] M.D. Shermis, J.C. Burstein, Handbook of Automated Essay Evaluation, Routledge, New York, 2013.

[104] M. Stevenson, A. Phakiti, The effects of computer-generated feedback on the quality of writing, Assess. Writ. 19 (2014) 51–65.

[105] G. Siemens, S. Dawson, G. Lynch, Improving the quality and productivity of the higher education sector. Policy and Strategy for Systems-Level Deployment of Learning Analytics. Canberra, Australia: Society for Learning Analytics Research for the Australian Office for Learning and Teaching, 2013, p. 31.

[106] J. Wilson, A. Czik, Automated essay evaluation software in English language arts classrooms: effects on teacher feedback, student motivation, and writing quality, Comput. Educ. 100 (2016) 94–109.

[107] J.R. Schoenherr, S. Millington, C.H. Lee, Efficiency and automaticity in the healthcare professions: operationalizing expertise development using the Speed-Accuracy Trade-Off, in: Fundamentals and Frontiers of Medical Education and Decision-Making, Taylor and Francis, 2023.

[108] A.R. Hakstian, W. Kansup, A comparison of several methods of assessing partial knowledge in multiple-choice tests: II. Testing procedures, J. Educ. Meas. (1975) 231–239.

[109] W. Kansup, A.R. Hakstian, A comparison of several methods of assessing partial knowledge in multiple-choice tests: I. Scoring procedures, J. Educ. Meas. (1975) 219–230.

[110] M.W. Wang, J.C. Stanley, Differential weighting: a review of methods and empirical studies, Rev. Educ. Res. 40 (1970) 663–705.

[111] D.A. Clement, K.D. Frandsen, On conceptual and empirical treatments of feedback in human communication, Commun. Monogr. 43 (1976) 11–28.

[112] J. McKendree, Effective feedback content for tutoring complex skills, Hum. Comput. Interact. 5 (1990) 381–413.

[113] V.J. Shute, Focus on formative feedback, Rev. Educ. Res. 78 (2008) 153–189.

[114] J.C. Archer, State of the science in health professional education: effective feedback, Med. Educ. 44 (2010) 101–108.

[115] J.M. Van De Ridder, K.M. Stokking, W.C. McGaghie, O.T. Ten Cate, What is feedback in clinical education? Med. Educ. 4 (2008) 189–197.

[116] E.K. Chiou, J.D. Lee, Trusting automation: designing for responsivity and resilience, Hum. Factors (2021). https://doi.org/10.1177/00187208211009.

[117] H.M. Truong, Integrating learning styles and adaptive e-learning system: current developments, problems and opportunities, Comput. Hum. Behav. 55 (2016) 1185–1193.

[118] J.R. Schoenherr, Adapting the zone of proximal development to the wicked environments of professional practice, in: International Conference on Human-Computer Interaction, Springer, 2020, pp. 394–410.

[119] C. Conati, V. Aleven, A. Mitrovic, Eye-tracking for student modelling in intelligent tutoring systems, in: Design Recommendations for Intelligent Tutoring Systems, vol. 1, 2013, pp. 227–236.

[120] R.W. Picard, Affective computing: challenges, Int. J. Hum. Comput. Stud. 59 (2003) 55–64.

[121] J. Tao, T. Tan, Affective computing: a review, in: International Conference on Affective Computing and Intelligent Interaction, Springer, Berlin, 2005, pp. 981–995.

[122] S. Greene, H. Thapliyal, A. Caban-Holt, A survey of affective computing for stress detection: evaluating technologies in stress detection for better health, IEEE Consum. Electron. Mag. 5 (2016) 44–56.

[123] R.R. Meijer, M.L. Nering, Computerized adaptive testing: overview and introduction, Appl. Psychol. Meas. 23 (3) (1999) 187–194.

[124] T. Davey, M.L. Nering, in: Controlling Item Exposure and Maintaining Item Security, Paper presented at the ETS-sponsored Colloquium entitled Computer-Based Testing: Building the Foundations for Future Assessments, Philadelphia PA, 1998.

[125] W. Celis, Computer admissions test found to be ripe for abuse, The New York Times, 1994. Retrieved June 17, 2022: https://www.nytimes.com/1994/12/16/us/computer-admissions-test-found-to-be-ripe-for-abuse.html.

[126] R. Sanchez, Computerized Graduate Exam called Easy Mark, The Washington Post, 1994. Retrieved June 17, from https://www.washingtonpost.com/archive/politics/1994/12/16/computerized-graduate-exam-called-easy-mark/d4069d3e-39f1-46b6-8d85-3d5f865d14fe/.

[127] T.Z. Keith, Validity and automated essay scoring systems, in: M.D. Shermis, J.C. Burstein (Eds.), Automated Essay Scoring: A Cross-Disciplinary Perspective, Lawrence Erlbaum Associates, Inc, Mahwah, NJ, 2003, pp. 147–167.

[128] Z.V. Zhang, K. Hyland, Student engagement with teacher and automated feedback on L2 writing, Assess. Writ. 36 (2018) 90–102.

[129] J. Buolamwini, T. Gebru, Gender shades: intersectional accuracy disparities in commercial gender classification, in: Conference on Fairness, Accountability and Transparency, PMLR, 2018, pp. 77–91.

[130] V. Eubanks, Automating Inequality: How High-Tech Tools Profile, Police, and Punish the Poor, St. Martin's Press, 2018.

[131] S.U. Noble, Algorithms of Oppression, New York University Press, 2018.

[132] R. Lippi-Green, English with an Accent: Language, Ideology, and Discrimination in the United States, Routledge, 2012.

[133] J.F. Dovidio, K. Kawakami, S.L. Gaertner, Implicit and explicit prejudice and interracial interaction, J. Pers. Soc. Psychol. 82 (2002) 62–68.

[134] A.G. Greenwald, L.H. Krieger, Implicit bias: scientific foundations, Calif. Law Rev. 94 (2006) 945–967.

[135] D.M. Quinn, Experimental evidence on teachers' racial bias in student evaluation: the role of grading scales, Educ. Eval. Policy Anal. 42 (3) (2020) 375–392.

[136] E.M. Bender, B. Friedman, Data statements for natural language processing: toward mitigating system bias and enabling better science, Trans. Assoc. Comput. Linguist. 6 (2018) 587–604.

[137] E.A. Hanushek, P.E. Peterson, L.M. Talpey, L. Woessmann, The unwavering SES achievement gap, in: Trends in US Student Performance, vol. No. w25648, National Bureau of Economic Research, 2019.

[138] P. Eckert, Variation and the indexical field 1, J. Socioling. 12 (4) (2008) 453–476.

[139] P. Garrett, N. Coupland, A. Williams, Investigating Language Attitudes: Social Meanings of Dialect, Ethnicity and Performance, University of Wales Press, 2003.

[140] C. O'Connor, 2019 Wallace Foundation distinguished lecture education research and the disruption of racialized distortions: establishing a wide angle view, Educ. Res. 49 (7) (2020) 470–481.

[141] B.J. Dietvorst, J.P. Simmons, C. Massey, Algorithm aversion: people erroneously avoid algorithms after seeing them err, J. Exp. Psychol. Gen. 144 (2015) 114.

[142] M. Itoh, K. Tanaka, Mathematical modeling of trust in automation: trust, distrust, and mistrust, in: Proceedings of the Human Factors and Ergonomics Society Annual Meeting, vol. 44, No. 1, SAGE Publications, Los Angeles, CA, 2000, pp. 9–12.

[143] M. Bouville, Plagiarism: words and ideas, Sci. Eng. Ethics 14 (2008) 311–322.

[144] N. Hayes, L.D. Introna, Cultural values, plagiarism, and fairness: when plagiarism gets in the way of learning, Ethics Behav. 15 (2005) 213–231.

[145] T. Bretag, S. Mahmud, Self-plagiarism or appropriate textual re-use? J. Acad. Ethics 7 (2009) 193–205.

[146] N.J. Auer, E.M. Krupar, Mouse click plagiarism: the role of technology in plagiarism and the librarian's role in combating it, Libr. Trends 49 (2001) 415–433.

[147] P.M. Scanlon, Student online plagiarism: how do we respond? Coll. Teach. 51 (2003) 161–165.

[148] C.R. Campbell, C.O. Swift, L. Denton, Cheating goes hi-tech: online term paper mills, J. Manag. Educ. 24 (2000) 726–740.

[149] L.P. Stavisky, Term paper" mills," academic plagiarism, and state regulation, Polit. Sci. Q. 88 (1973) 445–461.

[150] R.L. Genereux, B.A. McLeod, Circumstances surrounding cheating: a questionnaire study of college students, Res. High. Educ. 36 (1995) 687–704.

[151] D.L. McCabe, L.K. Trevino, Individual and contextual influences on academic dishonesty: a multi-campus investigation, Res. High. Educ. 38 (1997) 379–396.

[152] C.O. Swift, S. Nonis, When no one is watching: cheating behavior in projects and assignments, Mark. Educ. Rev. 8 (1998) 27–36.

[153] N. Wice, Copy & Paste: Term Paper Mills on the Web, Y-Life, 1997. Available: http://www.zdnet.com/yil/content/mag/9701/wice9701.html (6/10/98).

[154] A. Hern, AI Bot ChatGPT stuns Academics with Essay-Writing Skills and Usability, The Guardian, 2022. Retrieved December 16, 2022 from https://www.theguardian.com/technology/2022/dec/04/ai-bot-chatgpt-stuns-academics-with-essay-writing-skills-and-usability.

[155] A.M.E.T. Ali, H.M.D. Abdulla, V. Snasel, Overview and comparison of plagiarism detection tools, in: Dateso, 2011, pp. 161–172.

[156] A. Parker, J.O. Hamblen, Computer algorithms for plagiarism detection, IEEE Trans. Educ. 32 (1989) 94–99.

[157] A. Altheneyan, M.E.B. Menai, Evaluation of state-of-the-art paraphrase identification and its application to automatic plagiarism detection, Int. J. Pattern Recognit. Artif. Intell. 34 (2020) 2053004.

[158] A. Barrón-Cedeño, P. Rosso, On automatic plagiarism detection based on n-grams comparison, in: European Conference on Information Retrieval, Springer, Berlin, Heidelberg, 2009, April, pp. 696–700.

[159] A. Barrón-Cedeño, M. Vila, M.A. Martí, P. Rosso, Plagiarism meets paraphrasing: insights for the next generation in automatic plagiarism detection, Comput. Linguist. 39 (2013) 917–947.

[160] M. AlSallal, R. Iqbal, V. Palade, S. Amin, V. Chang, An integrated approach for intrinsic plagiarism detection, Futur. Gener. Comput. Syst. 96 (2019) 700–712.

[161] I.A. Penn, G.B. Pennix, Records Management Handbook, Routledge, 2017.

[162] J. Bailey, S.C. Carter, C. Schneider, T. Vander Ark, Data backpacks: portable records & learner profiles, in: Navigating The Digital Shift: Implementation Strategies for Blended and Online Learning, 2012, pp. 101–124.

[163] M. Alavi, D.E. Leidner, Knowledge management and knowledge management systems: conceptual foundations and research issues, MIS Q. (2001) 107–136.

[164] R.K. Rai, Knowledge management and organizational culture: a theoretical integrative framework, J. Knowl. Manag. 15 (2011) 779–801.

[165] S.N. Braxton, S. Carbonaro, N. Jankowski, Comprehensive learner record as a vehicle for assessment and learning transparency in a skills economy, in: Y. Huang (Ed.), Handbook of Research on Credential Innovations for Inclusive Pathways to Professions, 2022, pp. 214–233. IGI Global http://doi:10.4018/978-1-7998-3820-3.ch011.

[166] S. Carbonaro, Comprehensive learner record: exploring a new transcript for lifelong learning, in: IMS Global Learning Consortium edTech rEvolutions Leaders, 2020. Retrieved from https://www.imsglobal.org/article/edtech-leaders/clr.

[167] K. Bednar, S. Spiekermann, M. Langheinrich, Engineering privacy by design: are engineers ready to live up to the challenge? Inf. Soc. 35 (2019) 122–142.

[168] L. Sweeney, Achieving k-anonymity privacy protection using generalization and suppression, Int. J. Uncertain. Fuzziness Knowl. Based Syst. 10 (5) (2002) 571–588.

[169] Federal Bureau of Investigations, Privacy Industry Notification, 2022, Retrieved July 10, 2022 from https://www.ic3.gov/Media/News/2022/220526.pdf.

[170] S. Adam, The State of Ransomware in Education, 2021, 2021. Sophos. Retrieved June 17, 2022 from https://news.sophos.com/en-us/2021/07/13/the-state-of-ransomware-in-education-2021/.

[171] Stilgherrian, ANU Incident Report on Massive Data Breach is a Must-Read, ZDNet, 2019. Retrieved Oct 12, 2021 from https://www.zdnet.com/article/anu-incident-report-on-massive-data-breach-a-must-read/.

[172] C. Hayhurst, Universities Share Lessons Learned from Ransomware Attacks, EdTech Magazine, 2022. Retrieved July 10, 2022 from https://edtechmagazine.com/higher/article/2022/05/universities-share-lessons-learned-ransomware-attacks.

[173] L.P. Hollis, Ghost-students and the new wave of online cheating for community college students, New Directions for Community Colleges 2018 (2018) 25–34.

[174] S. Silva, J. Fernandes, P. Peres, V. Lima, C. Silva, Teachers' perceptions of remote learning during the pandemic: a case study, Educ. Sci. 12 (10) (2022) 698.

[175] B. Chen, M. West, C. Zilles, How much randomization is needed to deter collaborative cheating on asynchronous exams? in: Proceedings of the Fifth Annual ACM Conference on Learning at Scale, 2018, June, pp. 1–10.

[176] A. Lane, The impact of openness on bridging educational digital divides, Int. Rev. Res. Open Distance Learn 10 (5) (2009). https://doi.org/10.19173/irrodl.v10i5.637.

[177] C.J. Bonk, M.M. Lee, T.C. Reeves, T.H. Reynolds, MOOCs and Open Education Around the World, Routledge, 2015.

[178] X. Li, K.M. Chang, Y. Yuan, A. Hauptmann, Massive open online proctor: protecting the credibility of MOOCs certificates, in: Proceedings of the 18th ACM Conference on Computer Supported Cooperative Work & Social Computing, 2015, February, pp. 1129–1137.

[179] P. McAndrew, E. Scanlon, Open learning at a distance: lessons for struggling MOOCs, Science 342 (2013) 1450–1451.

[180] L. Yuan, S.J. Powell, MOOCs and Open Education: Implications for Higher Education, Centre for Educational Technology and Interoperability Standards, 2013.

[181] T. Staubitz, D. Petrick, M. Bauer, J. Renz, C. Meinel, Improving the peer assessment experience on MOOC platforms, in: Proceedings of the Third (2016) ACM Conference on Learning@ Scale, 2016, April, pp. 389–398.

[182] G. Alexandron, J.A. Ruipérez-Valiente, Z. Chen, P.J. Muñoz-Merino, D.E. Pritchard, Copying@ Scale: using harvesting accounts for collecting correct answers in a MOOC, Comput. Educ. 108 (2017) 96–114.

[183] Y. Bao, G. Chen, C. Hauff, On the prevalence of multiple-account cheating in massive open online learning: a replication study, in: Proceedings of the 10th International Conference on Educational Data Mining, EDM, 25–28 June 2017, Wuhan, Hubei, China, 2017.

[184] C.G. Northcutt, A.D. Ho, I.L. Chuang, Detecting and preventing "multiple-account" cheating in massive open online courses, Comput. Educ. 100 (2016) 71–80.

[185] J.A. Ruiperez-Valiente, P.J. Muñoz-Merino, G. Alexandron, D.E. Pritchard, Using machine learning to detect 'multiple-account' cheating and analyze the influence of student and problem features, IEEE Trans. Learn. Technol. 12 (1) (2017) 112–122.

[186] S. Bayne, M. Gallagher, J.M. Lamb, Being 'at' university: the social topologies of distance students, High. Educ. 67 (2014) 569–583.

[187] S. Cogdill, T.L. Fanderclai, J. Kilborn, M.G. Williams, Backchannel: whispering in digital conversation, in: Proceedings of the 34th Annual Hawaii International Conference on System Sciences, IEEE, 2001, January. pp. 8-.

[188] M. Eid, I. Al-Jabri, Social networking, knowledge sharing, and student learning: the case of university students, Comput. Educ. 99 (2016) 14–27.

[189] N. Jacobs, A. McFarlane, Conferences as learning communities: some early lessons in using 'backchannel' technologies at an academic conference – distributed intelligence or divided attention? J. Comput. Assist. Learn. 21 (5) (2005) 317–329, https://doi.org/10.1111/j.1365-2729.2005.00142.x.

[190] L.R. Kearns, B.A. Frey, Web 2.0 technologies and Back Channel communication in an online learning community, TechTrends 54 (2010) 41. https://doi.org/10.1007/s11528-010-0419-y.

[191] A. Purvis, H. Rodger, S. Beckingham, Engagement or distraction: the use of social media for learning in higher education, SEEJ 5 (1) (2016), https://doi.org/10.7190/seej.v5.i1.104. ISSN (online) 2047-9476.

[192] C. Toledo, S. Peters, Educators' perceptions of uses, constraints, and successful practices of backchanneling, Education 16 (2010) 1–15.

[193] S. Yardi, The role of the backchannel in collaborative learning environments, in: S.A. Barab, K.E. Hay, D.T. Hickey (Eds.), The International Conference of the Learning Sciences: Indiana University 2006. Proceedings of ICLS 2006, vol. 2, International Society of the Learning Sciences, Bloomington, Indiana, USA, 2006, pp. 852–858.

[194] D. Bouhnik, M. Deshen, WhatsApp goes to school: mobile instant messaging between teachers and students, J. Inf. Technol. Educ.:Res. 13 (2014) 217–231.

[195] D. Baron, A. Bestbier, J.M. Case, B.I. Collier-Reed, Investigating the effects of a backchannel on university classroom interactions: a mixed-method case study, Comput. Educ. 94 (2016) 61–76.

[196] S.Y. Harunasari, N. Halim, Digital backchannel: promoting Students' engagement in EFL large class, Int. J. Emerg. Technol. Learn. 14 (2019) 163–178.

[197] S. Stone, A. Logan, Exploring students' use of the social networking site WhatsApp to foster connectedness in the online learning experience, Ir. J. Technol. Enhanc. Learn. 3 (2018) 42–55.

[198] M. Crumb, Brief Notes on the Cheating Behavior Literature, 2022, Retrieved May 21, 2022 from https://crumplab.com/articles/blog/post_994_5_26_22_cheating/index.html.

[199] L.T. Hamilton, H. Daniels, C.M. Smith, C. Eaton, The Private Side of Public Universities: Third-Party Providers and Platform Capitalism, 2022.

[200] E. Peper, V. Wilson, M. Martin, E. Rosegard, R. Harvey, Avoid zoom fatigue, be present and learn, NeuroRegulation 8 (2021) 47, https://doi.org/10.15540/nr.8.1.47.

[201] A. Archie, C. Turner, U.S. Will Forgive $5.8 Billion of Loans to Corinthian Colleges Students, NPR, 2022. Retrieved January 4, 2023 from https://www.npr.org/2022/06/02/1101424651/corinthian-colleges-student-loan-forgiveness.

[202] R.S. Helderman, Trump Agrees to $25 Million Settlement in Trump University Fraud Cases, Washington Post, 2016. Retrieved January 4, 2023 from https://www.washingtonpost.com/politics/source-trump-nearing-settlement-in-trump-university-fraud-cases/2016/11/18/8dc047c0-ada0-11e6-a31b-4b6397e625d0_story.html.

[203] E.T. Chancey, J.P. Bliss, Y. Yamani, H.A.H. Handley, Trust and the compliance–reliance paradigm: the effects of risk, error bias, and reliability on trust and dependence, Hum. Factors 59 (2017) 333–345. https://doi.org/10.1177/0018720816682648.

[204] K.A. Hoff, M. Bashir, Trust in automation: integrating empirical evidence on factors that influence trust, Hum. Factors 57 (3) (2015) 407–434. https://doi.org/10.1177/0018720814547570.

[205] J.-Y. Jian, A.M. Bisantz, C.G. Drury, Foundations for an empirically determined scale of trust in automated systems, Int. J. Cogn. Ergon. 4 (1) (2000) 53–71. https://doi.org/10.1207/S15327566IJCE0401_04.

[206] A.D. Kaplan, T.T. Kessler, J.C. Brill, P.A. Hancock, Trust in artificial intelligence: meta-analytic findings, Hum. Factors (2021). 00187208211013988 https://doi.org/10.1177/00187208211013988.

[207] S.C. Kohn, E.J. de Visser, E. Wiese, Y.-C. Lee, T.H. Shaw, Measurement of trust in automation: a narrative review and reference guide, Front. Psychol. 12 (2021). https://doi.org/10.3389/fpsyg.2021.604977.

[208] J.D. Lee, K.A. See, Trust in automation: designing for appropriate reliance, Hum. Factors 31 (2004). https://doi.org/10.1518/hfes.46.1.50_30392.

[209] B.F. Malle, D. Ullman, A multidimensional conception and measure of human-robot trust, in: Trust in Human-Robot Interaction, Elsevier, 2021, pp. 3–25. https://doi.org/10.1016/B978-0-12-819472-0.00001-0.

[210] T.B. Sheridan, Individual differences in attributes of trust in automation: measurement and application to system design, Front. Psychol. 10 (2019) 1117. https://doi.org/10.3389/fpsyg.2019.01117.

[211] F.D. Davis, R.P. Bagozzi, P.R. Warshaw, User acceptance of computer technology: a comparison of two theoretical models, Manag. Sci. 35 (8) (1989) 982–1003.

[212] Venkatesh, Morris, Davis, Davis, User acceptance of information technology: toward a unified view, MIS Q. 27 (3) (2003) 425. https://doi.org/10.2307/30036540.

[213] E. Blasch, J. Sung, T. Nguyen, Multisource AI scorecard table for system evaluation, in: AAAI FSS20: Artificial Intelligence in Government and Public Sector, 2020. Washington, DC.

[214] L. Huang, N.J. Cooke, R.S. Gutzwiller, S. Berman, E.K. Chiou, M. Demir, W. Zhang, Distributed dynamic team trust in human, artificial intelligence, and robot teaming, in: C.S. Nam, J.B. Lyons (Eds.), Trust in Human-Robot Interaction, Academic Press, 2021, pp. 301–319. https://doi.org/10.1016/B978-0-12-819472-0.00013-7.

[215] National Academies of Sciences, Engineering, and medicine, in: Human-AI Teaming: State of the Art and Research Needs, The National Academies Press, 2021. https://doi.org/10.17226/26355.

[216] B. Shneiderman, Human-centered artificial intelligence: reliable, safe & trustworthy, Int. J. Hum.-Comput. Interact. 36 (6) (2020) 495–504. https://doi.org/10.1080/10447318.2020.1741118.

[217] J. Jordan, Portal Problems, Blackboard Blackout: Data Systems at CUNY Malfunction and Disrupt Classes, 2009, Retrieved June 28, 2022 from https://web.archive.org/web/20091206034013/http://media.www.thehunterenvoy.com/media/storage/paper1327/news/2009/02/18/News/Portal.Problems.Blackboard.Blackout-3635795.shtml.

[218] K. Miller, Blackboard Blackout, The Palm Beach Post, 2008. Retrieved June 28, 2022 from https://web.archive.org/web/20101013162312/http://blogs.palmbeachpost.com/extracredit/2008/04/18/blackboard-blackout/.

[219] McMaster University, A New Learning Management System, McMaster Daily News, 2010. Retrieved June 28, 2022 from https://web.archive.org/web/20100401112411/http://dailynews.mcmaster.ca/story.cfm?id=6675.

[220] J.R. Schoenherr, Adoption of surveillance technologies: data openness, privacy, and cultural tightness, IEEE Trans. Control Syst. Technol. 2 (2021) 122–127.

[221] Z. Chen, L. Mo, R. Honomichl, Having the memory of an elephant: long-term retrieval and the use of analogues in problem solving, J. Exp. Psychol. Gen. 133 (2004) 415–433.

[222] A.P. Fiske, Structures of Social Life: The Four Elementary Forms of Human Relations, Free Press, 1991.

[223] J. Graham, J. Haidt, S. Koleva, M. Motyl, R. Iyer, S.P. Wojcik, P.H. Ditto, Moral foundations theory: the pragmatic validity of moral pluralism, in: Advances in Experimental Social Psychology, vol. 47, Academic Press, 2013, pp. 55–130.

[224] J. Dewey, Democracy in education, Elem. Sch. Teach. 4 (1903) 193–204.

[225] E.L. Glaeser, G.A. Ponzetto, A. Shleifer, Why does democracy need education? J. Econ. Growth 12 (2) (2007) 77–99.

[226] A. Gutmann, Democratic Education, Princeton University Press, 1999.

[227] A. Spilimbergo, Democracy and foreign education, Am. Econ. Rev. 99 (2009) 528–543.

[228] K. Croke, G. Grossman, H.A. Larreguy, J. Marshall, Deliberate disengagement: how education can decrease political participation in electoral authoritarian regimes, Am. Political Sci. Rev. 110 (2016) 579–600.

[229] R. Breen, J.O. Jonsson, Inequality of opportunity in comparative perspective: recent research on educational attainment and social mobility, Annu. Rev. Sociol. (2005) 223–243.

[230] P. Brown, Education, opportunity and the prospects for social mobility, Br. J. Sociol. Educ. 34 (2013) 678–700.

[231] J. Sims, R. Vidgen, P. Powell, E-learning and the digital divide: perpetuating cultural and socio-economic elitism in higher education, Commun. Assoc. Inf. Syst. 22 (1) (2008). eprints.bbk.ac.uk/id/eprint/6276.

[232] J. Waycott, S. Bennett, G. Kennedy, B. Dalgarno, K. Gray, Digital divides? Student and staff perceptions of information and communication technologies, Comput. Educ. 54 (2010) 1202–1211.

Perceiving a humorous robot as a social partner

Haley N. Green, Md Mofijul Islam, Shahira Ali, and Tariq Iqbal

UNIVERSITY OF VIRGINIA, CHARLOTTESVILLE, VA, UNITED STATES

1 Introduction

With recent advancements in social robots, humans are more likely to encounter robots in their day-to-day lives. In particular, there has been an increase in robots working along-side humans in customer service, health care, and education [1–7]. The surplus in robots operating in close proximity to humans has led to increased efforts in improving the inter-active capabilities of these robots [8–20]. We are also seeing an increase in robots designed to model more human-like interactional strategies [21–25]. Specifically, many of these robots utilize a variety of communicatory approaches to facilitate conversations with human partners. Some of the more common tactics include: expressive emotions, explainability, and personal anecdotes [26–30]. However, there is still a need for robots that can employ dynamic, human-like conversational tactics that inspire trust and more enriching exchanges with humans.

For example, in human-human interactions, among many other conversational tech-niques, humans often wield humor to relieve tension, build relationships, cultivate trust, and recover from mistakes. Humor as a conversational tactic supplies an attractive alter-native to lengthy anecdotes and references. If used appropriately, humor can serve as a low-risk and easy opportunity to formulate connections with other humans. As a result, the use of humor is common in various workplace interactions. There are many reports on the benefits of humans utilizing humor in these high-stress and accident-prone set-tings [31–33].

While humor is a tool often employed by humans, there has been less examination in the context of how social robots can efficiently use humor in conversation. In previous literature, the joke capabilities of social robots consist of puns and "knock-knock" jokes that are not necessarily task-related [34, 35]. These particular forms of humor have been used to establish the robot as a more human-like and sociable interaction partner. Although people incorporate their past experiences, task awareness, and context to gen-erate more situationally relevant jokes, these robots cannot generate different types of jokes based on the conversational context. In consequence, the jokes generated by these robots are often misaligned and result in unpleasant interactions with humans. Therefore, along with combining various situational attributes to generate humor, it is essential to investigate how people perceive various types of humor delivered by a robot.

To our knowledge, there has not been a systematic study on human perceptions of a robot employing different humor types in a human-robot interaction (HRI) scenario. An arsenal of differing humor types would add dimension to a social robot's conversational capabilities. More broadly, understanding how humans perceive the various humor types will guide us to develop AI systems for robots that utilize humor more appropriately to achieve fluent interaction. Fluent interactions enable a better understanding and execution of expectations and interdependence between humans and robots, which can affect trust. Furthermore, accounting for trust in an interaction can enable robots to accurately calibrate a human agent's reliance on automation. Thus the implementation of different behaviors and techniques by robots in HRIs with the goal of analyzing the impact on trust is necessary to make advancements in this domain. We wish to observe if a social robot can also utilize more advanced, human-like humor styles. Consequently, in this work, we generate a set of jokes based on the preestablished set of humor types in which the robot capitalizes on its past experiences, task awareness, and situational context.

Specifically, in this study, we observe an NAO robot employing the four categories of humor widely used in human-human interaction: *affiliative*, *aggressive*, *self-defeating*, and *self-enhancing*. In our designed HRI scenario, the NAO robot engages in one of the humor types in response to a mistake made in a game called *iSpy*. We examine the ratings of perceived *funniness* for 20 jokes from these four humor categories. We then analyze the ratings against varying demographic factors in order to evaluate their effect on how the different humor types are perceived. Specifically, we report the effects of age, gender, and previous experience with robots on perceptions of the different humor types.

We performed a within-subjects study via Amazon Mechanical Turk ($n = 50$). The results suggest that prior experience with NAO robots and robots in general has a significant effect on perceived joke funniness. In particular, the results indicate that participants with more experience with NAO robots or robots in general found the jokes to be funnier than participants with less experience. Additionally, our results do not indicate a significant effect of humor type, gender, or age on perceptions of joke funniness. These results will enable the development of social robots and, in turn, AI systems capable of more efficient and fluent interactions.

2 Background

2.1 Humor in human interactions

Humor as a conflict resolution and stress-relief strategy has been examined thoroughly in human-human interactions [36–40]. For instance, Mesmer-Magnus et al. [32] reported on the stress relief and health benefits of positive humor in the workplace. Similarly, Kobel and Groeppel-Klein [31] found that humor could be used to reduce tension and frustration in cases of customer service failure. In addition to reducing conflict and stress, humor has also been used to enhance connections and office morale. Specifically, Warren et al. [33] examined how humor could be used to enhance relationships and achieve consumer goals. There have also been reports on how different factors can affect the interpretation of humor in human-human interactions. For example, Bippus [41] observed how internal versus external attributions of reasons for humor usage affected one's perceptions of humor when utilized in a conflict. Additionally, Norrick and Spitz [42] examined how factors such as the agent's relationship and the severity of the conflict can affect how humor is received in a human-human interaction.

Humor has been previously identified as an effective communication tactic in a variety of human-human interaction scenarios. We aim to expand on the findings of the aforementioned studies to evaluate perceptions of advanced, human-like humor in a HRI. Incorporating familiar communication patterns could provide human collaborative agents with a source of familiarity that could enable robots to be more accessible and usable in everyday environments. Furthermore, we could consider prior work on using humor to navigate delicate working relationships between humans and robot partners while strengthening bonds between these collaborators.

2.2 Humorous robots

There has also been recent work in utilizing humor as a social strategy in HRI. Robot humor has been observed in the context of stand-up comedy in order to examine the effects of timing and adaptivity on audience reception [43–45]. Weber et al. [46] examined how robots can analyze a person's reaction to humor and adapt their jokes to better align with the human's sense of humor. Social robots have used humor in order to optimize interactions with humans while achieving specific engagement goals. For example, in a review of humor in HRI, Oliveira et al. [47] highlighted the physical and psychological benefits that can be achieved through humor as well as how humor can be leveraged to reduce the effect of failure. Additionally, Adamson et al. [48] designed a robot-photographer that made use of humorous memes, GIFs, and sound effects to elicit spontaneous smiles. Sebo et al. [49] included a humor component in their study on how a social robot's behavior can be used to encourage human team members' expressions of trust-related behavior. Humor has also been examined as a conflict resolution strategy in HRI. Stoll et al. [50] examined perceptions of three humor forms in a robot versus human conflict mediator.

In human-human interactions, humor as a conversational tactic supplies an attractive alternative to lengthy anecdotes and references. If used appropriately, humor can serve as a low-risk and easy opportunity to formulate connections with other humans. In HRI to date, simple and often unrelated forms of humor have been utilized as a conversational tool to foster bonds and engage users. However, there has not been a systematic study investigating how people perceive various humor types employed by robots during HRIs. To address this gap, we are specifically interested in examining a robot that employs the more advanced, human-like humor strategies as a direct conversational partner.

2.3 Social recovery in HRI

Communicating and mitigating failures is another crucial domain in HRI research. Social robots have been designed to incorporate certain human-like social tactics to minimize the effects of task failure. These approaches can be grouped into the following categories: apologies, explanations, and preemptive attempts to set expectations. There have been previous studies on how humans perceive robots that utilize these more human-like social recovery approaches. Notably, Honig and Oron-Gilad [51] conducted a review on human perceptions of robot failure and mitigation strategies in HRI. In addition, Reig et al. [52] investigated human perceptions of different recovery and explainability tactics in a task failure scenario. In a different approach, Cameron et al. [53] observed how perceptions vary when a social robot explains its errors or merely apologizes. Similarly, Choi et al. [54] examined failure and recovery in the context of a service robot. Additionally, Lucas et al. [55] explored how social dialog was received when used in an attempt to recover from conversational errors.

Failure is possible in any interaction where agents are aiming to achieve a goal. However, failure can manifest in a myriad of ways and affect each agent differently. In addition, failure can range in severity and consequence. As a result, incorporating a broad spectrum of mitigation strategies for failure recovery can better minimize the negative effects of the different types of error that can occur. Specifically, by increasing the amount of social recovery strategies at the robot's disposal, we can equip robots with a greater arsenal of usable techniques. We aim to explore humor as a tactic with the ultimate goal of designing robots capable of generating the best response for recovery given the context and collaboration partner.

2.4 Trust in human-robot teams

The integration of robots into everyday life is dependent on whether human agents are willing to trust in the capabilities of robotic partners, aids, and assistants. As a result, there has been extensive exploration on how to design robots that embody, adapt to, and utilize trust [56–58]. Chen et al. [59] examined how behavioral patterns in a pick-and-place task could be modified to encourage trust between a human-robot pair. Habibian et al. [60] conducted research on robots capable of considering the human's point of view during their learning process in order to create more transparent and efficient human-robot collaborations. There has also been prior work on the examination of how a robot's performance affects trust [61, 62]. The attribution of human characteristics to robots and its effect on trust have been progressively studied [9]. In their study, van Pinxteren et al. [63] examined the effects of anthropomorphism on trust in a service robot. Additionally, in the context of the robot's physical characteristics, Bryant et al. [64] examined how robot gendering can affect how they are trusted and perceived by humans. Moreover, there has also been exploration of the measurement of trust in HRI. Chita-Tegmark et al. [65] examined how trust as a metric can be effectively used in HRI to develop more robust collaborative robot partners. Alternately, Desai et al. [66] evaluated the effectiveness of posttask evaluation on depictions of real-time trust in a HRI.

Understanding how robots impact, react to, and model trust in human-robot relationships is imperative for more fluent HRIs. Humans use trust to guide their lives, relationships, and decisions. Thus it is crucial to engineer robots that can also integrate trust into their design. By considering robots that utilize human-like communication strategies such as humor, we can further study the dynamic between trust, recovery, and reassurance in HRI. Moreover, humor could be used to facilitate more accessible and usable robot teammates. Specifically, humor supplies sentiment of inclusivity, which can transcend skill-based barriers and promote communication in agents with less robot experience.

3 Humor and trust

Trust is an essential component to fostering connections between agents. Moreover, a balance of trust is key for fluent collaborations. There are numerous strategies that can be used to minimize and maximize the trust between agents. Specifically, humans rely on intrinsic factors such as experience, mood, and personality when determining whether to trust in another agent. In addition to these internal characteristics, trust can also be influenced by certain external factors. For example, situational complexity, task risk, and collective goals can affect trust levels.

While trust is dynamic and changing with the human's intrinsic qualities and surrounding environment, it is also actively affected by the partnered agent. Factors such as the partner's appearance, dialog, performance, and familiarity can contribute to perceived trustworthiness. In our study, we focus on one specific component of trust: the dialog of the partnered agent. Specifically, we observe how humor can be used in dialog in a HRI as a communication strategy to foster trust.

We consider our study on humor in HRI as a natural extension of the prior work on the dynamic between humor and trust in human-human interactions. Particularly, in human-human interactions, humor is often used to engender trust in a interaction. Humor as a communication strategy can serve as a method for projecting one's personality, approachability, and even intelligence. Much can be deciphered from a humorous conversation without a need to express one's capabilities outright. Designing robots capable of utilizing contextually relevant, advanced humor could enable them to strengthen the trust between human-robot pairs. By using humor to encourage trust between agents, users would not need deep background or technical knowledge on the inter workings of their robot partners in order to collaborate effectively.

Our study on humorous robots creates an opportunity for a deeper dive into the autonomous and adaptable applications of humor in HRI. Specifically, we look to design robots capable of recognizing when humor is appropriate for the situation, interaction partner, and environment. By capitalizing on these different factors, we can design robots that use humor to effectively manage trust while promoting amicable interactions. In addition, these humorous robots can offer more human-like interactions and conversation patterns, which could aid in connecting human users that would otherwise struggle to adopt robot partners. Communication is an excellent method for building relationships and developing trust, and humor can elevate the experience by assuring users in a way that is inclusive and lighthearted.

4 Research questions

To develop our research questions, we referred to the aforementioned work on humor and social recovery in HRI. In our study, we also wanted to incorporate the humor types most widely used in human-human interactions. We aimed to examine the effects of the four different humor types through the following five research questions:

- **RQ1:** *What is the effect of humor type on joke rating?* In particular, we wish to examine if manipulating the robot's humor type will affect participants' perception of the joke funniness.
- **RQ2:** *What is the effect of gender on joke rating?* Specifically, we are interested in observing if participants of different genders differ in their funniness ratings for the different humor types.
- **RQ3:** *What is the effect of age on joke rating?* The third goal is to determine if participant age affects the perceptions of the different humor types.
- **RQ4:** *What is the effect of previous experience with NAO robots on joke rating?* We also wish to evaluate if participants with previous NAO experience differ in perceptions of the humor types when compared to participants with less experience.
- **RQ5:** *What is the effect of previous robot experience on joke rating?* Finally, we would like to observe the effects of previous participant experience with robots in general on their perceptions of the different humor types. Specifically, we are interested in determining if participants with more experience with robots will have different funniness ratings than participants with less experience.

5 Method

5.1 The humor types

In this study, we adopt the four humor types most commonly observed in human-human interactions [67]. These four humor categories adequately cover the conversational objectives that arise in an interaction. These objectives include: enhancing the self or the relationships with others by building up or tearing down oneself or others. An example joke from each humor category is presented in Table 9.1.

Affiliative humor is good-natured, nonhostile, and meant to amuse everyone. This low-stakes style is intended to affirm oneself and others.

Aggressive humor involves ridiculing or putting down others. This style is typically associated with sarcasm and teasing.

Table 9.1 Sample set of jokes for each of the four humor types.

Humor type	Joke example
Affiliative	"Uh oh, I hope they don't dock my pay."
Aggressive	"Interesting, I guess we're both bad at this game."
Self-defeating	"Well, I guess all I can do now is return to my box in shame."
Self-enhancing	"One for two, that's a pass on an engineering curve."

Self-defeating humor takes "laughing-at-oneself" a step further through excessive self-deprecation. In this style, the joke-teller becomes the target of derision in an attempt to amuse others in the process.

Self-enhancing humor is good-natured and playful like affiliative humor. However, unlike affiliative humor, this "laugh-at-oneself" style is more introspective and less focused on cultivating relationships.

5.2 Programming the NAO

In our study, we utilized SoftBank Robotics' NAO robot [68]. This humanoid robot is approximately 58 cm in height. For the robot's performance, we combined the expressive behavior modules in SoftBank's Choregraphe suite with Amazon Polly's text-to-speech platform [69]. We elected to use Amazon Polly over Choregraphe's text-to-speech option because of Amazon Polly's robust annunciation and timing capabilities. We use the "Ivy" voice as it was the most gender-neutral option to reduce any bias with perceptions of the robot's gender. Then, 20 sound clips of the jokes were added to the Choregraphe program and matched with some basic animation behaviors. In Choregraphe, we also utilized the speech recognition modules so that the robot could react accordingly to the three block colors upon instruction. The speech recognition modules were also used to trigger the humorous responses in these cases of block misidentification.

5.3 The *iSpy* testbed

The game consisted of the human partner prompting the NAO robot to identify colored blocks (see Fig. 9.1) and was loosely based on the children's game *iSpy*. The NAO robot was positioned to face a small stand holding the three, 1.5 in blocks (green, red, and blue). Since we were gathering participant perceptions of the different humor types, the robot was programmed to incorrectly identify the block and respond with a joke from one of the four humor categories. For this study, one of the experimenters sat opposite the NAO robot and played through the game for each joke condition. We recorded these interactions to use on Amazon Mechanical Turk. A sample frame of a recorded video is presented in Fig. 9.1, and a sample script of the interaction is included in Table 9.2.

FIG. 9.1 Image of the NAO robot playing the *iSpy* game.

Table 9.2 Sample script for one of the self-enhancing interactions.

Human: "I spy something blue."
NAO: "Is it this one?" [gestures to the green block]
Human: "No."
NAO: "Oh really? I must need to get my eyes checked."

6 Experiment

The study was conducted over Amazon Mechanical Turk to reach a large, diverse set of participants. In the study, participants were instructed to watch and rate 20 short clips of the NAO robot performing in the *iSpy* task. In each clip, the NAO robot misidentified one of the blocks and responded with a predetermined joke from one of the four humor categories. We also collected demographic information at the start of each study to examine how these factors affected the joke ratings.

6.1 Procedure

After reviewing the electronic study information document for consent and the task instructions, participants were asked to complete a brief demographic survey. Next, the participants were shown 20 videos in a randomized order. In each clip, the robot misidentified a block and followed the mistake with a joke from one of the four humor categories.

After viewing each interaction, the participants were asked to rate how funny the joke was on a Likert scale ranging from "not funny" (1) to "very funny" (5). For the recorded interactions, the longest video was 20 s, and the shortest was 12 s long ($M = 14$). At the conclusion of the study, participants were debriefed and compensated.

6.2 Measures

In addition to identifying the best humor types for robots to employ, we also wanted to examine how demographic differences could affect one's perception of humor. As a result, we asked participants to provide their gender, age, highest level of education, occupation, and nationality. Participants were also asked to rate their previous experience with both NAO robots and robots in general on a Likert scale from "no experience" (1) to "expert-level experience" (5).

6.3 Participants

A total of 50 adults participated in the study on Amazon Mechanical Turk (60% male [$n = 30$], 40% female [$n = 20$]). The mean age of participants was 33.76 years ($SD = 9.73$). All participants were based in the United States and required to be English speakers, at least 18 years of age or older, and have a Human Intelligence Task (HIT) approval rating of 99% or greater. In addition to the high approval rating requirement, we also added a timer to the video clips to ensure that participants watched the videos. The participants had a wide variety of occupations, and the highest level of education mainly consisted of Bachelor's ($n = 36$) and Master's ($n = 11$) degrees. During the study participants also reported their experience with both NAO robots ($M = 3.38$, $SD = 1.23$) and robots in general ($M = 3.70$, $SD = 1.06$). Participants were compensated $2 for participating in the study. Our study was approved by the Institutional Review Board.

7 Results

To address our research questions, we evaluated the effect of the four humor types on the funniness ratings. Additionally, we examined the effect of the humor types combined with other demographic factors such as gender, age, and previous robot experience.

7.1 RQ1: Effect of humor type

We conducted a repeated-measures ANOVA to examine the effect of humor type on the joke rating. Mauchly's test indicated that the assumption of sphericity had been violated, $\chi^2(5) = 14.25$, $P = .014$. Therefore, Huynh-Feldt estimates of sphericity are reported ($\epsilon = .87$). The results suggest that the joke ratings were not significantly affected by the type of joke, $F(2.62, 128.33) = 1.81$, $P = .156$. The mean ratings for the four humor categories are shown in Fig. 9.2.

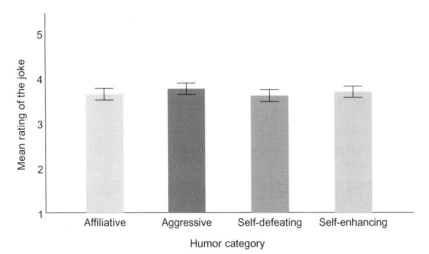

FIG. 9.2 The mean joke ratings on a scale from 1 (not funny) to 5 (very funny) for each of the humor categories: affiliative, aggressive, self-defeating, and self-enhancing.

7.2 RQ2: Effect of gender

We conducted a repeated-measures ANOVA to examine the effect of humor type and participant gender on the joke rating. Participants were able to self-describe gender; however, all participants selected either female or male. As a result, we analyzed results for these two groups. Mauchly's test indicated that the assumption of sphericity had been violated, $\chi^2(5) = .73$, $P = .011$. As a result, Huynh-Feldt estimates of sphericity are reported ($\epsilon = .88$). The results suggest that participant gender and the joke type did not significantly affect the joke rating, $F(2.65, 127.11) = 1.17$, $P = .32$. The mean ratings for joke type based on gender are shown in Fig. 9.3. In addition, the results suggest no significant effect of participant gender on the overall joke ratings $F(4, 45) = 1.24$, $P = .270$. The mean ratings for the overall joke ratings, categorized by gender, are shown in Fig. 9.4.

7.3 RQ3: Effect of age

We conducted a repeated-measures ANOVA to examine the effect of humor type and participant age on the joke rating. Participants were asked to provide their ages, and we grouped the ages before analysis. The ages were grouped as follows: 21–30, 31–40, 41–50, 51–60, and 61–70. Mauchly's test indicated that the assumption of sphericity had been violated, $\chi^2(5) = 11.65$, $P = .04$. Therefore, Huynh-Feldt estimates of sphericity are reported ($\epsilon = .97$). The results (Fig. 9.5) suggest that participant age and the joke type did not significantly affect the joke rating, $F(11.64, 130.92) = 1.37$, $P = .19$. Additionally, the results suggest no significant effect of participant age on the overall joke ratings $F(4, 45) = 1.36$, $P = .264$. The mean ratings for the overall joke ratings, categorized by the different age groups, are shown in Fig. 9.6.

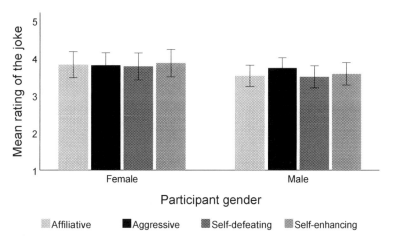

FIG. 9.3 The mean joke ratings on a scale from 1 (not funny) to 5 (very funny) for the four humor types based on participant gender.

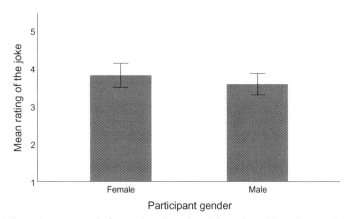

FIG. 9.4 The mean joke ratings on a scale from 1 (not funny) to 5 (very funny) based on participant gender.

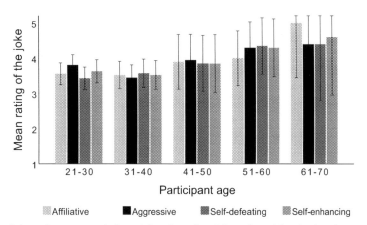

FIG. 9.5 The mean joke ratings on a scale from 1 (not funny) to 5 (very funny) for the four humor types based on participant age.

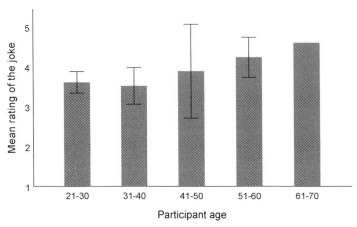

FIG. 9.6 The mean joke ratings on a scale from 1 (not funny) to 5 (very funny) based on participant age.

7.4 RQ4: Effect of previous NAO experience

We conducted a repeated-measures ANOVA to examine the effect of humor type and the participant's level of experience with NAO robots on the joke rating. Mauchly's test indicated that the assumption of sphericity had been violated, $\chi^2(5) = .67$, $P = .004$. Therefore, Huynh-Feldt estimates of sphericity are reported ($\epsilon = .91$). The results (Fig. 9.7) suggest that experience with NAO robots and the joke type did not significantly affect the joke rating, $F(11.00, 123.24) = 1.59$, $P = .11$. In other words, there was no significant difference in ratings of the four humor types for the varying NAO experience levels (1–5).

However, there was a significant main effect of participant experience with NAO robots on overall joke rating $F(4, 45) = 6.64$, $P < .001$. Participants with a higher level of experience with NAO robots found the jokes funnier than participants with less experience. The graph of mean joke rating and previous experience with NAO robots (Fig. 9.8) depicts that

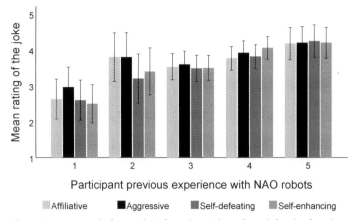

FIG. 9.7 The mean joke ratings on a scale from 1 (not funny) to 5 (very funny) for the four humor types based on previous experience with NAO robots on a scale from 1 (none) to 5 (expert-level).

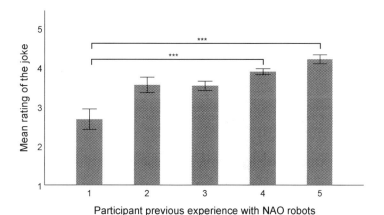

FIG. 9.8 The mean joke ratings on a scale from 1 (not funny) to 5 (very funny) based on previous experience with NAO robots on a scale from 1 (none) to 5 (expert-level). The significant difference in joke ratings between experience levels are shown in * (*$P < .05$, **$P < .01$, ***$P < .001$).

participants with a higher experience with NAO robots found the jokes to be funnier than participants with less experience. Specifically, participants with no NAO experience (1) rated the jokes significantly lower than participants with high and expert-level NAO experience (4 and 5), $P < .001$ and $P = .001$, respectively.

7.5 RQ5: Effect of previous robot experience

A repeated-measures ANOVA was conducted to examine the effect of humor type and the participant's level of experience with robots in general on the joke rating. Mauchly's test indicated that the assumption of sphericity had been met, $\chi^2(5) = .82$, $P = .11$. The results (Fig. 9.9) suggest that experience with robots in general and the joke type significantly affect the joke rating, $F(12, 135) = 2.20$, $P = .015$. In particular, for participants with no

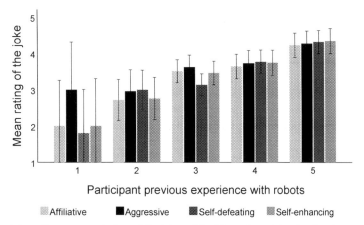

Affiliative ■ Aggressive Self-defeating Self-enhancing

FIG. 9.9 The mean joke ratings on a scale from 1 (not funny) to 5 (very funny) for the four humor types based on previous experience with robots in general on a scale from 1 (none) to 5 (expert-level).

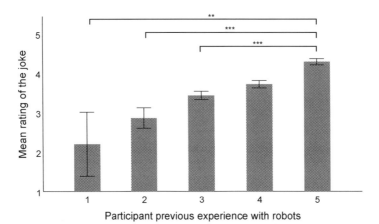

FIG. 9.10 The mean joke ratings on a scale from 1 (not funny) to 5 (very funny) based on previous experience with robots in general on a scale from 1 (none) to 5 (expert-level). The significant difference in joke ratings between experience levels are shown in * (*$P < .05$, **$P < .01$, ***$P < .001$).

robot experience (1), the aggressive humor type ratings were higher in comparison to the other categories (affiliative, self-defeating, and self-enhancing). Contrarily, participants with more previous robot experience did not have significantly different ratings for each of the humor types. In other words, participants with previous robot experience (2–5) had similar ratings of funniness for the different humor types.

In addition, there was a significant main effect of participant experience with robots in general on joke rating $F(4, 45) = 9.22$, $P < .001$. The graph of mean joke rating and previous experience with robots in general (Fig. 9.10) indicates that participants with a higher experience with robots found the jokes to be funnier than participants with less experience. Specifically, participants with expert-level robot experience (5) rated the jokes significantly higher than participants with low robot experience (2 and 3), $P < .001$ and $P = .001$, respectively. In addition, participants with expert-level experience (5) rated the jokes significantly higher than participants with no robot experience (1), $P = .008$.

8 Discussion and future work

In this study, we analyzed the human perceptions of funniness for an NAO robot that employed different humor types. We examined if social robots, like humans, can utilize advanced styles of humor. We collected a variety of demographic factors in order to see if they contributed to how the jokes were received. The results did not indicate that humor type alone affects the perceived funniness of the joke (**RQ1**). In turn, we reason that robots can utilize the same humor types as humans in simple interaction scenarios. Specifically, the results suggest that a social robot's jokes are perceived to be funny ($M = 3.67$), regardless of the humor type, in interactions involving a small number of agents. For **RQ2**, the results did not suggest that gender and humor type affect the joke ratings. Consequently, we reason that the four humor types can be adequately leveraged by a robot in interactions with both men and women. In the context of **RQ3**, we did not observe a significant effect of participant age on the joke ratings. We conclude from the results that the robot can utilize the various humor categories in HRIs without regard to age. Overall, we did not observe a significant effect of gender, age, and the joke type on perceptions of funniness for the jokes.

In response to **RQ4**, the results suggest that previous experience with NAO robots does not affect the ratings between the joke types. However, participants with more NAO robot experience did have significantly higher ratings of the jokes, in general. As a result, we can reason that participants who have more NAO robot experience are more receptive to robot humor in general. For **RQ5**, the mean ratings plotted in Fig. 9.9 suggest that the joke ratings for the different humor types are more similar when participants have a higher experience with robots. Contrarily, the joke ratings for the four humor types varied more when participants had less experience with robots. Additionally, the participant robot experience and joke rating results suggest that participants with more experience with robots tend to have more of an appreciation for robot humor in general.

Overall, the results suggest that previous robot experience has a stronger impact on how a social robot's jokes are perceived than demographic features like age and gender. This could be due to a familiarity effect. Specifically, it could be that humans are more likely to find a familiar or known conversation partner humorous. In addition, we note that the different jokes did not consist of any references or content that would alienate participants based on their demographic features. It could be that the humor style is less important than an overall understanding and appreciation of the joke's content.

The results indicate that participants' general experience with robots had a greater impact on joke reception than participant experience with NAO robots. It could be that people with less general robot experience have higher initial expectations of the more humanoid robot's joke-telling abilities and are consequently less amused by the performance. Conversely, participants with more general robot experience have a more accurate gauge of the robot's capabilities and, as a result, find the jokes funnier.

In this study, we were able to examine the perceived funniness of the different humor types. However, in the future, we would like to compare these cases against participants who were not exposed to humor. Thus, to extend the findings from this study, we wish to conduct another study that examines how the jokes are received in an in-person, dyadic setup. We conducted an additional study to examine the differences in failure reception [70].

Additionally, we are interested in exploring whether additional factors, such as propensity to trust, personality, novelty of the situation, and mistake severity, affect the reception of humor in response to task failure. In future work, we aim to examine how these additional factors affect how humor is perceived when used to recover from failure. Furthermore, we are interested in comparing the use of affiliative, aggressive, self-enhancing, and self-defeating humor against apologies, explanations, and preemptive attempts to set expectations.

9 Conclusion

Social robots are becoming increasingly present in day-to-day life. To improve the quality of HRIs and foster trust in human-robot teams, it is necessary to explore more dynamic, human-like conversational tactics such as humor. In human-human interactions, humor is a low-risk, simple social strategy for enriching a conversation. In our study, we attempted to examine if this tactic could be extended to social robots. The results suggest that previous experience with robots impacts the perception of a social robot's jokes more than demographic features like age and gender. Ultimately, our findings imply that robots can use advanced, human-like types of humor for more dynamic HRIs. We aim to expand upon our findings to further explore the dynamic between humor, failure recovery, and trust. With a robust understanding of how robots are perceived when they employ a myriad of failure mitigation tactics, we can create more fluent HRIs.

References

[1] M.C. Gombolay, X.J. Yang, B. Hayes, N. Seo, Z. Liu, S. Wadhwania, T. Yu, N. Shah, T. Golen, J.A. Shah, Robotic assistance in coordination of patient care, in: Robotics: Science and Systems, 2016.

[2] J. Wirtz, P.G. Patterson, W.H. Kunz, T. Gruber, V.N. Lu, S. Paluch, A. Martins, Brave new world: service robots in the frontline, J. Serv. Manag. 29 (5) (2018) 907–931.

[3] I. Leite, C. Martinho, A. Paiva, Social robots for long-term interaction: a survey, Int. J. Soc. Robot. 5 (2) (2013) 291–308.

[4] T. Belpaeme, J. Kennedy, A. Ramachandran, B. Scassellati, F. Tanaka, Social robots for education: a review, Sci. Rob. 3 (21) (2018) eaat5954.

[5] A. Kubota, T. Iqbal, J.A. Shah, L.D. Riek, Activity recognition in manufacturing: the roles of motion capture and sEMG+inertial wearables in detecting fine vs. gross motion, in: International Conference on Robotics and Automation (ICRA), 2019, pp. 6533–6539.

[6] S. Valencia, M. Luria, A. Pavel, J.P. Bigham, H. Admoni, Co-designing socially assistive sidekicks for motion-based AAC, in: HRI '21, Proceedings of the 2021 ACM/IEEE International Conference on Human-Robot Interaction (HRI), Association for Computing Machinery, 2021, pp. 24–33.

[7] T. Iqbal, L.D. Riek, Human robot teaming: approaches from joint action and dynamical systems, in: Humanoid Robotics: A Reference, 2017, pp. 2293–2312.

[8] S. Chernova, A.L. Thomaz, Robot Learning from Human Teachers, in: Synthesis Lectures on Artificial Intelligence and Machine Learning, vol. 8, 2014, pp. 1–121.

[9] M. Natarajan, M. Gombolay, Effects of anthropomorphism and accountability on trust in human robot interaction, in: HRI '20, Proceedings of the 2020 ACM/IEEE International Conference on Human-Robot Interaction (HRI), Association for Computing Machinery, 2020, pp. 33–42.

[10] M.F. Jung, Affective grounding in human-robot interaction, in: 2017 12th ACM/IEEE International Conference on Human-Robot Interaction (HRI), 2017, pp. 263–273.

[11] E.J de Visser, M.M. Peeters, M.F. Jung, S. Kohn, T.H. Shaw, R. Pak, M.A. Neerincx, Towards a theory of longitudinal trust calibration in human-robot teams, Int. J. Soc. Robot. 12 (2020) 459–478.

[12] T. Iqbal, L.D. Riek, A method for automatic detection of psychomotor entrainment, IEEE Trans. Affect. Comput. 7 (1) (2016) 3–16.

[13] T. Iqbal, S. Li, C. Fourie, B. Hayes, J.A. Shah, Fast online segmentation of activities from partial trajectories, in: IEEE International Conference on Robotics and Automation (ICRA), 2019.

[14] M.S. Yasar, T. Iqbal, A scalable approach to predict multi-agent motion for human-robot collaboration, IEEE Rob. Autom. Lett. (RA-L) 6 (2) (2021) 1686–1693.

[15] M.S. Yasar, T. Iqbal, Improving human motion prediction through continual learning, in: ACM/IEEE International Conference on Human-Robot Interaction (HRI), Lifelong Learning and Personalization in Long-Term Human-Robot Interaction (LEAP-HRI), 2021.

[16] M.M. Islam, T. Iqbal, HAMLET: a hierarchical multimodal attention-based human activity recognition algorithm, in: 2020 IEEE/RSJ International Conference on Intelligent Robots and Systems (IROS), 2020, pp. 10285–10292.

[17] M.M. Islam, T. Iqbal, Multi-GAT: a graphical attention-based hierarchical multimodal representation learning approach for human activity recognition, IEEE Rob. Autom. Lett. 6 (2) (2021) 1729–1736.

[18] M.M. Islam, R.M. Mirzaiee, A. Gladstone, H.N. Green, T. Iqbal, CAESAR: an embodied simulator for generating multimodal referring expression datasets, in: Thirty-Sixth Conference on Neural Information Processing Systems Datasets and Benchmarks Track, 2022.

[19] M.M. Islam, T. Iqbal, MuMu: cooperative multitask learning-based guided multimodal fusion, AAAI, 2022.

[20] M.M. Islam, M.S. Yasar, T. Iqbal, MAVEN: a memory augmented recurrent approach for multimodal fusion, IEEE Trans. Multimedia 25 (2023) 3694–3708.

[21] T. Iqbal, L.D. Riek, Coordination dynamics in multi-human multi-robot teams, IEEE Rob. Autom. Lett. 2 (3) (2017) 1712–1717.

[22] A. Kshirsagar, M. Lim, S. Christian, G. Hoffman, Robot gaze behaviors in human-to-robot handovers, IEEE Rob. Autom. Lett. 5 (4) (2020) 6552–6558.

[23] H. Admoni, B. Scassellati, Social eye gaze in human-robot interaction: a review, J. Hum.-Robot Interact. 6 (1) (2017) 25–63.

[24] T. Iqbal, S. Rack, L.D. Riek, Movement coordination in human-robot teams: a dynamical systems approach, IEEE Trans. Robot. 32 (4) (2016) 909–919.

[25] T. Iqbal, L.D. Riek, Temporal anticipation and adaptation methods for fluent human-robot teaming, in: IEEE International Conference on Robotics and Automation (ICRA), 2021.

[26] T. Williams, P. Briggs, M. Scheutz, Covert robot-robot communication: human perceptions and implications for human-robot interaction, J. Hum. Robot Interact. 4 (2) (2015) 24–49.

[27] A.S. Clair, M. Matarić, How robot verbal feedback can improve team performance in human-robot task collaborations, in: HRI '15, 2015 10th ACM/IEEE International Conference on Human-Robot Interaction (HRI), 2015, pp. 213–220.

[28] N. Martelaro, V.C. Nneji, W. Ju, P. Hinds, Tell me more designing HRI to encourage more trust, disclosure, and companionship, in: HRI '16, 2016 11th ACM/IEEE International Conference on Human-Robot Interaction (HRI), 2016, pp. 181–188.

[29] L. Takayama, V. Groom, C. Nass, I'm sorry, Dave: I'm afraid I won't do that: social aspects of human-agent conflict, in: Proceedings of the SIGCHI Conference on Human Factors in Computing Systems, Association for Computing Machinery, 2009, pp. 2099–2108.

[30] F. Correia, S. Mascarenhas, R. Prada, F.S. Melo, A. Paiva, Group-based emotions in teams of humans and robots, in: HRI '18, Proceedings of the 2018 ACM/IEEE International Conference on Human-Robot Interaction (HRI), Association for Computing Machinery, 2018, pp. 261–269.

[31] S. Kobel, A. Groeppel-Klein, No laughing matter, or a secret weapon? Exploring the effect of humor in service failure situations, J. Bus. Res. 132 (2021) 260–269.

[32] J. Mesmer-Magnus, D.J. Glew, C. Viswesvaran, A meta-analysis of positive humor in the workplace, J. Manag. Psychol. 27 (2) (2012) 155–190.

[33] C. Warren, A. Barsky, A.P. McGraw, Humor, comedy, and consumer behavior, J. Consum. Res. 45 (3) (2018) 529–552.

[34] N. Mirnig, G. Stollnberger, M. Giuliani, M. Tscheligi, Elements of humor: how humans perceive verbal and non-verbal aspects of humorous robot behavior, in: HRI '17, Proceedings of the Companion of the 2017 ACM/IEEE International Conference on Human-Robot Interaction (HRI), Association for Computing Machinery, 2017, pp. 211–212.

[35] I.M. Menne, B.P. Lange, D.C. Unz, My humorous robot: effects of a robot telling jokes on perceived intelligence and liking, in: HRI '18, Companion of the 2018 ACM/IEEE International Conference on Human-Robot Interaction (HRI), Association for Computing Machinery, 2018, pp. 193–194.

[36] W.J. Smith, K.V. Harrington, C.P. Neck, Resolving conflict with humor in a diversity context, J. Manag. Psychol. 15 (6) (2000) 606–625.

[37] S.F. Dziegielewski, Humor, Int. J. Ment. Health 32 (3) (2003) 74–90.

[38] M.H. Abel, Humor, stress, and coping strategies, Int. J. Humor Res. 15 (4) (2002) 365–381.

[39] J.-Y. Mao, J.T.-J. Chiang, Y. Zhang, M. Gao, Humor as a relationship lubricant: the implications of leader humor on transformational leadership perceptions and team performance, J. Leadersh. Org. Stud. 24 (4) (2017) 494–506.

[40] B.M. Savage, H.L. Lujan, R.R. Thipparthi, S.E. DiCarlo, Humor, laughter, learning, and health! A brief review, Adv. Physiol. Educ. 41 (3) (2017) 341–347.

[41] A.M. Bippus, Humor motives, qualities, and reactions in recalled conflict episodes, West. J. Commun. 67 (4) (2003) 413–426.

[42] N.R. Norrick, A. Spitz, Humor as a resource for mitigating conflict in interaction, J. Pragmat. 40 (10) (2008) 1661–1686.

[43] J. Vilk, N.T. Fitter, Comedians in cafes getting data: evaluating timing and adaptivity in real-world robot comedy performance, in: HRI '20, Proceedings of the 2020 ACM/IEEE International Conference on Human-Robot Interaction (HRI), Association for Computing Machinery, 2020, pp. 223–231.

[44] J. Vilk, N.T. Fitter, Comedy by Jon the robot, in: HRI '20, Companion of the 2020 ACM/IEEE International Conference on Human-Robot Interaction (HRI), Association for Computing Machinery, 2020, pp. 223–231.

[45] J. Vilk, N.T. Fitter, Jon the robot goes Hollywood, in: HRI '20, Companion of the 2020 ACM/IEEE International Conference on Human-Robot Interaction (HRI), Association for Computing Machinery, 2020, p. 644.

[46] K. Weber, H. Ritschel, I. Aslan, F. Lingenfelser, E. André, How to shape the humor of a robot - social behavior adaptation based on reinforcement learning, in: ICMI '18, Proceedings of the 20th ACM International Conference on Multimodal Interaction (ICMI), Association for Computing Machinery, 2018, pp. 154–162.

[47] R. Oliveira, P. Arriaga, M. Axelsson, A. Paiva, Humor-robot interaction: a scoping review of the literature and future directions, Int. J. Soc. Robot. 13 (2020) 1369–1383.

[48] T. Adamson, C.B. Lyng-Olsen, K. Umstattd, M. Vázquez, Designing social interactions with a humorous robot photographer, in: HRI '20, Proceedings of the 2020 ACM/IEEE International Conference on Human-Robot Interaction (HRI), Association for Computing Machinery, 2020, pp. 233–241.

[49] S.S. Sebo, M. Traeger, M. Jung, B. Scassellati, The ripple effects of vulnerability: the effects of a robot's vulnerable behavior on trust in human-robot teams, in: HRI '18, Proceedings of the 2018 ACM/IEEE International Conference on Human-Robot Interaction (HRI), Association for Computing Machinery, 2018, pp. 178–186.

[50] B. Stoll, M.F. Jung, S.R. Fussell, Keeping it light: perceptions of humor styles in robot-mediated conflict, in: HRI '18, Companion of the 2018 ACM/IEEE International Conference on Human-Robot Interaction, Association for Computing Machinery, 2018, pp. 247–248.

[51] S. Honig, T. Oron-Gilad, Understanding and resolving failures in human-robot interaction: literature review and model development, Front. Psychol. 9 (2018) 861.

[52] S. Reig, E.J. Carter, T. Fong, J. Forlizzi, A. Steinfeld, Flailing, hailing, prevailing: perceptions of multi-robot failure recovery strategies, in: HRI '21, Proceedings of the 2021 ACM/IEEE International Conference on Human-Robot Interaction (HRI), Association for Computing Machinery, 2021, pp. 158–167.

[53] D. Cameron, S de Saille, E.C. Collins, J.M. Aitken, H. Cheung, A. Chua, E.J. Loh, J. Law, The effect of social-cognitive recovery strategies on likability, capability and trust in social robots, Comput. Hum. Behav. 114 (2021) 106561.

[54] S. Choi, A.S. Mattila, L.E. Bolton, To err is human(-oid): how do consumers react to robot service failure and recovery? J. Serv. Res. 24 (3) (2021) 354–371.

[55] G.M. Lucas, J. Boberg, D. Traum, R. Artstein, J. Gratch, A. Gainer, E. Johnson, A. Leuski, M. Nakano, Getting to know each other: the role of social dialogue in recovery from errors in social robots, in: HRI '18, Proceedings of the 2018 ACM/IEEE International Conference on Human-Robot Interaction (HRI), Association for Computing Machinery, 2018, pp. 344–351.

[56] S. Herse, J. Vitale, B. Johnston, M.-A. Williams, Using trust to determine user decision making and task outcome during a human-agent collaborative task, in: HRI '21, Proceedings of the 2021 ACM/IEEE International Conference on Human-Robot Interaction, Association for Computing Machinery, 2021, pp. 73–82.

[57] A. Xu, G. Dudek, OPTIMo: online probabilistic trust inference model for asymmetric human-robot collaborations, in: 2015 10th ACM/IEEE International Conference on Human-Robot Interaction (HRI), 2015, pp. 221–228.

[58] P.A. Hancock, D.R. Billings, K.E. Schaefer, J.Y.C. Chen, E.J de Visser, R. Parasuraman, A meta-analysis of factors affecting trust in human-robot interaction, Hum. Factors 53 (5) (2011) 517–527.

[59] M. Chen, S. Nikolaidis, H. Soh, D. Hsu, S. Srinivasa, Planning with trust for human-robot collaboration, in: HRI '18, Proceedings of the 2018 ACM/IEEE International Conference on Human-Robot Interaction, Association for Computing Machinery, 2018, pp. 307–315.

[60] S. Habibian, A. Jonnavittula, D.P. Losey, Here's what I've learned: asking questions that reveal reward learning, CoRR 11 (4) (2022) 1–28.

[61] P. Robinette, W. Li, R. Allen, A.M. Howard, A.R. Wagner, Overtrust of robots in emergency evacuation scenarios, in: 2016 11th ACM/IEEE International Conference on Human-Robot Interaction (HRI), 2016, pp. 101–108.

[62] P. Robinette, A.M. Howard, A.R. Wagner, Effect of robot performance on human-robot trust in time-critical situations, IEEE Trans. Hum.-Mach. Syst. 47 (4) (2017) 425–436.

[63] M. van Pinxteren, R. Wetzels, J. Rüger, M. Pluymaekers, M. Wetzels, Trust in humanoid robots: implications for services marketing, J. Serv. Mark. (2019).

[64] D. Bryant, J. Borenstein, A. Howard, Why should we gender? The effect of robot gendering and occupational stereotypes on human trust and perceived competency, in: Proceedings of the 2020 ACM/IEEE International Conference on Human-Robot Interaction, Association for Computing Machinery, 2020, pp. 13–21.

[65] M. Chita-Tegmark, T. Law, N. Rabb, M. Scheutz, Can you trust your trust measure? in: HRI '21, Proceedings of the 2021 ACM/IEEE International Conference on Human-Robot Interaction, Association for Computing Machinery, 2021, pp. 92–100.

[66] M. Desai, P. Kaniarasu, M. Medvedev, A. Steinfeld, H. Yanco, Impact of robot failures and feedback on real-time trust, in: 2013 8th ACM/IEEE International Conference on Human-Robot Interaction (HRI), 2013, pp. 251–258.

[67] R.A. Martin, P. Puhlik-Doris, G. Larsen, J. Gray, K. Weir, Individual differences in uses of humor and their relation to psychological well-being: development of the humor styles questionnaire, J. Res. Pers. 37 (1) (2003) 48–75.

[68] SoftBank Robotics, NAO, 2021. Retrieved 29 July 2021 https://www.softbankrobotics.com/emea/en/nao.

[69] Amazon Web Services, Inc., Amazon Polly, 2021. Retrieved June 2021 https://aws.amazon.com/polly/.

[70] H.N. Green, M.M. Islam, S. Ali, T. Iqbal, Who's laughing NAO? Examining perceptions of failure in a humorous robot partner, in: HRI '22, Proceedings of the 2022 ACM/IEEE International Conference on Human-Robot Interaction, 2022.

10

Real-time AI: Using AI on the tactical edge

Hesham Y. Fouad[a], Oliver Broadrick[b], Benjamin Harvey[b], Charles Peeke[b], and Bhagi Narahari[b]

[a]INFORMATION TECHNOLOGY DIVISION, US NAVAL RESEARCH LABORATORY, WASHINGTON, DC, UNITED STATES [b]DEPARTMENT OF COMPUTER SCIENCE, THE GEORGE WASHINGTON UNIVERSITY, WASHINGTON, DC, UNITED STATES

1 Introduction

Current machine learning models are used to answer increasingly more complex questions, with the caveat that each more complex model requires more computational resources or more time to process the data and produce an accurate result. In resource constrained environments where results need to be calculated in real time, more computational power and more time are not always available. This work is motivated by the growing importance and deployment of mission critical edge artificial intelligence (AI) applications that must meet real-time requirements. For example, in the domain of mission critical military operations, the use of AI in mission planning and execution imposes temporal constraints on results produced by machine learning models. Current AI implementations have nondeterministic execution times, which poses a serious problem when used in tactical military settings because the results of AI computations, with desired precision, may not be available when needed. The problem is compounded when considering realistic, multimission scenarios that must be carried out concurrently. In some situations, the system should adapt to provide additional predictive power for high-interest events that are detected. The imprecise computation model provides a framework in which to study this problem. Rather than focus on a computational model that results in the highest possible accuracy, this chapter considers the imprecise computation model, which considers scheduling the tasks to meet timing constraints while trading off accuracy. It proposes a novel infrastructure to schedule AI tasks as imprecise computations in hard real time that can be implemented on edge architectures.

In the infrastructure proposed in this research, the output of each intermediate layer in the neural network may be extracted, allowing very fine-grained tuning of the trade-off between time and accuracy. Each intermediate output is optimized to achieve both approximate answers in less time as well as more accurate answers with more computational time. A fast scheduling algorithm is presented, which accounts for task priority to

Putting AI in the Critical Loop. https://doi.org/10.1016/B978-0-443-15988-6.00004-2

generate schedules that provide higher accuracy for more important tasks, while running significantly faster than past solutions to the problem of scheduling neural networks as imprecise computations [1].

1.1 Imprecise computation in AI

The design of scheduling algorithms that can effectively manage the real-time execution of AI applications in military and civilian applications is a particularly challenging problem. AI algorithms are computationally intensive and exhibit a highly dynamic and non-deterministic execution time due to the nature of the computations. In hard real-time systems, each task has a deadline that must be met and if a single task does not produce a result within the deadline, the entire system fails as well. A promising approach to achieving real-time AI is to use graceful degradation when the available computing resources are not sufficient. The imprecise computations model [2, 3] is based on the premise that, in some cases, producing less-than-perfect results on time is better than producing no results at all. It provides a rigorous model for graceful degradation where the accuracy of the result is tuned to meet time constraints. This approach prevents total system failure by scheduling and executing the tasks such that an approximate result is produced within the deadline, and if time remains, the result may use more time within the deadline to refine the result to achieve a more precise result. This imprecise model is used in Ref. [1] to design a real-time deep learning system.

Consider, for example, a track prediction algorithm that will necessarily have some degree of error. As long as that error is bound within an acceptable range, the predicted tracks provide value by limiting the search space. In the imprecise computation model, a number of track prediction algorithms, executing concurrently, would all be given sufficient execution time to produce results at a minimum precision (i.e., accuracy)—this is referred to as the *mandatory* part of the tasks [2]. Higher priority tracks, referred to as the *optional* tasks [2], would be given additional execution time to produce more precise results. If the computational resources do not allow all tracks to produce results at a minimum precision, the system would fail predictably and alert the user.

1.2 AI on edge devices

Advances in technology have led to more capability in edge devices, including the ability to perform various AI tasks such as facial recognition and other image processing tasks. Systems that deploy edge and IoT devices, such as drones and other IoT devices with sensors, are becoming ubiquitous. In combination with the hardware capabilities, advantages such as reduced latency, lower demands on network bandwidth, and potential for improved security and privacy have seen an increase in the development of edge AI systems that partition the computing across both the edge device and the cloud, which may provide high performance GPUs. These systems have been used in diverse applications such as mission critical object recognition, autonomous cars [4, 5], agriculture [6], wildlife tracking [7], and animal behavior [8].

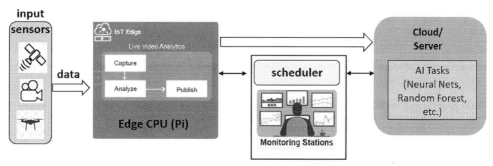

FIG. 10.1 System architecture.

The overall system view of our edge architecture is shown in Fig. 10.1. The edge devices, such as a Raspberry Pi, collect the data from different sensors such as cameras. We assume a periodic task scheduling model wherein data is sent for processing at periodic intervals, but our system allows the period to change based on user input. Since edge devices are constrained in computational power, we assume these cannot support a complete deep learning network that will meet the timing constraints. However, such devices are capable of running lean machine learning algorithms in a time-efficient manner [6]. We utilize these edge devices to provide the approximate results using our proposed multitask neural network (MNN) model. These results are then sent to the cloud server to run more complex machine learning computations on the tasks that are identified as requiring more predictive power. A key component in our system is the monitoring station, which includes the scheduler, and also includes the option of a *user in the loop*. Based on the approximate results from the edge device and the user-specified task priorities (i.e., a ranking of the importance of each task), the scheduler determines the amount of execution time for each task to optimize the accuracy metric while meeting the time deadline. One option for implementation of our system is to view the mandatory part of the computation, the MNN, as taking place on the edge device, and the optional tasks requiring more detailed analysis and accuracy are executed on the server. An alternate option is to compute some stages of the MNN on the edge device and the remaining mandatory tasks and the optional tasks on the server. Both options can be handled by our scheduler.

1.3 Contributions

Many of the assumptions used in the scheduling algorithms in the literature, in terms of both partitioning a job set across processing platforms as well as scheduling jobs on a single processor, assume fixed job priorities and a fixed job set [1, 9–13]. We consider a system wherein task priorities as well as the scheduling period could change dynamically (automated or user-chosen in the monitoring station module of our system). Going back to our track prediction example, we would expect track priorities to change dynamically based on the type, location, and velocity of a track. Additionally, we expect the job set to vary dynamically depending on objects recognized in the current period and on the user in the loop input (i.e., the decisions made in the monitoring station, if any). Finally, unlike the typical approach in the literature, the neural network itself needs to be trained for a model where not all stages of the network may be executed.

This chapter proposes a fine-grained infrastructure where (a) the output of each intermediate neural network layer is accessible, allowing for a trade-off between accuracy and computational time and (b) tasks are prioritized. The network is trained for use in the imprecise computation model, so each intermediate layer is optimized. In order to create a network that will more reliably run on the edge in a short amount of time using fewer computational resources, we created a modified version of this neural network to reduce computational power, improve the timing, while minimizing the expected decrease in overall accuracy that comes from fewer computations. We also propose a scheduling algorithm that takes into account task priority in this model. Our experimental results demonstrate that (a) the outputs of the intermediate layers are more accurate than intermediate results of networks trained using standard approaches, thus greatly improving the time/accuracy trade-off; (b) the scheduling algorithm runs much faster than past work while providing comparable quality for the average accuracy performance metric and provides better quality when the system objective is to optimize the accuracy of high-priority tasks.

Section 2 discusses the imprecise computation model and how neural network design and training can be adapted to this model. Related work is discussed in Section 3. Section 4 presents our approach for designing and training neural networks to support the imprecise computation model, and presents experiments and results. Section 5 presents the scheduling algorithm, and the experiments and results, and Section 6 concludes the chapter.

2 Problem definition

This section presents the imprecise computation model for our system. In order to integrate complex AI models with real-time systems and overcome the challenges with this process [14], the real-time system must be developed around the capabilities of the AI model: (i) the AI models must be constrained to meet the requirements of an existing real-time system, or (ii) AI models and real-time systems must work collaboratively to improve functionality of the system as a whole without a direct dependency. Our approach falls under the first category—we develop a model for training neural networks for object recognition based on the imprecise computation model to meet real-time constraints.

2.1 Imprecise computations model

One approach to prevent total system failure is to create tasks such that an approximate result is produced within the deadline, and if time remains, the result may use more time within the deadline to refine the result to achieve a more precise result. This approach to meeting deadlines is most commonly referred to as an imprecise computation; each task is defined to consist of a mandatory subtask and a collection of optional subtasks [2, 3, 9, 10]. In this chapter, the precision of a task is defined as the accuracy of the neural network prediction scores. Henceforth, we use the terms task precision and task accuracy interchangeably.

For our system, we consider the hard real-time, periodic workload model for scheduling the execution of a set of tasks. In this workload model, we have to schedule a set of n tasks at each period. We allow the duration of the period to change, either based on user input or automates based on some parameter in the scheduler, from one period to the next. This model applies to cases where the input data is collected from the sensors at regular intervals and the AI task needs to be completed before the next set of input data is collected. For example, the system could be identifying the different objects in the images being received at each time interval. If the system and user detect objects that could pose a threat, then the sampling interval could be decreased, thus leading to a smaller period in which to execute the tasks. The different tasks correspond to the different objects: for example, the objects to be recognized could be buildings, cars, trucks, and airplanes, and the recognition algorithm for each is viewed as a task. If identifying whether an object in the image is an (fast moving) airplane is of the highest importance (in the use case), then this task is assigned a high priority. Furthermore, if the object recognition algorithm detects the possibility of an airplane, then the system could, in addition to allocating more computing resources to this task, reduce the sampling period to track such fast-moving objects.

The classical imprecise scheduling problem [2, 9] specifies the input as a set of tasks or job set $\{\tau_1, \tau_2, ..., \tau_n\}$ where each task τ_j must be completed within a deadline time D_j and all tasks must be completed within the period deadline D. To capture the relative importance of the different tasks, each task τ_j is assigned a priority V_j. To model the mandatory and optional parts of the task set, each task τ_j consists of a mandatory subtask m_j and an optional subtask o_j. In the case where each task is a neural network, we can define the mandatory subtask to be the first k stages of the network and the optional subtask to be the remaining l stages. For a schedule to be valid, all the mandatory subtasks must be executed to meet the requirement to produce acceptable results with a minimum quality for the overall task set. For example, we want the system to spend some amount of time on detecting all objects in the input image. After all mandatory tasks have been scheduled, if there is time remaining to meet the period deadline D then the optional subtasks can be executed; for example, once the k mandatory stages in the neural network have been executed then we can execute two more stages if their execution time falls within the deadline. Under this model, the more time we allocate to the optional subtask the higher the quality of the output, that is, higher accuracy of the task. In other words, the precision, or accuracy in our case, is monotonic as a function of the execution time. We note that in our problem, where the tasks are neural networks, we can assume this condition holds without any loss of generality since we can simply ignore the cases where more stages in a neural network will result in lower accuracy and thus never schedule to execute the network with k layers if $k - 1$ layers have a better accuracy.

The scheduling of the mandatory subtasks reduces to the classical scheduling problem [2]. To determine how much time is needed to execute the optional subtasks, the priority of the task is taken into account. Intuitively, the system must generate higher accuracy for the more important tasks (which have a higher priority V_j). Furthermore, we assume that the priority of the tasks could change between one period to the next based on the user monitoring station and the results produced in the previous period. The goal is to increase the overall accuracy of the system, as defined by some performance metric by allocating more time to the appropriate set of optional subtasks. The quality of the final schedule should reflect the overall accuracy of the system. There have been several different metrics proposed to define this overall quality [2, 3, 9], and we consider two such metrics in the design of our scheduling algorithm. The first is a weighted average (or total) accuracy [2] and the second metric is the accuracy of the highest priority task. Section 5 presents the details of the scheduling model in terms of scheduling MNNs as imprecise computations.

The AI tasks, in this chapter and in related work, are neural networks and the accuracy of the network is considered as a function of the number of stages in the network [1, 10, 11]. The distinguishing feature of our approach is that we train the neural networks assuming the system scheduler uses the imprecise computation model. The starting point is to view a neural network as consisting of several subtasks where each subtask is simply a single stage in the network. The mandatory subtask itself could be more than one layer depending on the minimum accuracy specified for each task. For example, if the minimum accuracy for all tasks is set to 0.5, then this could result in a specific neural network requiring two stages to be executed to meet this minimum accuracy. Next, since we allow a subset of the stages of the network to be selected for execution, we approach the training process accordingly; that is, rather than train the entire n layer network and assign identical weights we train a k stage network using the training for $k-1$ stage network. Section 4 presents the details of our approach—the MNN model.

3 Related work

Considerable work in the literature has addressed various aspects of real-time AI systems. These have included new architectures [15], techniques for partitioning the AI task across the processing units in the edge architecture [11, 13], scheduling under power and network constraints [11, 12, 16], and the work closely related to ours that discussed scheduling deep neural networks as imprecise computations [1].

In Ref. [14], the different challenges of real-time systems for AI are discussed, but they do not consider the salient aspects of applying the imprecise computations model to AI tasks. Microsoft's BrainWave system for real-time AI is a configurable cloud-scale architecture, using field programmable gate arrays (FPGAs), which provides a highly parallel implementation of deep neural network models. This system is observed to achieve an order of magnitude improvement in latency over state-of-the-art GPUs. The use of FPGA enables configuration of the architecture to the models at compile time [15]. While the performance gains afforded by this system are substantial, it does not address the lack of determinism in the time complexity of AI computations.

In Ref. [17], deliberative real-time AI systems are considered with fixed priority scheduling using anytime scheduling algorithms that are special cases of imprecise computations. However, their model assumes fixed priority among the tasks and cannot consider the dynamic adaptive priorities that our system aims to model. Researchers have also explored real-time scheduling of neural network models using GPUs, including approximation aware neural networks. These approaches [4, 10] are motivated by energy and network availability constraints and are not modeled as a scheduling problem that would explicitly allow us to trade off accuracy to meet strict time constraints.

Resource constraints, such as power and network connectivity, play an important role in the implementation of edge AI systems and several projects have addressed this problem and provided solutions. In Ref. [18], the authors proposed compiling the neural network computations directly into functionally equivalent C leading to more predictable timing behavior. Samplawski et al. [13] consider the problem of partitioning computations, that is, the neural network, for object detection across a heterogeneous network with a specific instance of an edge device communicating with a cloud server that has to account for dynamically varying the network bandwidth. Their solution first partitions a deep neural network model at a given layer and then applies progressive transmission of intermediate convolutional filter maps. In Ref. [11], the authors designed a continuously tunable method for leveraging both local and remote resources to optimize performance of a deep learning model. Their system, CLIO, splits machine learning models between an IoT device and the cloud in a progressive manner that adapts to wireless dynamics. This enables graceful degradation when resources are constrained. Snigdha et al. [12] consider trading off accuracy for energy consumption by tuning the parameters in the neural network implementation, such as by bitwidth reduction and activating subsets of the neurons while keeping the same number of stages in the network. Embedded neural networks for animal behavior studies, examined in Ref. [16], considered the full implementation of a reconfigurable neural network embedded into an animal collar device. While all of these solutions trade off accuracy to meet a resource constraint, they are not geared to explicitly consider trading off accuracy to meet the real-time constraint. Thus their neural networks are not trained for the event in which only a subset of the network stages is executed.

The research that comes closest to this chapter considers scheduling deep neural networks as imprecise computations [1]. In Ref. [1], the authors developed a technique to schedule deep neural networks. The process was similar to the first strategy listed previously; develop an AI model, the deep neural networks, then create a real-time system that could schedule the tasks while considering the capabilities of the model. The scheduling algorithm and neural networks in this chapter reverse the strategy in this chapter to develop a similar scheduling technique with a novel approach to training and implementing neural networks in order to constrain an AI model to the hard real-time scheduling system. Our approach uses training techniques and AI models that are embedded within a real-time system that expands to implementation on the edge. With the system proposed in this chapter, devices on the edge can collect and input data to the system, and serve as a modular implementation of AI models to increase the efficiency of the real-time system [19].

The key difference in our approach is that the AI algorithms, in particular the neural network model, are designed in collaboration with the imprecise computing model. The concept of imprecise computation is considered during the training of the neural network, with the assumption that only a subset of the network layers may be computed. This MNN model provides us with the option of deploying it on the edge devices and using the cloud/server to focus on the more important tasks. A secondary difference is that our scheduling algorithm addresses different priorities among the tasks and, since the scheduler could be invoked during execution, we provide an algorithm that runs much faster than the solution in Ref. [1].

4 Multitask neural network model

This section presents our neural network model, the MNN, which is designed for imprecise computations. After a brief review of neural networks design, the key idea of training individual layers to better match an imprecise model and experimental results are presented. Finally, this approach is then applied to design a neural network that could run more reliably on an edge device in a shorter amount of time and require fewer computational resources.

Machine learning, and neural networks in particular, have become used more frequently and across wider applications in recent years. By attempting to mimic the processes of a human brain, neural networks have performed increasingly close to human levels for many complex tasks, including computer vision and natural language processing. In order to perform at this level, however, state-of-the-art neural networks require a substantial amount of computational power and data. And so, if a single model was trained with a stream of the same input data across various analytics, it is possible that not all analytics would be needed to achieve accurate results.

4.1 Neural networks

The development of artificial neural networks began from the analysis of the network of neurons that exist within the human brain [20]. In general, neural networks have a layered structure and a structured flow from input layer, to computational layer(s) (also referred to as hidden layers), and finally, an output layer such that the output from the layer serves as the input for the next layer. External inputs to the network, such as images from a camera, are applied to the first layer. The output from this initial layer is fed into the following computational layer(s). Each layer itself is a set of computations (nodes or neurons), each computing a different feature from the input to that layer. As more layers are added to this flow, the more computations are needed before a final "output" layer receives input to produce a result. This output layer performs a set of computations similar to other layers, except the size of this layer is defined based on the domain of the problem. In this application, the output layer produces a set of classification results based on the results of the total previous computations.

After training a network to classify contents of an image, a single image, for example a picture of a dog, could be input to the neural network. The expectation is that the neural network processes the images through all of the layers and completes the computations of all nodes, and then produces a final set of results that identifies a dog to be the most likely object within an image compared to other objects the network can classify, such as a cat or airplane.

In a traditional machine learning scenario, only a single set of weights for all nodes would be trained for the network. The entirety of a network is trained on one output layer and the parameters of each node's computation are created from the results. In order to identify an object, the same number of computations would need to be completed before a single set of results was calculated.

This concept of training a neural network with a singular output layer to create a single set of weights for all layers will be referred to as the "traditional" model and is most similar to the training schedule used for a majority of neural networks to date.

4.2 Our multitask model with sequential training

Take, for example, a computer vision model trained to be deployed to a security system camera. Such a model could be trained to identify multiple different objects, such as individuals and vehicles. If a malicious or potentially dangerous event is identified, however, such as an individual carrying a weapon or vandalism, it would potentially be valuable for the model to initiate some kind of focus on those events, spending more computational resources on identifying those events and prioritizing the computations on the individual as compared to identifying vehicles. In this chapter, we define and show results for a method to create this kind of focus in deep neural networks on the CIFAR-10 dataset [21]. By training the model in such a way that each intermediate hidden layer is simultaneously configured and trained to pass its activations to an output layer and to the next hidden layer, we create a single model that can be scaled up or down in terms of computational complexity and time to predict. By creating a main embedding network and additional smaller networks attached to this main network, it would then be possible to reshape the network and focus on individual tasks at inference time.

To continue, every image passed into this system would have a mandatory computation to classify the contents of the image. This stage would identify a human in an image, versus a dog in an image, versus a vehicle in an image. From here the results of this one computer vision model, in combination with priorities from the user, could schedule a new stage to further process the image. The new task from these results could select a model that identifies specific attributes of a human, breeds of dogs, or models of vehicles as each are identified in images.

We propose a new training technique to not only vary the size and computational resources of a neural network, but to increase accuracy for each intermediate output of the network. If we send the output of each hidden layer to a classification layer, then we can create various size tasks, which have different computational and time requirements, and thus will allow our system to create a schedule with tasks that will meet different deadlines and have different levels of accuracy depending on the requirements of the system.

4.2.1 Experiment

In order to refine training time, and achieve accurate intermediate results for these tasks, we trained each hidden layer of our neural network model to more precisely calculate the set of weights used within the network. At training time, each output layer is trained in sequence. After one output layer has been trained, this layer and all layers leading into it are frozen such that training the next output layer only requires training the next hidden layer and the appropriate output layer. This training schedule is designed to ensure that the intermediate output layers provide increasing accurate predictions and computational power with respect to the task at hand. We hypothesize that training all layers together will not significantly hinder overall performance compared to the traditional model, and will also provide accurate results at each intermediate output to refine the time needed for intermediate results.

To test our hypothesis that a single model can be created with multiple options for complexity and predictive power, we ran an experiment by creating a model with multiple intermediate outputs on the CIFAR-10 dataset, a well-known color image classification dataset containing 10 classes. Suppose that the perfect mapping from input pixels (X) to output class values (Y) is some function $\theta(X) = Y$. We seek to create approximations of this function using the available subset of data (x) and the available subset of labels (y) such that we minimize the categorical cross entropy loss from predicted class labels (y') to the true class labels. To do this, we create a neural network $\theta_\sigma(X) = Y'$ where σ corresponds to the weights in the network.

The first iteration of training would provide input to the network, perform the convolutional embedding, pass those results to Layer 1, and from there classify results within the output layer labeled L1 (Fig. 10.3).

The second iteration of training freezes the weights of the first layer. This ensures that the training from the first iteration is not lost to the second set of training. Input is then provided to the network and the convolutional embedding occurs as previously mentioned. Here, the computations in Layer 1 still occur, although the output is passed as input to Layer 2 rather than output L1. After the computation of Layer 2, the output of Layer 2 is passed to output layer L2 to be classified. This process then continues for each subsequent layer.

4.2.2 Neural network specifications

Consider the networks presented in Figs. 10.2 and 10.3. The Convolutional Embedding section of this network contains five "stacks" of two convolutional layers followed by a single max pooling layer and finalized with a dropout layer that randomly removes 20% of the

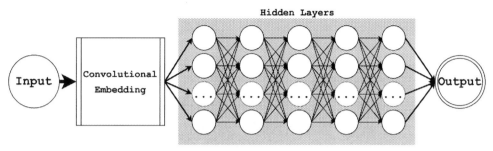

FIG. 10.2 Traditional NN model—five hidden layers—one output layer. Note: The single background color indicates the single set of weights created during traditional training.

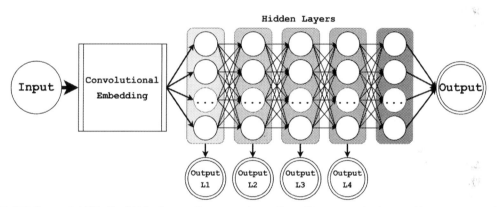

FIG. 10.3 Sequential NN—five hidden layers—five output layers. Note: The separate background boxes represent the individual sets of weights created during the sequential training.

weights during training. Each of these five stacks has a static number of convolutional filters, starting with 16 for the first stack and doubling at each stack until the final stack has 512.

After this Convolutional Embedding section of the network, each hidden layer contains 512 neurons and utilizes ReLU activation. Each output layer contains 10 hidden neurons, one for each of the classes in the dataset, and utilizes softmax activation. If no decrease greater than 0.01 is detected after five epochs, early stopping is implemented and the best weights with respect to validation loss are restored.

For each training, 20% of the training data is reserved for validation, and the model is tested against this validation data for decreases in loss. Batch sizes of 512 were used for training in all instances.

4.2.3 Experimental results

After the overall network had been trained successfully, predictions for the hold out test data were created using each of the individual output layers. We then calculated the overall accuracy and F1 score for each class for each of these output layers. The 10 classes or tasks in the CIFAR-10 dataset correspond to recognizing airplane (C1), automobile (C2), bird (C3), cat (C4), deer (C5), dog (C6), frog (C7), horse (C8), ship (C9), and truck (C10).

A traditional neural network was trained as a control group for the experiment (Fig. 10.2). After a total of runs, the average accuracy and results were calculated based on the total accuracy of each run as well as the accuracy to classify images into each of the 10 classes.

The results of this model and training technique are provided in Table 10.1.

This traditional network, described earlier, does not have any intermediate outputs at each layer, only a single layer after all computations are finished. We introduce the intermediate outputs to the traditional model during the testing phase. The traditional neural network model is still only trained from the final output layer, L5, but during the execution, each intermediate output is also extracted.

The network was similarly trained using the CIFAR-10 dataset with the same convolutional embedding processing as described earlier. The intermediate outputs were then introduced for the testing set and were not part of the training. The results of this technique are listed in Table 10.2.

The new model (Fig. 10.3) was trained layer by layer independently and sequentially using the same dataset and the same convolutional embedding processes as previously. Different from the traditional training method, once one layer was trained for the output layer, the set of weights for that layer was maintained for the training of each subsequent layer. The results of this technique are provided in Table 10.3.

During the execution of each prediction set, the time was recorded for the total set of predictions and was divided to achieve the time taken for a single result. Here, the mean time and standard deviation of each layer were used over the course of 10 executions, as shown in Table 10.4.

Table 10.1 Traditional neural network model: F1 scores—Average after five runs (Fig. 10.2).

Overall accuracy	C1	C2	C3	C4	C5	C6	C7	C8	C9	C10
0.75	0.78	0.86	0.65	0.55	0.72	0.63	0.81	0.79	0.85	0.82

Table 10.2 Traditional neural network (Fig 10.4): Intermediate output F1 scores.

Layers	Overall accuracy	C1	C2	C3	C4	C5	C6	C7	C8	C9	C10
1	0.05	0.00	0.08	0.12	0.04	0.06	0.13	0.01	0.02	0.00	0.03
2	0.09	0.01	0.00	0.19	0.01	0.04	0.05	0.01	0.02	0.02	0.53
3	0.11	0.01	0.29	0.02	0.26	0.01	0.10	0.10	0.01	0.00	0.00
4	0.07	0.02	0.03	0.06	0.06	0.37	0.06	0.09	0.01	0.04	0.01
5	0.74	0.77	0.85	0.65	0.52	0.71	0.60	0.82	0.79	0.83	0.82

Table 10.3   Sequential neural network (Fig 10.3): Intermediate output F1 scores.

Layers	Overall accuracy	C1	C2	C3	C4	C5	C6	C7	C8	C9	C10
1	0.76	0.78	0.87	0.68	0.54	0.73	0.66	0.79	0.79	0.85	0.84
2	0.75	0.78	0.86	0.65	0.55	0.71	0.64	0.78	0.79	0.85	0.84
3	0.75	0.79	0.85	0.66	0.50	0.71	0.67	0.79	0.78	0.86	0.83
4	0.76	0.79	0.84	0.67	0.57	0.72	0.64	0.79	0.79	0.86	0.82
5	0.76	0.80	0.86	0.66	0.56	0.72	0.66	0.80	0.79	0.86	0.84

Table 10.4   Sequential neural network (Fig 10.3): Prediction times.

Layers	Time per prediction (mean ±std. dev. of 10 runs)
Layer 1	6.67 ± 0.07 ms
Layer 2	7.28 ± 0.27 ms
Layer 3	8.25 ± 0.12 ms
Layer 4	8.69 ± 0.18 ms
Layer 5	9.32 ± 0.16 ms

4.2.4 Observations and discussion

The traditional network (Fig. 10.2), as expected, performed consistently and accurately. With overall results of about 75% accuracy, this network serves as the baseline of the experiment.

The results of the second experiment extracting intermediate outputs from a traditional neural network (Fig. 10.4, Table 10.2) show an overwhelming lack of accuracy of all layers except for the final layer. This shows that training a network traditionally and then observing intermediate outputs does not yield accurate intermediate results. The

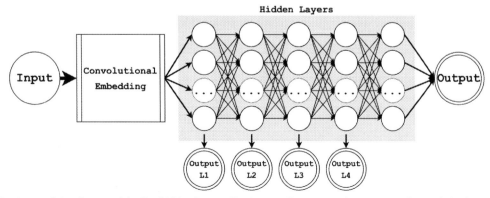

FIG. 10.4 Traditional NN model—five hidden layers—five intermediate output layers. Note: The single background color indicates the single set of weights created during traditional training.

most accurate results from this approach to training a network require that all layers and all computations be completed before accurate results are achieved. This set of results also serves as the metrics for which the new sequential neural network can be compared against to ensure that not only final layer accuracy is maintained compared to the traditional network, but also the results from intermediate output layers are improved.

The intermediate results in Table 10.3 reinforce the new approach to train neural networks sequentially to allow for accurate intermediate results. This new method of training neural networks maintains a similar level of overall accuracy and improves the intermediate results compared to the traditional neural network.

It appears that most of the initial hypothesis of the experiment holds. The overall accuracy of the network holds with the new training technique for neural networks. However, there is a variation in accuracy (i.e., the F1 score) across specific classes at each of the output layers. While overall accuracy and F1 score of each intermediate output have increased compared to the traditional training, the same cannot be interpreted within the results of each class. The changes in performance from class to class (i.e., tasks) are much more intricate, and clearly not monotone in nature. For example, for class 4 (the task of recognizing cats in an image), intermediate outputs are monumentally better than the intermediate outputs of the traditionally trained neural network. However, even in the novel approach, relatively equal performance can be achieved from utilizing either the fourth or fifth output layer, but there is a slight accuracy dip while using only three layers. These results indicate that accuracy varies with the number of layers in the neural network and the context of the problem.

Although our hypothesis is reinforced and the new method of training neural networks is shown to be decently effective, this new approach must also better satisfy the requirements of imprecise calculations compared to a traditional neural network. The traditional neural network requires a fixed number of layers, thus a set computational power requirement and a fixed time requirement for any input data that enters the network before a single result is calculated. As more layers, and thus more computational power, were used to predict a result, we see a slight trend toward more accuracy of the network, but not as significant as expected. There is only one class, C3, where the accuracy after more layers and computational resources was lower than the first layer by more than 1%. Since all neural networks here use the same convolutional embedding, and layers were trained individually, there is no surprise that each layer provides a level of accuracy similar to the others. The accuracy was not consistently diminished on a lower number of layers, but there is a possibility of fewer layers providing a less accurate result in other applications if the model cannot properly fit to the data.

This new training technique starts to solve a problem of embedding AI tasks into a hard real-time environment. A neural network configured and trained as discussed shows that there are other applications and situations in which "shortening" a network in certain contexts may improve the speed of each prediction while also using less computation and still achieve accurate results. This ability to more finely control the computational power used by a network would also allow other cases to use the full, robust network when more computational power is available, or for more complex applications.

It is also apparent that accuracy and time do not linearly scale. The results do not reflect the monotonic increase in accuracy expected when calculating results with more power and time. That is, the more time a network spends to create the results does not guarantee that the result has a higher level of accuracy. This observation implies that other approaches to embed AI tasks in real-time environments may need to examine tasks more critically in order to get the best results in the most efficient time.

In order to deploy a large size neural network across many devices, a lot of storage or computational power may be needed, and not all devices, on the edge for example, have the computational resources. To combat this, the network design could be extended to create a modular version of this network to handle the various size and computational power requirements of these devices. With one more central computational resource, smaller and more lightweight networks and computations can be deployed on a wider scale to achieve fast and accurate results.

In the current system, the complete convolutional neural network trained and tested as shown earlier would serve as the central computation of this structure.

On one side of the infrastructure, each edge device could use the network and process each image through an instance of the network. Each image, based on time and computational resources of the edge device, could be analyzed with a different number of layers to meet scheduling needs. If this step is mandatory for every image, then the main computational resource could serve as the additional, or optional, task for each image to get more specifics about the image, improve results from further computation, or take other actions from these primary results.

On the other side of the infrastructure, after processing each image with the mandatory computation, using this classification network, the central computational resource may take action to create a new task to identify specific features about an image. In addition to the edge devices processing input data, the total infrastructure may support other modular tasks that contain a model trained for each specific use case. Each smaller, modular device could be configured with a machine learning model, such as a neural network like YOLO v3, where deploying the network on smaller devices with limited power and computational resources yields effective results [6]. These individual networks would serve to identify specifics about each object. One module would be trained and responsible for identifying key features about airplanes (C1), such as a model of the aircraft, company, FAA registration numbers, etc., and another module would serve to identify the specific breed of dogs (C6), for example, and other key features such as injury or trauma. The results from these modules would accurately assess situations so a user can interpret the results, and take appropriate actions.

Given one of the earlier use cases of a field camera system, suppose that the central computational network contained two tasks, one that identified people, and the other that identified wildlife. By default, this network could balance the utilization of both tasks at an intermediate level. If, for example, the central network detected some form of wildlife, for example, elephants, with a confidence >70%, then a task would be created to identify the specifics about the elephant(s) classified in the image. The central module would continue to process images to identify people and wildlife, while a separate module would further process the image.

A module trained to recognition specifics of elephants would take the image and identify the species, breed, how many are in the image, if there are any with injuries, etc. The results of this module could be monitored by authorities near the location of the specific camera that captured the image, to aid the wildlife and monitor the animals more closely. While the first module is processing the image, the same central network processes the next image from a camera in a similar location. From the new image, the central network detects a person in the image and thus a smaller module processes the image to identify the number of people in the image, if they are carrying anything like hunting equipment, and other useful characteristics. This module responsible for object detection/identification in the image could be integrated with existing innovations like SPOT [7] to take action and alert authorities. The same authorities that examined the previous image are taking action to help the wildlife and would also be alerted about the hunters and could take additional actions to not only help the wildlife, but also respond to the identified hunters.

4.3 Edge focused model

The preceding results show a promising technique to use a neural network model that is computationally expensive and train it in a new way to extract accurate intermediate results in order to refine the time and improve the application of the neural network. The design itself is not focused for edge devices, and so for an application of this network to expect the same results in the same amount of time, when deployed on an edge device, is not reasonable. In order to create a network that will more reliably run on the edge in a short amount of time using fewer computational resources, we created a modified version of the previous neural network to reduce computational power and improve the timing of it, while minimizing the expected decrease in overall accuracy that comes from fewer computations.

The new, edge-focused, model starts by decreasing the convolutional embedding computations of the network. Previously, there were five stacks of convolutions, each of increasing size in a sequential fashion, before that data was fed into the hidden layers of the network. Now, the convolutional embedding process is reduced to one convolutional stack using the smallest size of the previous experiment.

This modification itself reduces computations needed for every single image that is introduced to the system. Since the convolutional steps occur every time no matter where the intermediate output is taken out, the overall time should decrease to process any results in the network. On a machine with lots of computational resources, this step may not matter, but to deploy this on an edge device, reducing computations for every image will create a significant decrease in overall time and computational power required to get accurate results.

As previously seen, the most accurate results for training a network come from the traditional model to send all the data through the input, and to train the weights of the

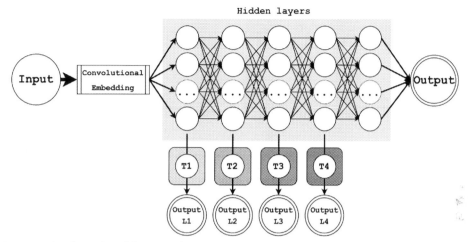

FIG. 10.5 New edge-focused model: intermediate translation and output layers. Note: The single background color indicates the single set of weights created during traditional training.

network based off the final output layer. Since this network does less processing on an image to begin with, this lighter-weight model is trained traditionally to ensure that if an edge device has the time to process all computations in the network, the model produces the most accurate results of the network.

Once the network is trained traditionally, the intermediate outputs do not provide accurate results, as expected from results seen in Table 10.2. To improve the intermediate output results without affecting total accuracy, we introduce "translation" layers (Fig. 10.5, T1–T4) to serve as smaller layers that take intermediate results from the hidden layers and are trained to provide accurate results to output classification layers without modifying the weights of the main network.

After training the network traditionally, the hidden layers are frozen so the weights continue to provide the most accurate results when all five layers are processed. Each translation layer is then trained individually to increase accuracy of the intermediate results.

4.3.1 Experiment

The convolutional embedding step contains one stack of size 16 before the data enters the hidden layers. Each of the five hidden layers contains 512 neurons and each translation layer contains 64 neurons. This size ensures that every intermediate output extracted from the hidden layers is guaranteed to use less computational resources to get an output than processing an additional hidden layer.

Parallel to the previous experiment, the new network was trained and tested using the CIFAR-10 dataset. For each training, 20% of the training data is reserved for validation, and the model is tested against this validation data for decreases in loss. After the model is trained traditionally, the five hidden layers of the network are frozen to ensure further training does not have an effect on the results.

We hypothesize that this model will achieve accurate results in less time due to the reduced computations, while improving intermediate results compared to the previous traditional training techniques.

4.3.2 Experimental results

Since this network was trained traditionally and the weights were frozen for additional training, the results shown in Table 10.1 were calculated for this network and included in Table 10.5 in the last row rather than in a separate table.

During the execution of each prediction set, the time was recorded for the total set of predictions and was divided to achieve the time taken for a single result. Here, the mean time and standard deviation of each layer were used over the course of 10 executions, as shown in Table 10.6.

4.3.3 Observations and discussion

As hypothesized, reducing the convolutional embedding stage of the network does reduce the overall computations for every image in the system, and shows a noticeable time improvement across every layer. Getting results from this smaller, more lightweight network can reduce the time to get results by more than 10%. To the average human, a millisecond or two improvement may not seem like that much time. When this infrastructure scales, and these networks are deployed on edge devices, the difference in processing time compounds and becomes more significant to the overall efficiency. Over time, the longer the system is running, the more important and more noticeable this improvement will be.

Table 10.5 New edge-focused neural network: Intermediate output F1 scores (Fig. 10.5).

Layers	Overall accuracy	C1	C2	C3	C4	C5	C6	C7	C8	C9	C10
1	0.48	0.55	0.56	0.36	0.28	0.41	0.39	0.53	0.53	0.58	0.54
2	0.59	0.64	0.73	0.45	0.39	0.52	0.49	0.66	0.66	0.71	0.65
3	0.62	0.69	0.75	0.5	0.41	0.56	0.51	0.69	0.7	0.75	0.69
4	0.64	0.7	0.76	0.51	0.42	0.57	0.54	0.7	0.71	0.76	0.71
5	0.64	0.7	0.75	0.55	0.35	0.59	0.54	0.69	0.7	0.77	0.71

Table 10.6 New edge-focused neural network: Prediction times (Fig. 10.5).

Layers	Time per prediction (mean ±std. dev. of 10 runs)
Layer 1	5.87 ± 0.26 ms
Layer 2	7.02 ± 0.16 ms
Layer 3	7.86 ± 0.26 ms
Layer 4	8.41 ± 0.13 ms
Layer 5	8.91 ± 0.24 ms

The decrease in accuracy from this network is also expected. As proposed earlier, the expectation is that more computational power, more time, and more computations will result in a more accurate output. The same is true in the inverse process. When computational power is limited and the number of computations are reduced, the time and accuracy will also be reduced.

In order to more effectively design a modular system, as described in the previous discussion, this lightweight network was designed to be deployed on edge devices. These edge devices serve as the first step between input devices such as cameras, sensors, etc., and the central computational resource, as seen from Fig. 10.1.

Continued from the earlier example in which this system is deployed in a field camera system, this new network, as described previously, contains two tasks that would identify people or wildlife and instead is deployed on the edge device, not in the central computational resource. Each time an edge device collects an image from a camera, the device has only 6 ms to process each image. Once processed, the results must be sent to the central computational system. The first network, although seemingly more accurate, cannot produce an answer in less than 6 ms to begin with, regardless of the number of layers used. In the new network, results can be achieved in less than 6 ms when using the intermediate results from the first hidden layer.

For example, in the field camera system, an edge device receives an image from a camera. The result is processed within the scheduled period, and the network finds a horse (C8) to be the most likely object in the image with a confidence of 53%. When the result of this edge-focused network is published to the central server, the image is sent as well as the results of the network and is accessible by the user(s) at the monitoring station. Each time an image with results is received in the server, the next task must be scheduled. The system is configured such that results with a confidence over the threshold of 50% are automatically scheduled for further processing. If the users responsible for the monitoring station have an incident, then the priority of the task may increase, and thus the image of a horse collected from the edge is now the first task to be scheduled. Without the initial processing done quickly on the edge device, the server would have no way of knowing how to prioritize the image and corresponding task, and would have to spend extra time, computational resources, and scheduling bandwidth of the server to get the results of the classification network before creating a new task for the image.

The benefit of a network like this comes from the design and usability in edge applications. It is less computationally expensive and more closely designed to the type of networks that will be efficient when deployed on edge devices. Since edge devices are so computationally limited, a network that has a reduced level of accuracy is shown to output results in a much faster period of time.

4.4 Discussion

The current configuration and training of these neural networks imply that the input is the same across every application (i.e., images input into the network are the same resolution). A future extension of this work could also examine aspects of neural networks that could scale up or down on multiple tasks with input size rather than number of layers. With a higher resolution image, for example, there is more data for the network to use and this may yield more accurate results with extra time or computational power.

An AI task, like neural networks, with this new training technique, edge application, and proposed architecture, which can create accurate results based on additional dynamic constraints of accuracy, timing, etc., provides a foundational step toward reevaluating other AI tasks and their applications/implementations, as well as other constraints and assumptions about AI algorithms, in order to solve other hard real-time problems.

This work shows promising results for the ability to create a self-focusing and self-constructing network that is able to scale up or down computational power and predictive performance on multiple tasks.

5 Scheduling

5.1 Scheduling model

We consider the hard real-time, periodic workload model for scheduling the execution of a set of tasks. In this workload model, a task set, $\tau = \{\tau_j\}$ where $j \in \{1, ..., n\}$, consists of a set of independent tasks, each making periodic requests for the same execution. Each task generates a set of stages $L_{j,k}$ for $k = 1, 2, 3, ..., l_j$. Thus l_j is the number of stages for task τ_j. While the tasks are independent of each other, the stages within a task must execute sequentially. That is, stage $L_{j,k+1}$ cannot begin execution until stage $L_{j,k}$ has completed execution. We denote the set of all tasks' stages as $L = \{L_{j,k}\}$. We denote the size of L by $N = |L|$. For task τ_j, the first ω_j stages are mandatory and the remaining are optional. Each task τ_j has a deadline, D_j. Mandatory stages must be scheduled to complete execution before their deadline, whereas optional tasks can be scheduled for execution before the deadline or not scheduled at all. Estimates of expected worst-case run time and expected precision of each stage of each task are known in advance of scheduling. That is, it is known a priori that stage $L_{j,k}$ has expected worst-case run time $t_{j,k}$ and expected cumulative precision $P_{j,k}$. Finally, each task has priority V_j.

In this work, we consider a periodic scheduling problem, and so we will assume that the deadlines of all tasks are equal (at the end of the period). So $D_i = D_j$ for all i, j, and we denote this single deadline D. Thus the scheduling problem is not only to choose the order of task execution, but more importantly to choose the number of stages to execute before the deadline for each task. Note that while we consider algorithms that fit our model of a single deadline for all the tasks, they can be extended in natural ways to account for distinct deadlines, as will be discussed.

A schedule S for this model then is a determination, for each task, of the number of stages to execute. The execution implied by such a schedule, for a single processor, is simply executing the chosen number of stages for each task, in increasing order of deadline. A correct schedule is one in which the mandatory stages of each task are expected to complete execution before their deadlines. If there is excess time for execution after completion of the mandatory stages of all tasks, the remaining time can be allocated to optional stages. The choice of how to allocate time among the optional stages comes down to a balancing of expected precision, task priority, and expected run time. A number of objective functions can be employed to measure the quality of proposed schedules. Such objective functions can be used to inform design of scheduling algorithms, and these functions can be used as metrics for benchmarking algorithms' performance.

For a single processor, the scheduling problem for this model can be stated simply as follows: Given task set τ with associated priorities V_j and stage set L with expected runtimes $t_{j,\,k}$ and precisions $P_{j,\,k}$, specify the order in which stages are run to maximize objective function $\mathcal{C}$. In this work, we consider various objective functions—or equivalently, performance metrics—$\mathcal{C}$.

Objective functions are used in designing scheduling algorithms and measuring the quality of their outputs. An objective function takes a schedule $\mathcal{S}$ as input and outputs a value $m \in \mathbb{R}$, giving a measure of the quality or utility of the schedule. We consider two main objective functions in this chapter. First, it may be useful to maximize the quality of the results for the highest priority task above all others. Suppose the tasks $\tau_1, \tau_2, \dots, \tau_n$ are numbered in order of priority such that $V_1 \geq V_2 \geq \cdots \geq V_n$. Then our first objective function can be written:

$$\mathcal{C}_{max}(\mathcal{S}) = P_{1,k},$$

where $k = \arg\max_S L_{1,\,k}$. In other words, $\mathcal{C}_{max} = max_{V_j}\{P_{j,k}\}$.

On the other hand, there may be multiple tasks of high importance, and so also allowing lower priority tasks to contribute to the objective function leads us to our second natural choice, which is one of the classic metrics considered in the literature [3, 9]. We now consider a weighted sum of the expected precisions of all tasks where the weights are given by the priority of the tasks. Thus our second objective function is

$$\mathcal{C}_{sum}(\mathcal{S}) = \sum_{j=1}^{n} V_j \cdot P_{j,k_j},$$

where $k_j = \arg\max_S L_{j,\,k}$.

Both objective functions can be extended to accommodate applications in which certain a priori information is available, for example, regarding the distribution of task priorities.

Finally, we consider a classic metric from the literature [3, 9] that considers the maximum weighted precision of the schedule $\mathcal{S}$ and is defined as

$$\mathcal{C}_{weight}(\mathcal{S}) = max_j\{V_j P_{j,k}\},$$

where $k_j = \arg\max_S L_{j,\,k}$.

The mandatory stages of each task comprise a hard real-time scheduling problem, while the optional stages constitute the imprecise scheduling side of the problem. We consider two broad approaches to the scheduling problem: a dynamic programming approach and a greedy algorithm approach. We compare both approaches subject to each of the specified objective functions, $\mathcal{C}_{max}$ and $\mathcal{C}_{sum}$, by implementing variations of each and running Monte Carlo simulations. We observe, as expected, that the dynamic programming approach performs well for the weighted sum metric C_{sum}, whereas the greedy algorithms perform well for the simpler max priority task metric C_{max}. We suggest that scheduling algorithms for this model can be tailored to their specific application, depending on resources and desired objective functions.

Scheduling algorithms for this problem model essentially simplify to a choice of the number of stages to execute for each task. If the a priori expected runtimes are accurate, execution order matters minimally because the period is the deadline for all tasks. Consequently, we describe algorithms that assign the number of stages (or depth) to be executed for each task, and we assume earliest deadline first (EDF) is used when actually dispatching tasks for execution.

5.2 Dynamic programming approach

We first consider the dynamic programming approach presented in Ref. [1] for this scheduling problem. This approach relies on a notion of reward. The reward is used to describe the quality or utility of particular scheduling decisions and thus is used to lead the scheduling algorithm toward a solution with high value by some objective functions. In Ref. [1], as the model has expected precision but no priority, expected precision is used as the reward function. The primary scheduling problem is reduced to subproblems $S(j, r)$, which pose the question: for tasks $\tau_1, \tau_2, \ldots, \tau_j$, what schedule, if any, achieves exactly reward r in minimal time? For each solution found to $S(i, r)$, the corresponding minimum execution time is recorded in $S_t(i, r)$. Such a problem is simple to solve for $j = 1$, and for $j > 1$ solutions are computed inductively. The remaining detail to be specified involves the reward space $[0, r_{max}]$ where r_{max} is the maximum possible achievable reward. This space is discretized by dividing it into intervals of constant size Δ, a parameter of the algorithm. Thus the solutions to the subproblems $S(j, r)$ make up a table with rows $j = 1, 2, 3, \ldots$ and columns $r = 0, \Delta, 2\Delta, 3\Delta, \ldots$ from which the optimal schedule considered can be easily derived; the furthest right cell from row n corresponds to the schedule of all tasks with the greatest reward.

Inductively solving the subproblems involves computing the function in Eqs. (10.1), (10.2). For ease of notation, denote $T_{i,l} \triangleq \sum_{k=1}^{l} t_{i+1,k}$. We also use shorthand to deal with rewards. Let $R_{i,j}$ be the cumulative reward for scheduling the first l stages of task i. Then for shorthand, for any r, denote $r'_{i,l} \triangleq r - \lfloor R_{i,l} \rfloor$. Then the dynamic programming formulation is as follows:

$$S(i+1, r) = \begin{cases} \arg\min_{l \in \{\omega_{i+1}, \ldots, l_{i+1}\}} \{T_{i+1,l} + S_t(i, r'_{i+1}), S_t(i, r)\} & \text{if } T_{i+1,l} + S_t(i, r'_{i+1}) \leq D_{i+1} \\ \varnothing & \text{otherwise} \end{cases} \quad (10.1)$$

$$S_t(i+1, r) = \begin{cases} \min_{l \in \{\omega_{i+1}, \ldots, l_{i+1}\}} \{T_{i+1,l} + S_t(i, r'_{i+1}), S_t(i, r)\} & \text{if } T_{i+1,l} + S_t(i, r'_{i+1}) \leq D_{i+1} \\ \infty & \text{otherwise} \end{cases} \quad (10.2)$$

This approach fundamentally relies on discretization of the search space, and thus the parameter Δ influences not just the accuracy but the runtime of the algorithm. For near optimal solutions, small Δ should be used and higher runtime is the consequence.

We study an unmodified implementation of the dynamic programming algorithm [1], and we also implemented a modified version that also considers the priority of tasks. We do so by letting the reward measure be the objective function that weights precisions by priority, $r = C_{sum}$. That is, $R_{i,j} = V_i \cdot P_{i,j}$. This is a natural extension of the dynamic programming algorithm to account for task priority, an important component of our problem model.

This algorithm has time complexity highly dependent on the quantization of the reward space. Assume without loss of generality that the reward function r maps to the interval $[0, 1] \subseteq \mathbb{R}$. Then for N total stages, the dynamic programming algorithm has pseudopolynomial time complexity $O(N \log N + \Delta^{-1} N)$.

5.3 Greedy algorithm

In a resource constrained computing environment, the temporal complexity of the scheduler itself needs to be considered. In our problem, the period of computation (i.e., the deadline time in the periodic schedule) includes the computation performed by the scheduler. Thus the time taken by the scheduler contributes to the overall execution time within each schedule period. Therefore, we explore greedy approaches to our scheduling problem to offer a faster alternative to the dynamic programming formulation.

We considered a variety of greedy approaches. Implementations of all algorithms mentioned are provided in our public software repository [22]. Ultimately, a single, simple formulation proved sufficient, and so our experiments use that single method to exemplify the greedy alternative to the dynamic programming approach of Section 5.2.

One greedy approach to this scheduling problem is to (after scheduling all the mandatory stages) sort the stages in order of priority and then add as many stages as possible beginning with the highest priority task. This approach is optimal with respect to the C_{max} objective function in the sense that no algorithm, for any input, can achieve a higher value of C_{max}. As an alternative greedy approach, after all mandatory stages are added and the tasks are sorted by priority, a single stage can be added per task in a "sweep" through the tasks beginning with the highest priority task. The "sweep" concludes when some heuristic indicates that starting a new sweep (i.e., adding the second stage of the highest priority task) is better than continuing the current sweep (i.e., adding the first stage of the next task by priority). Such "sweeps" continue until no time remains, that is, the deadline is reached. This method, while not optimal, performs better with respect to the C_{sum} objective function. The heuristic, among other reasonable options, could be the product of expected precision and priority divided by expected time, giving a measure of the return on investment in time. Implementations of these algorithms are given in Ref. [22].

Ultimately, we found that a simple greedy algorithm G is sufficient to achieve strong performance for both metrics C_{max} and C_{sum}. For the remainder of this work we focus

```
heuristicVals ← empty 1D list of 3-tuples
for j ∈ {1, 2, 3, ...} do
    for k ∈ {1, 2, 3, ..., l_j} do
        heuristicVals.append((j, k, h(L_{j,k})))
    end for
end for
sortedHeuristicVals ← sort(heuristicVals) //ascending, by third tuple entry
timeScheduled ← 0
for i ∈ {1, 2, 3, ...} do
    j ←sortedHeuristicVals(i, 1)
    k ←sortedHeuristicVals(i, 2)
    if timeScheduled +t_{j,k} > D then
        break loop
    end if
    schedule.add(L_{j,k})
    timeScheduled ← timeScheduled +t_{j,k}
end for
```

FIG. 10.6 Pseudocode description of greedy algorithm G.

on G. In G, a heuristic is computed for all stages of every task, the stages are sorted according to these heuristic values, and then stages are added until there is not enough remaining excess time to schedule any more stages. A pseudocode description of G is given in Fig. 10.6. The heuristic in G is a map $h : L_{j,k} \to \mathbb{R}$ given by

$$h(L_{j,k}) = \frac{V_j \cdot P_{j,k}}{T_{j,k}}.$$

Intuitively, $h(L_{j,k})$ measures the contribution of task τ_j toward the overall objective function C_{sum} per unit time allocated to the task by the schedule when the first k stages of task τ_j are scheduled. That is, the heuristic h gives a measure of the return on investment of time for a given task and number of stages. While we use h for the heuristic in G, other heuristics could be designed for other objective functions. We use h because of its simplicity and effectiveness for both the functions C_{max} and C_{sum}.

Importantly, the time complexity of G is polynomial in the total number of stages, N. It is straightforward to implement G (i.e., using an efficient sorting algorithm) to achieve time complexity $O(N\log N)$.

Note that while we describe G for the periodic model (with single deadline D), the greedy algorithm can be extended to the per-task deadline model as well by checking a task's own deadline when considering adding one of its stages to the schedule.

5.4 Experiments

We run Monte Carlo simulations to test and compare our implemented algorithms. Short of a fully implemented system for comparison within a specific application, the simulations allow us to compare average performance of the algorithms but also to have direct control over the input spaces. That is, we can explore the performance of the algorithms by the various metrics (objective functions) subject to various underlying probability distributions of the inputs.

Each trial in our simulations consists of

(1) creating a complete scheduling problem (using a pseudorandom number generator for the at-random parameters);
(2) then solving the scheduling problem with each of the algorithms we seek to compare; and finally
(3) applying both objective functions (C_{sum} and C_{max}) to the schedules produced and recording the values.

Between algorithm parameters and scheduling problem inputs, there are many dimensions of data that can be varied and analyzed. For initial tests, we set the parameters of our simulations according to Table 10.7. Notice that the probability distribution for task priority is not specified in Table 10.7 because we run simulations in which we use various distributions for task priority to gain insight into algorithm behavior. Note that we set only the first of six stages for each task to be mandatory. This allows more flexibility in the scheduling because more optional stages means more valid schedules, but it also is a reasonable decision in practice. Whichever components of an AI task (layers of a neural network) are to be made mandatory can be abstracted into a single, first stage.

We report measures of both objective functions C_{max} and C_{sum}. We begin by confirming that our augmentation to the dynamic programming approach behaves as predicted.

We focus the remainder of our simulations on comparing the greedy and dynamic approaches. For these comparisons, we select the modified dynamic programming approach and the greedy algorithm G. We also track the run time of these algorithms. All experiments are run on a 2020 MacBook Pro with M1 chip. Relative times, rather than absolute times, are considered since we care only to compare the approaches. We investigate the effect of several different probability distributions for priority. We begin with a uniform distribution and a Gaussian distribution, as shown in Figs. 10.10 and 10.11.

We also consider a distribution each with high mean and low mean (skew-left and skew-right, respectively) using the Beta distribution. Results for these skewed distributions are shown in Figs. 10.7 and 10.8.

Table 10.7   Parameter values for our Monte Carlo simulations.

Number of tasks	2–30, equal number of trials for each value
Number of stages per task	6 (1 mandatory, 5 optional)
Expected runtimes	Uniform over (0, 1)
Expected precisions	Uniform over (0, 1)
Deadline	Average of expected mandatory time and total expected time
Number of trials	$10{,}000 = 10^4$

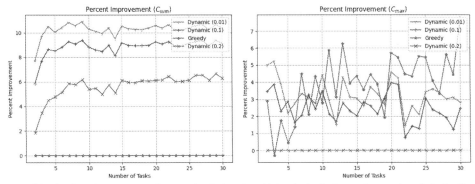

FIG. 10.7 Objective function measures for our simulations in which priority is drawn from a skew-left distribution. (Beta distribution with $\alpha = 8$ and $\beta = 2$.)

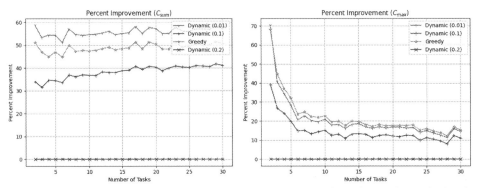

FIG. 10.8 Objective function measures for our simulations in which priority is drawn from a skew-right distribution. (Beta distribution with $\alpha = 2$ and $\beta = 8$.)

5.5 Observations and discussion

First, taking into account priority in the reward function of the dynamic programming algorithm improves performance for both C_{max} and C_{sum}, as shown in Fig. 10.9. The improvement for C_{max} is greater, which makes sense since only the highest priority task is considered, and so high precision for low-priority tasks does not contribute to the metric like it does for C_{sum}.

Then, we observe the ordering of performance for the uniform and Gaussian priorities in Figs. 10.10 and 10.11. Note that the dynamic programming approach with $\Delta = 0.2$ has the lowest average performance by both metrics and both priority distributions, and so it is used as the baseline for measuring percent improvement, hence having percent improvement equal to 0. For both priority probability distributions, by the G_{sum} metric, the dynamic programming approach with $\Delta = 0.01$ exhibits the best performance, seconded by the same approach with $\Delta = 0.1$. Then the greedy algorithm is an improvement over the worst-performing dynamic approach with $\Delta = 0.2$. As expected, as Δ

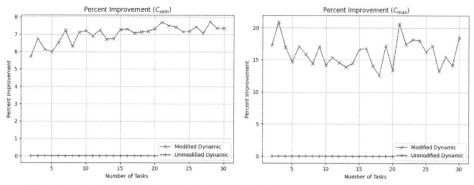

FIG. 10.9 Objective function measures for our simulations of the dynamic programming approaches both unmodified and modified to consider priority. Task priorities are drawn from a uniform distribution.

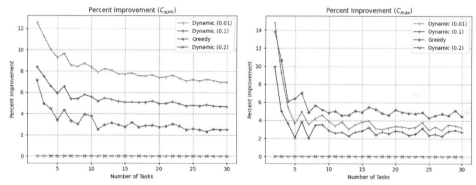

FIG. 10.10 Objective function measures for our simulations in which priority is drawn from a uniform distribution.

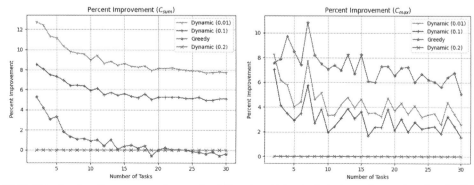

FIG. 10.11 Objective function measures for our simulations in which priority is drawn from a Gaussian distribution.

decreases, performance of the dynamic programming algorithm increases. We observe that the greedy algorithm performs comparably to the dynamic programming, depending on Δ. Finally, we observe that for the Gaussian distributed priority the percent improvements are on average greater than those in the experiments with the uniform distribution. On the right-hand side of Figs. 10.10 and 10.11, we see the percent improvement for the G_{max} objective function. For this metric, the greedy algorithm outperforms the dynamic programming approach, and the same trend is seen among the dynamic programming cases with respect to Δ.

For the experiments with uniform priority, the average run time per scheduling problem of each of the algorithms is shown in Fig. 10.12. Note that the scale of the vertical axis is logarithmic. The greedy algorithm G outperforms all three instances of the dynamic programming approach and by roughly an order of magnitude. This confirms the expectation that the greedy approach offers a significantly faster alternative to the dynamic programming approach. Among the dynamic programming algorithm instances, run time increases as the quantization parameter Δ decreases, with the $\Delta = 0.01$ instance running on average almost two orders of magnitude longer than the $\Delta = 0.2$ instance. This reflects the pseudopolynomial time complexity of the dynamic programming approach and its high sensitivity to the quantization parameter Δ. Relative run times for each of the other priority probability distributions we considered was similar.

The skewed distributions yield interesting results. The skew-left distribution has mean 0.82, unlike the 0.5 mean of both the uniform and Gaussian distributions from before. The skew-right distribution has mean 0.18. By the metric C_{sum}, we observe that for the skew-

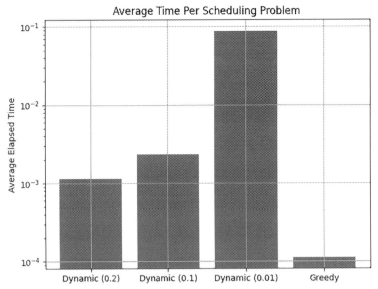

FIG. 10.12 Average run time per scheduling problem of the dynamic programming algorithm ($\Delta = 0.01, 0.1, 0.2$) and the greedy algorithm G for the experiments with uniform priority.

left distribution the greedy algorithm performs worse than all three dynamic algorithms. For the skew-right distribution, on the other hand, the greedy algorithm outperforms even the dynamic algorithm with $\Delta = 0.1$. Further, we see that the percent improvements for the skew-right distribution are much larger (approximately between 10% and 70%).

From our experiments, we learn that a simple greedy algorithm is sufficient to gain comparable performance by both the C_{sum} and C_{max} metrics. The greedy algorithm solves scheduling problems roughly 10 times faster than the dynamic approach and thus offers a faster alternative when time constraints are significant. We observe that by the C_{sum} metric, the dynamic approach with small quantization parameter $\Delta = 0.01$ gives the best performance, while the greedy algorithm tends to perform best using the C_{max} metric. Finally, we observe that algorithm performance is sensitive to the distribution of task priorities. Consequently, we conclude that choice of scheduling algorithm can be tailored according to several factors, including time/resource constraints, choice of objective functions/metrics, and expected priority distribution.

Note that these algorithms are naturally extended to the multiprocessor model. As is a standard approach to multiprocessor scheduling, an assignment algorithm would be sufficient to still utilize the advantages of the greedy and dynamic programming methods. Another useful direction for this work is similar problem models but with intertask dependencies. Similarly to multiprocessing, the algorithms discussed in this work can be used in models with task dependencies by decomposing dependent tasks into stages, fitting this model. One way to accomplish this is to specify different start and deadline times for each task, and thus the tasks become independent again.

6 Conclusions

This chapter used the imprecise computing model to propose an approach for executing real-time AI tasks on an edge architecture. Our premise was that trading off of accuracy to meet time constraints allows for graceful degradation. To achieve this, the neural network itself was designed to support the imprecise computing model by training the network assuming only a subset of the layers would be executed. Experiments showed that this method results in higher accuracy when intermediate results, at each layer, are used. In order to create a network that will more reliably run on the edge in a short amount of time using less computational resources, we created a modified version of this neural network to reduce computational power and improve the timing of it, while minimizing the expected decrease in overall accuracy that comes from fewer computations. Finally, we then provided a fast greedy scheduling algorithm that accounts for task priority. Experiments demonstrated that our scheduler, when compared to past methods, takes less time to execute and provides better quality of output for the performance metric of maximizing accuracy of high-priority tasks, while providing comparable quality for the metric of average accuracy across all tasks. Future directions include taking a more comprehensive approach to the problem of scheduling imprecise real-time AI tasks on edge architectures by including power and network constraints. This would include new neural network models, both training and implementations, geared toward resource constrained systems and scheduling algorithms that explicitly address partitioning the tasks across multiple compute resources.

Acknowledgments

The authors thank Jacob Renn at AI Squared for providing computational cloud services support and valuable discussions. The authors additionally thank Kyle Vitale for his helpful comments.

References

[1] S. Yao, Y. Hao, Y. Zhao, H. Shao, D. Liu, S. Liu, T. Wang, J. Li, T. Abdelzaher, Scheduling real-time deep learning services as imprecise computations, in: 2020 IEEE 26th International Conference on Embedded and Real-Time Computing Systems and Applications (RTCSA), Gangnueng, Korea (South), 2020, pp. 1–10. https://doi.org/10.1109/RTCSA50079.2020.9203676.

[2] J.-Y. Chung, J.W.S. Liu, K.-J. Lin, Scheduling periodic jobs that allow imprecise results, IEEE Trans. Comput. 39 (9) (1990) 1156–1174, https://doi.org/10.1109/12.57057.

[3] J.W. Liu, K.J. Lin, W.K. Shih, C.C. Yu, J.Y. Chung, W. Zhao, Algorithms for scheduling imprecise computations, IEEE Comput. 24 (1991) 58–68.

[4] S. Bateni, H. Zhou, Y. Zhu, C. Liu, PredJoule: a timing-predictable energy optimization framework for deep neural networks, in: 2018 IEEE Real-Time Systems Symposium (RTSS), 2018, pp. 107–118, https://doi.org/10.1109/RTSS.2018.00020.

[5] J. Kim, H. Kim, K. Lakshmanan, R. Rajkumar, Parallel scheduling for cyber-physical systems: analysis and case study on a self-driving car, in: 2013 ACM/IEEE International Conference on Cyber-Physical Systems (ICCPS), 2013, pp. 31–40.

[6] V. Mazzia, A. Khaliq, F. Salvetti, M. Chiaberge, Real-time apple detection system using embedded systems with hardware accelerators: an edge AI application, IEEE Access 8 (2020) 9102–9114, https://doi.org/10.1109/ACCESS.2020.2964608.

[7] E. Bondi, F. Fang, M. Hamilton, D. Kar, D. Dmello, J. Choi, R. Hannaford, A. Iyer, L. Joppa, M. Tambe, R. Nevatia, SPOT poachers in action: augmenting conservation drones with automatic detection in near real time, in: Proceedings of the AAAI Conference on Artificial Intelligence, vol. 32, 2018, https://doi.org/10.1609/aaai.v32i1.11414.

[8] R. Arablouei, L. Wang, L. Currie, J. Yates, F.A.P. Alvarenga, G.J. Bishop-Hurley, Animal behavior classification via deep learning on embedded systems, Comput. Electron. Agric. 207 (Apr) (2023), https://doi.org/10.1016/j.compag.2023.107707.

[9] J.W. Liu, W.K. Shih, K.J. Lin, R. Bettati, J.Y. Chung, Imprecise computations, Proc. IEEE 82 (1) (1994) 83–94.

[10] S. Bateni, C. Liu, ApNet: approximation-aware real-time neural network, in: RTSS'2018, IEEE Real-Time Systems Symposium, IEEE, 2018.

[11] J. Huang, C. Samplawski, D. Ganesan, B.M. Marlin, H. Kwon, CLIO: enabling automatic compilation of deep learning pipelines across IoT and Cloud, in: Proceedings of the 26th Annual International Conference on Mobile Computing and Networking, 2020.

[12] F.S. Snigdha, I. Ahmed, S.D. Manasi, M.G. Mankalale, J. Hu, S.S. Sapatnekar, SeFAct: selective feature activation and early classification for CNNs, in: Proceedings of the 24th Asia and South Pacific Design Automation Conference, 2019.

[13] C. Samplawski, J. Huang, D. Ganesan, B.M. Marlin, Towards objection detection under IoT resource constraints: combining partitioning, slicing and compression, in: Proceedings of the 2nd International Workshop on Challenges in Artificial Intelligence and Machine Learning for Internet of Things, 2020.

[14] D.J. Musliner, J.A. Hendler, A.K. Agrawala, E.H. Durfee, J.K. Strosnider, C.J. Paul, The challenges of real-time AI, Computer 28 (1) (1995) 58–66, https://doi.org/10.1109/2.362628.

[15] J. Fowers, K. Ovtcharov, M. Papamichael, T. Massengill, M. Liu, D. Lo, S. Alkalay, M. Haselman, L. Adams, M. Ghandi, S. Heil, P. Patel, A. Sapek, G. Weisz, L. Woods, S. Lanka, S.K. Reinhardt, A.M. Caulfield, E.S. Chung, D. Burger, A configurable cloud-scale DNN processor for real-time AI, in: 2018 ACM/IEEE 45th Annual International Symposium on Computer Architecture (ISCA), 2018, pp. 1–14, https://doi.org/10.1109/ISCA.2018.00012.

[16] D. Gutierrez-Galan, J.P. Dominguez-Morales, E. Cerezuela-Escudero, A. Rios-Navarro, R. Tapiador-Morales, M.R. Pérez, M.J. Domínguez-Morales, A. Jiménez-Fernandez, A. Linares-Barranco, Embedded neural network for real-time animal behavior classification, Neurocomputing 272 (2018) 17–26.

[17] Y. Chu, A. Burns, Supporting deliberative real-time AI systems: a fixed priority scheduling approach, in: 19th Euromicro Conference on Real-Time Systems (ECRTS'07), 2007, pp. 259–268, https://doi.org/10.1109/ECRTS.2007.32.

[18] H.A. Pearce, X. Yang, P.S. Roop, M. Katzef, T.B. Strøm, Designing neural networks for real-time systems, CoRR abs/2008.11830 (2020). https://arxiv.org/abs/2008.11830.

[19] S. Yao, J. Li, D. Liu, T. Wang, S. Liu, H. Shao, T. Abdelzaher, Deep compressive offloading: speeding up neural network inference by trading edge computation for network latency, in: SenSys '20, Proceedings of the 18th Conference on Embedded Networked Sensor Systems, Association for Computing Machinery, 2020, pp. 476–488.

[20] W.S. McCulloch, W. Pitts, A logical calculus of the ideas immanent in nervous activity, Bull. Math. Biophys. 5 (4) (1943) 115–133, https://doi.org/10.1007/bf02478259.

[21] The CIFAR-10 Dataset, 2009. https://www.cs.toronto.edu/kriz/cifar.html.

[22] Software Repository, 2022. https://github.com/obroadrick/imprecise.

11

Building a trustworthy digital twin: A brave new world of human machine teams and autonomous biological internet of things (BIoT)

Michael Mylrea

UNIVERSITY OF MIAMI, INSTITUTE OF DATA SCIENCE AND COMPUTING, CORAL GABLES, FL, UNITED STATES

1 Introduction

The digital transformation of biopharma and other advanced manufacturing has fundamentally altered the contract of trust between humans and machines. The 5Vs (velocity, volume, variety, veracity, and value) of data [1] required in the biopharma manufacturing process increases human reliance on automation and machines. Artificial intelligence (AI) and machine learning (ML) algorithms help translate insight from these prodigious datasets into automated processes that govern robotics. To improve quality, integrity and productivity, scientists must understand the stochastic nature of the biologic interactions as well as analyze, aggregate, and automate processes to improve related manufacturing processes. These challenges require technical innovation to create new efficiencies and value for modern manufacturers, from scaling up production to improving analytics and control. However, as information technology (IT) and operational technology (OT) increasingly converge in manufacturing environments and the 5Vs grow exponentially, trust between humans and machines becomes more challenging. The critical systems that underpin modern manufacturing—primarily industrial control systems (ICSs) and industrial internet of things (IIoT)—are inherently vulnerable to human error and cyberexploitation. These systems lack basic encryption, authentication, and other security and integrity requirements for high assurance and trust [2]. The last two decades of security research have highlighted related challenges in monitoring, securing, and controlling these cyber-physical systems [3]. However, there is a major gap in examining the increasing convergence of BIoT or cyber-physical-biological systems. This presents a brave new world with few precedents to safely guide our journey.

Putting AI in the Critical Loop. https://doi.org/10.1016/B978-0-443-15988-6.00005-4

These challenges have been exacerbated by sophisticated exploitation of these vulnerabilities by nation states that has led to revolution in military affairs, where cyberweapons are designed to cause kinetic damage. Various nations and advanced persistent threats have used stealthy cyberattacks to cause kinetic damage, from the Stuxnet attacks that derailed a rogue nuclear reactor in Natanz, Iran, to cyberpayloads that shut down and destroyed electricity in infrastructure in Ukraine. In addition, there are many attempted attacks, such as the malware that could have caused a mass poisoning event at a water utility in Oldsmar, Florida [4]. The global significance of these cyberattacks crossing the kinetic barrier by exploiting known vulnerabilities in cyber and physical systems has been explored in academic and industry literature. These works highlight how the proliferation of cyberweapons available in the global marketplace could lead to mass destruction [5]. In this context, AI/ML algorithms automating cyber-physical-biological systems introduce a brave new world of both promise and peril. This raises grand challenges around trust as stakeholders explore the opportunities to improve computational modeling and control of AI driven digital twins with new challenges around manipulating and controlling biological interactions that are difficult to explain and, for that matter, control. If trust is underpinned by a contract of assumed behavior that is random, then autonomy presents new unexplored risks for complex adaptive systems (CAS).

The cyber-physical-biological world that is producing an array of devices in the biological internet of things (BIoT) has not been adequately explored by current literature through the lens of trust between humans and AI driven machines. This chapter explores that dynamic by highlighting current security gaps and proposing technical remediations that foster an improve framework of trust. The dynamic of trust and autonomy in human-machine teams is explored through the use case of applied research in building an AI driven industrial immune system to secure and optimize the biological internet of things (BIoT). This exploration is timely, as the Covid-19 pandemic has reaffirmed that all modern infrastructures require a resilient health ecosystem to function. The proposed BioSecure industrial immune system is part of the response to build a more resilient biopharmaceutical manufacturing capacity and supply chain to rapidly respond to man-made and naturally occurring biological-cyber-physical threats.

"We know that the first human cases that were detected were detected in Wuhan in December 2019," he said. "We also know that this virus belongs to a group of viruses that have their original niche in bat populations. In between these two points, we don't know much. (Peter Ben Embarek quoted from video recording in January 2021 before heading off to China as leader of a World Health Organization (WHO) fact-finding mission into the origins of the SARS-CoV-2 coronavirus, the source of the COVID-19 pandemic.)"

Peter Ben Embarek, leader of a World Health Organization fact-finding mission into the origins of the Covid-19 pandemic.

2 Examination of the current state of biosecurity: What does assured trust in BIoT look like? What happens when it breaks down?

Biology is random and data intensive. Application of AI/ML technology can help humans better understand the billions of molecular transactions that are happening inside bioreactors. These machines, and the AI/ML algorithms that will increasingly control and optimize them, are complex adaptive systems (CAS) or "systems in which a perfect understanding of the individual parts does not automatically convey a perfect understanding of the whole system's behavior" [6,7]. Building trust between human-machine teams is a critical factor, especially where value is created in increasing the levels of autonomy in rapid decision-making. Seminal work by Lawless et al. [8] further highlights that trust is often determined by the effectiveness of communications between teams. But the BIoT speaks different protocols that rarely translate from machine to machine or between human-machine teams. What are the technical specification requirements to foster trust? How can they be translated into a framework that translates into improved machine state integrity? What happens to trust when communications break down?

The Covid-19 pandemic is a grim reminder of what happens when data is not shared and communications fail. There was a trust contract in place or an assumption that humans, machines, and corresponding teams at institutions—like the WHO, National Institute of Health, etc.—were in place to provide early warning and response to biological events. Those assumptions proved to be false, highlighting that we are both vulnerable and ill-prepared to rapidly detect, respond to, and recover from biological events. Lack of timely communications and data sharing is part of the reason why the origin of the virus remains unknown. This also reaffirms the importance of creating trustworthy systems and security methods to protect the machines and AI/ML models that increasingly interact, measure, and control biological processes and manufacturing. The lack of explainability for the current origins of the pandemic is what happens when the trust contract between humans, machines, and the biological processes in between breaks down. This chapter explores the gaps and mitigations needed to build a trust framework and improve the predictability of our assumptions, risks, and processes in between AI/ML applications and human-machine teams. The lack of understanding of the origins of Covid-19 highlights the importance of establishing a trust framework for AI applications and digital twins that model, analyze, optimize, control, and secure biopharma manufacturing.

The implications of this framework applied to various use cases, such as understanding the origins of the pandemic, and highlighting that it is timely and necessary to save lives by curtailing the existing and preventing the next mutations of this deadly virus. Moreover, Covid-19 is an important use case in understanding how AI applications are critical to building a more resilient biopharma manufacturing capacity and responding to man-made and naturally occurring biologic events. Before diving into *how* AI driven digital twins create new opportunities in improving modeling, analytics, and anomaly detection in a stochastic space, it is also important to explore the impact and potential danger of the many unknowns and randomness in manufacturing biological products.

2.1 Use case: Digital trust in AI driven BIoT in an era of pandemic

A quick summary of the open-source intelligence published to date suggests there is little or no trust that we understand the origins of one of the world's most deadly pandemics. This alarming revelation was echoed in an open letter [9] published by distinguished scientists calling for an independent investigation on the grounds that the WHO "did not have the mandate, the independence, or the necessary accesses to carry out a full and unrestricted investigation into all the relevant SARS-CoV-2 origin hypotheses." [10] Governments of 14 countries subsequently expressed concern that the WHO "lacked access to complete, original data and samples." [11]. After examining all available intelligence community reporting and other information, though, the US intelligence remains divided on the most likely origin of COVID-19. Recent unclassified findings published by the US Office of the Directorate of National Intelligence (ODNI) highlight the lack of data defined by the 5Vs is in part because we don't have any trustworthy or conclusive findings. Thus we are left with plausible hypotheses that continue to be inconclusive fodder for media talking points and increasingly weaponized in various disinformation campaigns (source). The following hypotheses shared are important to mention as we explore the potential impact and implications for AI/ML manipulating biological processes as well as the trust framework needed to help secure these systems and processes. Moreover, it reaffirms the importance of building a more resilient ecosystem of biological security, safety, and monitoring. A recent unclassified ODNI assessment concluded:

i. All agencies assess those two hypotheses as plausible: natural exposure to an infected animal and a laboratory-associated incident. Four IC elements and the National Intelligence Council assess with low confidence that the initial SARS-CoV-2 infection was most likely caused by natural exposure to an animal infected with it or a close progenitor virus—a virus that probably would be more than 99% similar to SARS-CoV-2.
   a. These analysts give weight to China's officials' lack of foreknowledge, the numerous vectors for natural exposure, and other factors.
ii. One IC element assesses with moderate confidence that the first human infection with SARS-CoV-2 most likely was the result of a laboratory-associated incident, probably involving experimentation, animal handling, or sampling by the Wuhan Institute of Virology.
   a. These analysts give weight to the inherently risky nature of work on coronaviruses.

The discovery of the origins of SARS-CoV-1 took a decade and was largely unimpeded by geopolitics, so it is not completely surprising we lack any trustworthy conclusions. However, the lack of any definitive conclusion highlights major gaps in the trustworthiness, communications, and situational awareness between organizations funding research into deadly pathogens and disease. The US National Institutes of Health (NIH) cofunded research at the Wuhan Institute of Virology (WIV) that deserves further examination under the hypothesis of a lab-lead of the virus [12]. But what safeguards and monitoring were used to validate and verify that they had the proper safeguards in place? What would those safeguards look like from a biological, cyber, and physical perspective? In considering these questions in defining a framework to empower human-machine teams applying AI applications to this space, it is important to note that trust is also not static. An effective trust contract or framework must respond to real-time assessment in machine state, given an operator's perception that is governed by a summation of their experience [13]. Factors that define the social contract of trust are beyond the scope of this study. Instead, this chapter focuses on the data confidentiality, integrity, and availability requirements to establish a trust framework between human-machine teams operating AI driven systems.

3 Security maturity of cyber-physical-biological systems in the biopharma sector

Trust was lost in the data sharing and communications process needed to identify Covid-19 origins and respond. The event also highlighted that the current cyber-physical security for systems that govern and manipulate the molecular structure and biological processes is lacking. Various lab-leak hypotheses suggest Covid-19 originated in a lab, such as the Wuhan Institute of Virology (WIV), where the first major cluster of infections occurred. The lab has a history of analyzing bat coronaviruses and the hypothesis suggests pathogens were modified in a vulnerable facility without understanding their lethality. An effective trust framework for human-machine teams would help answer requirements for questions, such as: What is the current cybersecurity maturity of laboratories working on biological agents? Of those labs, how many contain systems and agents that could potentially be weaponized or escape from an accidental lab release and so-called zoonotic spillover? How have labs handling deadly pathogens adapted to the digital transformation accompanied by new cyber-physical threats?

Two years before the novel coronavirus caused a deadly pandemic, the State Department cables warned of safety issues at the Wuhan Institute [14]. In 2015, WIV became China's first laboratory to achieve the highest level of international bioresearch safety (known as BSL-4). WIV is just one of 59 maximum containment labs with biosafety level 4 (BSL4) labs [15]. What defines the trust contract that enables humans and machines and algorithms to predict behavior when working with some of the most dangerous pathogens on the planet? Do machine learning algorithms and computational modeling software and analytics have to be explainable? What about the sensors and actuators that measure deviation in bioreactors? At what time, frequency, and accuracy must they read biological reactions that could potentially cause the next deadly pandemic and for which no treatment exists? (Fig. 11.1).

The origins of Covid 19 will likely remain unknown as long as critical data is being withheld [16]. What is known is that, despite the subsequent deadly pandemic, the security of biopharma manufacturing systems and facilities remains antiquated. Trust in the cybersecurity of biopharma manufacturing—defined as the confidentiality, integrity, and

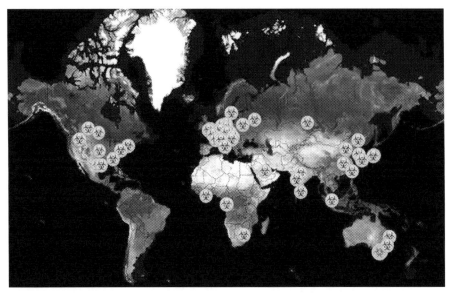

FIG. 11.1 Location of BSL4 labs that are spread over 23 countries; the largest concentration of BSL4 labs is in Europe, with 25 labs. North America and Asia have roughly equal numbers, with 14 and 13, respectively [15].

availability of data—remains at low levels of maturity as defined by cybersecurity standards and best practices (e.g., NIST Cybersecurity Framework and 800 series of standards). Data integrity, audit and monitoring regulations, requirements, and safeguards remain antiquated. Despite the millions of lives lost in the pandemic, we have *not* fundamentally altered how we regulate biopharma and life sciences systems. The current form of FDA data requirements as defined by CFR 21 Part 11 provides little protection for innovative digital approaches to biopharma manufacturing and data exchange that are currently being applied. Add that these systems are increasingly controlled by AI/mL algorithms that are not completely explainable, and systems that lack basic monitoring are troubling.

We may never understand the biological process that occurs in a bioreactor in a way that lets us explain the millions of interactions. A bioreactor is the core of manufacturing and the biological processes. Digital transformation of the biopharma sector has introduced AI/ML driven automation and robotic technologies that are increasingly dictating the manipulation of the molecular process in bioreactors. Machine learning has helped us advance computational fluid dynamics and design. This has advanced bioreactor systems and analytics of biological systems and processes, such as genetic manipulation, cell growth, metabolism, and protein expression. These advances provide valuable insight into the physical and chemical environments needed to optimize productivity and control to realize cost-effective large-scale manufacturing [17]. But if we don't fully understand the biologic interaction, then how do we gain confidence and trust in the digital systems that are integral to that process? What does this mean for human-machine teams that design, deploy, and manage the code and learners that define neural networks? How do we trust

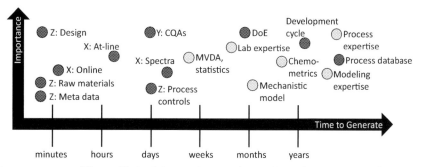

FIG. 11.2 Time to generate different kinds of data analyses, and expertise vs their importance for decision-making in bioprocessing [18].

our boundary conditions and level of acceptable risk as we determine what data is being fed to our algorithms and how they are classified to dictate critical biological processes? What should be included in a trust framework for improving trust in digital twins and other AI driven solutions for high-assurance biopharma systems?

In developing a trust framework for AI/ML driven biopharma manufacturing, it is important to understand the complexity of the data and corresponding processes. Fig. 11.2 highlights "the current landscape of different data (in red; dark gray in print version) and expertise (in yellow; gray in print version) available in bioprocessing together with the duration to generate these and a tendency of their utilization importance for decision-making [18]."

Fig. 11.2 highlights how the Data 5Vs create prodigious datasets and complexity of process to increase human reliance on machines to derive insight. Deep domain expertise gained over years is not always translated into machine learning algorithms and corresponding computational models in a way that is predictable. In his article "Decision Making and Risk Management in Biopharmaceutical Engineering," Michael Sokolov highlights this gap in the trust contract between humans, machines, and the data being exchanged under time constraints between parties:

> *"Because of significant time pressure in development and risk mitigation pressure in manufacturing, decisions are often made on an ad hoc basis involving expert meetings where all readily available data, analysis results, and experience sources are taken into account without ensuring consideration of all possible available information hidden in the databases or inside the potential of (not automatedly retrained or connected) predictive models."*

These trust constraints raise serious questions in the context of building an AI driven digital twin that acts as an industrial immune system. Yet, the introduction of new modalities such as gene therapy to manipulate the anomalies and markers of genomic code to give sight to the blind, hearing to the deaf, and cures for deadly disease seems to be worthy of the risks in manipulating randomness. One promising example is AlphaFold, an innovative artificial intelligence (AI) tool that can predict the structure of nearly the entire human proteome (the full complement of proteins expressed by an organism). As of 2022, it had more than 200 million protein structures uploaded from 1 million species, available through a public database, which vary in their accuracy [19]. In building an AI-human trust framework around this technology it is imperative to determine an acceptable level of accuracy for humans and machines and their teams. What is an acceptable level of risk that the algorithm is wrong? What is an acceptable level of false positives and negatives and how will they be validated and verified in a timely way so as not to be prohibitive to discovery and operations? How should speed be balanced with accuracy? AI-human trust is largely based on a contract that a behavior we assume will happen occurs. Thus if the AI platform has predicted almost complete proteomes for various other organisms, ranging from mice and corn to the malaria parasite, what is not to trust when AlphaFold provide answers to technical questions humans could not previously solve? What is the right balance between XAI explainability and velocity? If we don't fully understand the biological interactions at molecular level, what are the potential cascading failures of getting the algorithm wrong?

4 Antiquated biosafety and security net

The framework established in this paper helps define the security and monitoring requirements to improve trust in AI platforms in the life science space with a focus on biopharma. Data requirements for trust in explainable AI or XAI are out of scope for this study. However, this will be increasingly important with the rapid advances in AI that revolutionize the life sciences. These questions highlight the prospective and potential peril at the nexus of cyber-physical-biological systems. They also help solidify a design of an experiment to build an industrial immune system in response to these gaps. Biopharma investments, which include multiple modalities that range from gene editing to advanced manipulation of bioreactors, continue to increase with new urgency in the age of the pandemic. Yet, the regulations and safeguards for securing these systems remain antiquated: relics of a bygone era where validating quality was a paper exercise and verifying the integrity of systems didn't require consideration of their exposure to the internet's Wild West of complex, nonlinear adversaries [2]. At unprecedented speed we have moved forward with investments in human-brain interfaces, editing of genomic data, and digital bioreactors, creating new complex, nonlinear, and evolving challenges that will fundamentally alter humans' relationship with machines, especially with issues related to autonomy, cybersecurity, and trust explored in this chapter.

4.1 BIoT gap analysis

Several major security challenges exist at the nexus of cyber-physical-biological manufacturing. Solving any of these presents the opportunity to give impetus to a new wave of biological innovation. The *form* in which our human-machine teams orchestrate the digital fabric that underpins Industry 4.0 will determine the success of its *function* of resilience. Currently, there is no framework that governs how AI/ML platforms are trained, learn, and interact with biological systems. This brave new world requires an improved computational modeling framework or digital twin to measure the trustworthiness of the digital fabric that is increasingly autonomous and vulnerable to supply chain manipulation and stealthy, persistent cyber adversaries. Current cybersecurity measures of trust are found in best practices and controls that are measured around confidentiality, integrity, and availability of systems and networks. The science of cybersecurity, however, lacks the metrics of evaluation required to define trustworthiness [20]. This is true for everything from machine state integrity of bioreactors to the systems that track critical biological materials through their chain of custody, to the integrity of code that drives our machine learning algorithms in various life sciences platforms.

There is little to no trust in AI driven systems manipulating biological processes we don't fully understand. Thus we turn to a zero-trust security paradigm, which essentially takes a defense-in-depth approach to systems engineering and combines it with a wholistic people, process, and technology approach. However, zero-trust security solutions offered up by industry present a false panacea for cybersecurity without offering any verifications that they reduce inherent risk of critical systems and networks in which they are applied [21]. Similarly, there are few measures of trust to assess the integrity of the digital fabric (systems, networks, and data exchange) that underpins biopharma manufacturing. Current cybersecurity measures that underpin it are nonorthogonal and require an improved framework to measure trust. Filling this gap will require advances in measuring trust in applications of everything from artificial intelligence to privacy preserving zero-knowledge proofs found in some blockchain technology. Some of these advances are being applied to improve measures of trust defined by confidentiality and integrity in data exchange. This is important, as there are current major vulnerabilities or gaps in the exchange of data in the biopharma sector.

Applying homomorphic encryption to sensitive data can help ensure *confidentiality* of sensitive data. But then the privacy preserving function of cryptography that underpins it now lacks availability for machine learning applications to illuminate insights. Similarly, blockchain technology helps overcome availability of data in privacy preserving ways; however, proof-of-work consensus algorithms are susceptible to various attacks that undermine integrity and trust. Quantum computing will increase the number of vulnerable solutions. This paper examines these critical gaps in monitoring advances in AI driven biological internet of things (BIoT) and proposes a digital twin framework to help improve trust, integrity, and control of the systems and processes manipulating these complex systems. Some of these critical gaps and major challenges explored through this applied study include but are not limited to:

Measuring trust: If we can't define trust, zero trust doesn't mean anything more than buzzword marketing. The same can be said about confidentiality, integrity, and availability of systems and data in the BIoT space.

Monitoring, visibility, and attestation challenge: Antiquated technology and processes create gaps in telemetry, monitoring, and control of biomanufacturing. The inability to protect the machine state integrity of our operational technology (OT) creates security, safety, and quality gaps. The confidentiality challenge requires receiving, storing, and performing computational operations on encrypted batch recipes and other sensitive info without seeing or storing the data in its unencrypted form. Current solutions often trade scale and efficiency for security. The integrity challenge requires formal methods to prove to the owner of the batch recipe or sensitive data that their privacy was preserved while retaining granular data provenance and nonrepudiation through chain of custody. The availability challenge requires a cloud-based manufacturing as a service architecture that can support trusted execution environments on modern processors without sacrificing speed for security, or vice versa.

Cybersecurity and biopharma economics: Cyber-physical-biological systems are increasingly woven together in a digital fabric of manufacturing. The current cybersecurity economics encourages allocation of the least number of resources to buy down the most risk to the enterprise. The challenge in all of these is capital expenditures, and resources allocation is static when compared to the complex, nonlinear, evolving cyberthreat that continues to exploit the expanding attack surface from digital transformations. As Moore's Law plays out with drug discovery, falling costs of discovery due to ubiquitous data and technology, such as AlphaFold, enable small teams to make potential breakthrough discoveries. However, high capital costs to manufacture and the regulatory burdens make it difficult for small upstream discovery operations to compete with major vertically integrated discovery and contract manufacturers. This puts pressure on to prioritize investment in connectivity and productivity over cybersecurity, which, if effective, can be difficult to measure. Unfortunately, the same regulatory envelope that limits biopharma innovation doesn't limit adversary ingenuity in exploiting the asymmetry of cyber-vulnerabilities in manufacturing. Thus there are few incentives for large manufacturers to make the investments needed to provide trustworthy security commensurate to the risk at the nexus of BIoT.

Supply chain challenges in tracking and tracing the quality and integrity of critical materials and processes through their lifecycle is another major challenge, given the global supply chain and the many third and fourth parties involved in fulfilling this supply.

Mitigating these gaps is imperative in solidifying the trust contract between humans and machines and the AI/ML algorithms that underpin the processes that will drive competitiveness, integrity, and visibility of the manufacturing process via high-fidelity BioSecure industrial immune systems. The AI driven digital twin that underpins it will help advance secure, resilient, and agile capabilities to respond to the current and future pandemic biological, military, health, and economic threats. As a critical national capability for bioterror events, the current state of cyberphysical security in biopharma manufacturing requires a more resilient approach to improve the security, visibility, and control of the rapid disease response, therapeutic resilience, and economic opportunities created by modern digitally supported biological systems.

5 BioSecure digital twin response

The AI/ML driven BioSecure Digital Twin highlights features of an industrial immune system that will both optimize and secure the biopharma process by rapidly identifying anomalies and mitigating the behaviors that deviate from the norm. This will include inputs from all sources of sensing to include acoustic, network, infrared, and process-oriented monitoring, detection, and mitigation. It creates a modular digital infrastructure that is portable and can be deployed as a virtualized testbed for both cyber wargaming and enhanced workforce development. This will help transform biopharma cybersecurity resilience, integrity, and monitoring of the supply chain and production lifecycle, while improving resiliency. This helps fill a major gap identified earlier for biopharma security as well as other advanced manufacturing sectors that lack the required security. Unlike most security solutions, this will also advance computational modeling by improving simulation and analytic capabilities. To realize this goal, the digital twin integrates the physical plant, data collection, data analysis, and system control in a secure digital environment, which can assist in product development, process prediction, decision-making, and advancing security, safety, and quality. While most cybersecurity solutions create new functionality and interoperability challenges, this architecture and solution help optimize processes. The competitiveness of the US bioeconomy on the global stage depends on improving real-time visibility and analytic capabilities of the biopharma production lifecycle. However, moving from improved monitoring and response to predictive to real time neutralization requires additional exploration of the trust relationship between human-machine teams extending the state of the art of autonomy to BIoT [2] (Fig. 11.3.).

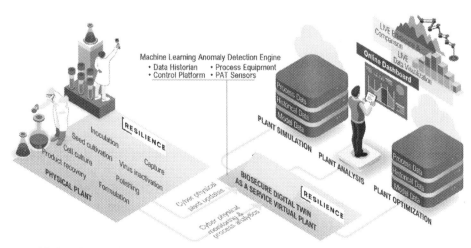

FIG. 11.3 Highlights of AI/ML driven biosecure digital twin features.

6 Trust between human-machine teams deploying AI driven digital twins

A large catalogue of literature explores various applications of AI/ML advances to advanced manufacturing. Surprisingly absent in the age of pandemic is an examination of the technical specification requirements to improve trust in AI driven biodefenses. The current gaps in applied research and solutions were reaffirmed as the United States and other nations proved to be unprepared to respond with the manufacturing of a pandemic response. This reaffirmed the need for improved velocity in decision-making [22] in the context of how to respond to adversarial advances in weapons technology [23]. Part of the challenge is that commercial interests often blind the need for appropriate safeguards and guardrails in the technology. The pandemic created that fog of war where ill intent in the application of AI/ML was abstracted when examined in the context of modern conveniences, such as protein folding in AlphaFold or bioreactor scale-up and optimization. However, to improve trust in AI-humans, we need to acknowledge these vulnerabilities along with the many opportunities afforded by AI advances in self-driving cars, robotics, and other areas where AI automation is creating value [24]. This chapter helps overcome this bias by exploring both the opportunities and challenges in the application of machine learning in building the BioSecure Digital Twin or industrial immune systems. This chapter builds on Digital Twin research conducted on active defense and defending against stealthy zero-day exploits. One of the first demonstrations of the self-healing capability to defend against zero days was demonstrated by GE Research on the world's largest combined cycle powerplant. Note that a bioreactor has a small sensor footprint but is more stochastic in its molecular interactions. This study found that horizontal digital twin technology can be applied to create a self-healing autonomous defense of biopharma manufacturing [25]. The applied research underway is exploring both the opportunities as well as limitations of humans in the loop, especially around active defense and neutralization applications that can help mitigate threats and attacks in real time.

Trust between human-machine teams is a critical factor of the study, especially where value is created in increasing the levels of autonomy in rapid decision-making. Seminal work by Lawless et al. [8] highlights that the delta between velocity and trust is often determined by the effectiveness of communications between the team. Their work emphasized the importance of both the availability of data to communicate as well as the importance of preserving privacy of some information [26]. That finding was reaffirmed in the process of designing an experiment to test the efficacy of an industrial immune system [2]. For one, the processes that are mirrored on the digital twin include biopharmaceutical intellectual property that ranges from some of the most valuable vaccine information to batch recipes of emergency medical countermeasures to respond to a bioattack. Protecting the confidentiality, integrity and availability of this data is paramount. However, the insight gained in scaling up production, predictive maintenance, and tuning a Pareto optimum also comes from applying machine learners on top of this valuable data. The digitization, networking, and automation of critical manufacturing processes are also increasing the potential attack surface that can be exploited by an adversary. Thus a wholistic response is needed to create resiliency.

7 Zero-trust approach to biopharma cybersecurity

One popular approach to increasing resiliency is a zero-trust approach. This is important as internet connectivity increasingly connects to biopharma systems and manufacturing lines that are inherently vulnerable to cyberattacks. Zero trust has become a popular approach to apply cybersecurity people, process, and technology best practices to create a defense in-depth strategy. Improving trust via zero trust will strengthen our national security by developing next-generation defensive capabilities for the critical healthcare and bioeconomy sectors. Biopharma and other advanced manufacturing platforms are vulnerable to catastrophic attacks, supply chain shortages, human error, and other naturally occurring disasters capable of wiping out most of the US bioeconomy, currently estimated at more than 5% of US GDP ($950B) annually [27]. The ability to respond to pandemic events, retain economic competitiveness, and realize the full potential of the bioeconomy depends on our ability to monitor, predict, and protect its underlying cyber-physical infrastructure.

Digital twinning facilitates a zero-trust approach by creating a high-fidelity representation of the machine state, enabling computational modeling and analytics to verify the product integrity through its entire chain of custody, and compare machine and workflow states across geographies and history. This reaffirms the importance of the BioSecure Digital Twin for critical biopharma processes and supply chains that employ a zero-trust approach to security cyber-physical-biological systems. However, with improved fidelity or availability of data and processes for modeling comes the challenges of protecting the confidentiality or privacy of that valuable information and association intellectual property. Privacy and other ethical assurances are increasingly part of a trust contract in deploying AI/ML to critical systems, but oftentimes speed, functionality, and profitability are prioritized over security and privacy in these systems. One of the novel findings of digital twins for anomaly detection is they have the potential to optimize performance, reduce maintenance costs and downtime, as well as improve overall resilience [25]. These gains are in stark contrast to most security controls and tools that limit functionality, ease use, and impose new costs.

8 Trust framework for biological internet of things (BIoT)

These systems include an array of vulnerable Biological Internet of Things (BIoT) devices. Biopharma infrastructure is vulnerable to a host of adversaries and naturally occurring hazards, from cyberthreats to supply chain shortages, from access issues directly related to the pandemic, to simple human errors. The current state of biopharma manufacturing system and process vulnerabilities exacerbates these challenges and undermines the preparedness and competitiveness of the US bioeconomy. Table 11.1 highlights security gaps

Table 11.1 Current and future state of trust in AI driven digital twins [2].

Low level of trust in current state	High level of trust in future state
Critical production systems are not encrypted, authenticated	Data is encrypted in transit and at rest, trusted execution environments combine with multi-factor authentication
Low levels of visibility into machine state integrity and thus when an anomaly occurs it is hard to determine the cause or mitigation for the anomaly	The BioSecure Digital Twin has high fidelity telemetry monitoring critical production systems. This includes not only networking level measurements, but monitoring the sensors and actuators, acoustic and infrared, frequency and voltage to understand the behavior behind the anomaly
Currently architectures focus on compensating controls and segmentation	With BioPharma systems increasingly orchestrated from the cloud, architectures will increasingly leverage advances in software defined networking and perimeters to ensure that the BioSecure digital twin, and the underlying physical processes, are cybersecure without losing functionality and control
Current Systems Lack Security Controls and Contain Multiple Vulnerabilities	Zero trust Architectures introduce security controls while detecting and mitigating cyber vulnerabilities
Systems lack basic cybersecurity analysis tools	Digital twining of critical systems advances approaches to continuous monitoring and detection as well as computational modeling
Systems have NO traceable Bio Integrity	Blockchain and distributed ledger technology improve data provenance through chain of custody, improved scaling of software bill of materials provides new insight into software supply chain integrity

and vulnerabilities that create low levels of trust in the current state and the advances needed to realize a more trustworthy future state. This applied research is timely as the frequency, severity, and sophistication of attacks targeting life sciences and other critical sectors have increased during the pandemic (Source). Current digital biosecurity and monitoring deficiencies allow offenders to carry out low-cost and deniable cyberattacks with a huge impact on public health. The COVID-19 crisis has accelerated the need for onshore critical manufacturing that includes high levels of autonomy.

AI/ML insights continue to help advance the resilience and autonomy of US manufacturing. However, as AI advances its level of general artificial intelligence and contextual reasoning it will impact autonomy and nature will have to be reassessed through a lens of trust. What is required to form a trusting relationship between human-machine teams? The performance of more autonomous manufacturing of critical systems teams will be led by human-machine teams. Trust must be optimized via an understanding of the nature of autonomy as well as the dynamic between teams [28]. Optimized trust yields optimal benefits.

Designing systems that include human-machine partnerships requires an understanding of the rationale of any such relationship, the balance of control, and the nature of autonomy, The convergence of these two fields coincides with the emergence of high impact and persistent digital biosecurity attacks. Even relatively simple attacks can cause daily losses in the millions of dollars and can easily cause months of downtime due to circumstances and constraints unique to the bioeconomy. Unfortunately, the entire bioeconomy is currently vulnerable to a wide range of attacks that include targeted ransomware, biological industrial control system vulnerabilities, and bioeconomy-specific laboratory instrumentation vulnerabilities. Unchecked, our adversaries can hold US citizens hostage to future pandemics, crippling our economy.

The Covid-19 pandemic highlighted how dependent critical infrastructures are on a stable health ecosystem. In the last 2 years, critical ports and transportation ground to a halt, hospitals and essential services were overwhelmed, and critical supply chains and manufacturing disrupted. While researchers worked to develop a vaccine for a complex, nonlinear, evolving biothreat, several sophisticated cyber adversaries took advantage of the fog of war against Covid to weaken their opponents. An ongoing cyberwar fueled by nation state adversaries and well-funded criminal gangs took on new meaning as malware targeting operational technology became kill ware, meant to degrade, destroy, and kill physical things. During the pandemic, cyberattacks increasingly targeted healthcare, pharma and medical device manufacturing, and device supply chains; ransomware spread through hospital networks; stealthy cyberweapons were launched to steal intellectual property; and misinformation campaigns questioned the viruses' source and efficacy of its cure.

The promise for digital transformations to disrupt many sectors via advances in automation driven by AI/ML algorithms and other autonomic systems is an uncertain path in the absence of our ability to measure trust. To help fill this gap, this chapter explores: how trust, as it relates to information assurance, can be measured in digital twinning or essentially autonomous computation modeling of physical systems? What is an acceptable level of risk for digital twins? What if those twins remove humans from the decision loop and are increasingly autonomous—how does that change the equation if their algorithms are governed machine learning (ML) applications responsible for identifying, detecting, and responding to anomalies in biological processes and instilling trust of our cyber-physical-biological systems? The following paper helps answer these questions by establishing a blockchain trust framework to measure the ability of these technologies to protect the confidentiality, integrity, and availability of digital data.

These questions are timely, as digital twin technologies are being applied now to secure cyber-physical systems in critical infrastructures, from bioreactors to the power grid to industrial IoT, medical devices, and supply chains. Cyber-physical systems are being rapidly extended to biological processes; as we decode our genome and alter our own genetic code, we will need a distributed ledger to track and trace their manipulations, preserve privacy, and trust that we can apply machine learning on top of encrypted data. Yet, even while existing blockchain frameworks [29] are starting to establish common taxonomies to compare trust afforded by their consensus algorithms that underpin them, there is a lack of quantitative metrics to measure trust in these applications. Filling in these gaps is imperative as the digital transformations underway in everything from finance to health and defense to critical manufacturing have fundamentally altered the interactions and trust between humans and machines [24].

The pandemic has accelerated these impacts. Examinations of autonomy and trust have explored how these challenges impact cyber-physical systems [24]. For the distributed form of organizations to function in the modern digital age, an improved trust framework is needed to quantify trust as defined by its application to the science of information assurance, or the confidentiality, integrity, and availability of digital data at rest or in transit in computer systems and networks. Current cybersecurity measures of trust are found in best practices and controls, such as the National Institute of Standards and Technology (NIST) Cybersecurity Frameworks, Cybersecurity Maturity Models, and the necessary other models used to create a baseline for organizations and systems. The science of cybersecurity, however, lacks the necessary metrics to evaluate trust in everything from machine-state integrity to supply chains, machine learning algorithms, and from there to blockchain and distributed ledger technology.

9 Digital twin trust framework for human-machine teams

Digital twins act as a high-fidelity, real-time representation of a physical system or network to advance computational modeling and analytics. When applied to the problem of anomaly detection in stochastic biological systems, validating and verifying that the inputs and outputs are representative of the machine state integrity is imperative. This requires formalization of a trust contract framework guided by guidelines that help define assumptions, documentation, methods, and analysis to guide the behavior that is expected to occur. Jacovi et al. state that standardizing contracts enables us to clarify the goal of anticipation between human-machine teams and autonomous AI driven systems for the following reasons:

> *"(1) it has, though recent, precedence in sociology; (2) it opens a general view of trust as a multi-dimensional transaction, for which all relevant dimensions should be explored before integration in society, and importantly, (3) the term implies an obligation by the AI developer to carry out a prior or expected agreement, even in the case of a social contract [30]."*

The following AI driven Digital Twin Trust Framework for human-machine teams builds on and adapts European requirements for trustworthy AI for the purpose of this study by defining the methods, analyses, and documentation needed to sustain trust in human-machine teams (Table 11.2).

Table 11.2   AI driven Digital Twin trust framework for human-machine teams [30].

Key requirements	Trust factors and assumptions	Documentation	Explanatory methods and analysis
Human agency and oversight	Human rights oversight & fundamentals Aligns with users' agency Humans in the loop for oversight	Human rights requirement checklists	Explainable AI (XAI) required to show how algorithms were trained and arrived at their conclusions Humble AI where algorithm alerts humans in the loop when outside of decision manifold Ethical and humble AI to guide the lifecycle of training, learning and AI/ML inference
Technical robustness and safety	Resilient to attacks on confidentiality, integrity, and availability Redundancy, agility, and response plan is documented & tested Accuracy levels high and reproducible Reliable and explainable results	Penetration testing results and maturity level assessment Response to all hazards tested and assessed against reproducibility checklists	Continuous penetration testing via chaos engineering, application of extreme learning to test boundary conditions Formal method proof of how the algorithm arrived at its conclusion backed by proof from telemetry monitoring that further validates the results. This provides an additional method of formal verification beyond what Dodge et al. (2019) demonstrated with "Show your work"
Privacy and data governance	Confidentiality and privacy preserved Validated and verify quality and integrity of data Data management plan that defines how data is collected, shared, and stored in a way that preserves confidentiality, integrity	Requirements checklist Requirements checklist Data management plan that can be tested, validated, and verified	Formal methods proof that data was neither accessed nor manipulated Data artifacts and documentation Attestation of policies of data management plan through continuous testing
Transparency	Telemetry and monitoring of all systems and networks including physical (frequency and voltage) measurements of the sensors responsible for measuring	Anomaly detection logs of deviations and Checklist and factsheets Factsheets (explainability) Human-machine teams inventory list	Heuristic documentation on normal as well as defined boundary conditions Combine and correlate defined heuristic with methods documented

Continued

Table 11.2 AI driven Digital Twin trust framework for human-machine teams [30]—cont'd

Key requirements	Trust factors and assumptions	Documentation	Explanatory methods and analysis
	Formal methods enable explainability Adaptable user-centered explainability Identify human and machine inputs and outputs to classify data transactions		by Jacovi et al. (2019) "Saliency maps [30], self-attention patterns [31], influence functions [32], probing [33] Counterfactual [34], contrastive [35], free text [36], concept-level explanations [37]"
AI ethics in data use and design	Document, define and limit bias Use diverse training data that optimizes accessibility and universal design Include human feedback loop to	AI Ethics checklists included in requirements checklist through design and implementation of experiment lifecycle	Bias testing and defining classifiers and boundary conditions throughout ML algorithm training, learning and implementation lifecycle
Societal well-being	Assess and limit adverse impact on individuals, groups, and society	AI Ethics checklists and constraints on neutralization and self-healing algorithms that could potentially be poisoned to cause adverse results	Validate and verify efficacy of training and results through project lifecycle, including application of chaos engineering and full monitoring of telemetry withing heuristic of "normal" as well as boundary conditions
Accountability	Validation and verification of algorithms, data, design through lifecycle from training to application Examine and document bias, assumptions, trade-offs in accuracy vs speed, etc.	Factsheet, checklists, and technical specification requirements that can be audited and explained	Monitoring and logging of deviations from "normal" heuristic and boundary conditions and assumptions. Audibility that confirms data provenance and non-repudiation through project lifecycle from design, training, and implementation

10 Digital twin opportunities and challenges to improve trust in human-machine teams

The AI/ML driven BioSecure Digital Twin efforts underway are in the process of validating and verifying the preceding trust framework requirements. Initial findings suggest these guidelines will help solidify trust between AI-human-machine teams. Moreover, these also provide a roadmap for a more effective biopharma manufacturing

Table 11.3 Highlights gaps that need to be solved to improve trust between human machine teams applying AI applications such as digital twin to biopharma manufacturing [2].

Monitoring, visibility & data integrity of critical	Confidentiality	Integrity	Availability
Gap: Antiquated tech and processes create gaps in telemetry, monitoring, and control of biomanufacturing. The inability to protect the machine state integrity or our operational technology (OT) creates major security, safety, and quality gaps. **Improvement:** Biosecure digital twin detects faults, human error and machine state degradation, and stealthy attacks. Normal behavior and boundary conditions establish pathway for optimization and operating systems at their pareto optimum	**Gap:** Confidentiality of sensitive data and processes can't be maintained in way that protects privacy, integrity, and availability of the data at scale. **Improvement:** Application of privacy preserving applications to improve trust in receiving, storing, and performing computational operations on encrypted batch recipes and other sensitive info without seeing or storing the data in its unencrypted form. Additional improvements need to be made to add speed without sacrificing security and visibility of data	**Gap:** Critical cyber-physical systems in the biopharma manufacturing process lack monitoring and check to ensure integrity through their lifecycle. **Improvement:** Improving trust requires formal methods to prove to the owner of the batch recipe or sensitive data that their privacy was preserved while retaining granular data provenance and non-repudiation through chain of custody. Provides provenance and non-repudiation. Provides cryptographically auditable trail designed to preserve data lineage and provenance for both physical and cyber properties in corona virus therapeutics	**Gap:** Cyber-physical systems lack interoperability, speak different protocol, and present other challenges to data availability. Data 5Vs add additional challenges when considering stochastic nature of biological data. **Improvement:** Advancement of AI/ML applications for trust include sustainability/ethical guardrails such as explainable / humble AI algorithms as well application of privacy preserving and confidential computing to make data available without sacrificing confidentiality and integrity of data

preparedness to respond to biological events. This combines with improved trust in the processes needed to ensure integrity and cybersecurity for the biopharma industry through the pandemic detection and response lifecycle. Table 11.3 describes initial gap findings and associated improvements needed to improve trust between human-machine teams.

Monitoring, visibility, and data integrity: Trust between human-machine teams can be solidified via identification of shared goals. Both parties require high assurance and integrity of process and data, which requires trusting the data as well as the systems that measure and output data. Digital twinning helps realize that goal via attestation of data and integrity of the machine state. This is done by making critical production systems visible in both the biological and cybersecurity dimensions, both cyber and physical. In the biological dimension, it enables a continuous quality approach that is data-driven and real-time linked to high-fidelity process models. In the cyber dimension, it provides strong logging, dramatically improves attacker detection based on physical behaviors, and makes it harder for unsophisticated attackers to have an impact [2].

Currently, operators are blind and lack the telemetry, visibility, and integrity checks to attest to the state of the machine. That includes machine degradation from overutilization and sensor drift as well as vulnerabilities in the firmware or software that controls the monitoring of the system. Today, monitoring in the biomanufacturing sector largely follows historical definitions of pharmaceutical quality, with an emphasis on paper records and strong change management processes. Digital transformation is improving connectivity and autonomy, providing efficiency gains; however, very little is being done to monitor the integrity of the measurements. Sensor drift, machine state degradation, and human error as well as malicious cyberattacks can all create false measurements. McDermott highlights that understanding these boundary conditions and integrity gaps must be achieved in the design as part of the systems engineering (SE) approach to improve resilience, noting:

> *"Improved resilience from design errors and malicious attacks is a concern for use of AI/ML in critical applications. Protection from adversarial attacks and general robustness cannot be provided by add-on applications. It must be designed into the learning process. McDermott et al. [39] provide an overview of this research area and some possible defensive techniques. In the long term, adaptation and contextual learning in AI/ML systems across long system lifecycles, and the resilience of these systems to changing contexts (environment, use, etc.) will be an active area of research and development in the engineering community. Cody, Adams, and Beling [40] provide an example of the need and possible approaches to make an AI/ML application more robust to changes in a physical system over time. This article provides a good example of the challenges of ML in operational environments. Eventually, the broad use of learning applications for multiple interconnected functions in complex systems will arrive. At some point, the SE community will no longer be able to rely primarily on decompositional approaches to system design and must adopt new, more holistic approaches [28]."*

This also has significant business and biological SE design implications, such as long lead times (often several days) to identify root cases for batch deviations, difficulty in comparing real-time conditions with all historical batches, and near-total reliance on individual staff's historical experience and memory. In the cyberdomain, this lack of high-fidelity monitoring represents an enormous advantage for adversaries: it makes detection difficult; attribution is significantly more complex; lateral movement by attackers within networks is simplified; and adversaries can use less sophisticated methods. This high-fidelity cyber-physical visibility improves trust in systems and processes that underpin manufacturing.

11 Future research and conclusion

Previous research explored how to apply an AI driven digital twin to act as an industrial immune system [25]. An applied study showed how this horizontal technology could be applied to improve biopharma security and optimize productivity in advanced manufacturing [2]. This study built on both efforts by filling gaps in human-AI trust. Future research needs to bolster the quantification of this trust framework. For example, how can digital twin applications add autonomy to high assurance cyber-physical systems in a way that both parties feel confident that there is reasonable assumption on how risk will be dealt with and decisions made. That research should focus on the contextual dynamics of measuring trust as social construct and trust anchor in an era of uncertainty and change. The pandemic reaffirms the timeliness of this research and risk, highlighting the potential devastation and damage when our trust contract breaks. The next pandemic or Covid strain might be more virulent and deadly. We may not have another opportunity to harness the power of AI driven cyber-physical-biological systems to get it right. The world remains ill prepared to respond to the next pandemic, which could be even more deadly. As AI driven twins improve our fidelity of data for computational modeling, and as AlphaFold and other advances in machine learning improve our predictability, we need to redefine our social contract and relationship between human-machine teams to shape how they advance their contextual awareness and apply predictability in an ethical and sustainable way.

Future research needs to extend the state of the art of confidential computing and privacy preserving solutions. Combining these advances with the trust framework established in this chapter could help move BIoT towards resilience and better prepare our next response. This would improve the scalability of the Digital Twin for anomaly detection and optimization by enabling federated learning that leverages use of AI and ML based algorithms (combined with cryptography such as homomorphic encryption and trusted execution environment (TEE)) to securely generate actionable insights from complex datasets and mathematical models. While application of these solutions is being applied to the research underway, critical questions remained unanswered:

(1) How do we quantify trust and explainability of AI driven solutions?
(2) How do we quantify zero trust and apply it to a trust framework to build trust in the loop of human-machine teams?

Answering these questions in a timely way is part of harnessing AI advances in an ethical, sustainable way. Individual freedom is predicated on the ability to preserve the individual's privacy. Thus privacy preserving solutions that afford protections to human without curtailing the advances of AI present an exciting new lens through which to view this study of AI driven digital twins. The lens of confidentiality introduces new philosophical and ideological challenges among humans. How will that conversation and trust contract change as the discourse starts to include sentient machines that are fueled by data, not an emotional connection to protect its privacy? There are numerous challenges with sending, receiving, and storing of sensitive data that are exacerbated by the prodigious amounts of sensitive data required for the life sciences space. This is especially true when combined with the stochastic random nature of biology that is data intensive but limited in its ability to correlate complex interactions that occur in the biologic process.

Examples of privacy preserving solutions that would help solidify trust between human-machines teams include, but are not limited to:

○ **Zero-knowledge proofs**: Enables improved trust by enabling one party to prove a statement regarding certain information without revealing the information itself, using a secret key that is generated before the transaction happens. Various applications compromise between scalability and the security of the privacy solution. However, advances in zk-ConSNARKs provide a near-constant proof size while removing some of the challenges and security issues with implementation. Applied examples can be found in various decentralized finance platforms and cryptocurrencies, such as Z-Cash and Monero. Additional research should explore application to BIoT and life sciences where sensitive data needs to be exchanged between parties.

○ **Hardware root of trust**: Extending trusted execution environments into the privacy preserving space should be explored with future research, including novel private key development using the entropy of the boot. Each TEE-capable manufacturing system must have a unique identity, cryptographically secured with a hardware secret. "This secret may, for example, be sampled and recorded in fuse banks within the device at the end of its manufacturing process, and the corresponding public key may be harvested by the manufacturer to issue the platform certificate" [41].

○ **Confidential computing and ledgers**: Enables improved trust between humans and machines in autonomous multiparty computation environments where users have competing interests and/or lack of trust. Consider this application in the context of Covid-19, where one nation needs to share data, but may not want to be linked to it or is mutually distrusting in the provenance of data. Future studies into confidential computing should examine its application to run joint computations and share their results without revealing their sensitive inputs "to one another or to anyone with physical or logical access to the hardware on which the computations execute" [41,42].

○ **Confidential cloud computing**: Biopharma production systems are increasingly orchestrated from the cloud. While this improves visibility and control, future research needs to examine how these processes can preserve functionality and visibility without increasing risk that confidentiality data, such as batch recipes and processes for vaccines, will be visible to unauthorized users. This is especially true in considering a digital twin use case that applies federated learning. Confidentiality cloud computing may provide a new trust anchor by running digital twins and other critical processes inside trusted execution environments that provide verifiable proofs inside a distributed ledger [41].

These advances, however, often trade off speed and security, scale, and efficiency. Future research can help overcome these limitations while advancing trust between untrusted parties and systems leading AI-human-machine teams. Trust can also be advanced by ensuring a more holistic cybersecurity (data confidentiality, integrity, availability) or zero-trust approach through multiple layers of cryptography, such as homomorphic encryption and trusted execution environment (TEE) for compute over encrypted data. This may increase trust in the cyber-physical resiliency and fault tolerance of high-assurance systems through redundant computing and threshold cryptography. For the form of these technical specification requirements to function as a trust contract, it must be guided by the core tenets of the trust framework presented in this research. Realizing these goals then improves trust between human-machine teams applying AI advances to high-assurance systems because it can enable sharing of data and communications while retaining confidentiality, integrity, and availability of sensitive data.

References

[1] A. Jain, A. Jain, The 5 V's of big data-Watson health perspectives, Retrieved January, 25 (2016) 2018.

[2] M. Mylrea, C. Fracchia, H. Grimes, W. Austad, G. Shannon, B. Reid, N. Case, BioSecure digital twin: manufacturing innovation and cybersecurity resilience, in: Engineering Artificially Intelligent Systems, Springer, Cham, 2021, pp. 53–72.

[3] T. Miller, A. Staves, S. Maesschalck, M. Sturdee, B. Green, Looking back to look forward: lessons learnt from cyber-attacks on industrial control systems, Int. J. Crit. Infrastruct. Prot. 35 (2021) 100464.

[4] F. Robles, N. Perlroth, Dangerous Stuff': Hackers Tried to Poison Water Supply of Florida Town, The New York Times, 2021.

[5] N. Perlroth, This Is how they Tell me the World Ends: The Cyberweapons Arms Race, Bloomsbury Publishing USA, 2021.

[6] A.K. Raz, J. Llinas, R. Mittu, W. Lawless, Engineering for Emergence in Information Fusion Systems: A Review of some Challenges, Fusion 2019, Ottawa, Canada | July 2–5, 2019, 2019.

[7] W.F. Lawless, R. Mittu, D.A. Sofge, T.M. Shortell, McDermott, T. A. (Eds.), Systems Engineering and Artificial Intelligence, Springer, 2021.

[8] W.F. Lawless, R. Mittu, D. Sofge, S. Russell (Eds.), Autonomy and Artificial Intelligence: A Threat or Savior? Springer, 2017.

[9] Open letter, New York Times. Call for a Full and Unrestricted International Forensic Investigation into the Origins of COVID-19, 2021, Accessed on March 2021 at https://int.nyt.com/data/documenttools/covid-origins-letter/5c9743168205f926/full.pdf.

[10] G. Lawton, Did covid-19 come from a lab? New Scientist 250 (3337) (2021) 10–11. ISSN 0262-4079, Accessed on July 2022 at https://doi.org/10.1016/S0262-4079(21)00938-6. https://www.sciencedirect.com/science/article/pii/S0262407921009386.

[11] Joint Statement on the WHO-Convened COVID-19 Origins Study accessed on March 2022 at https://www.state.gov/joint-statement-on-the-who-convened-covid-19-origins-study/.

[12] P.D. Thacker, Covid-19: Lancet Investigation into Origin of Pandemic Shuts Down Over Bias Risk, 2021.

[13] Department of Defense Research & Engineering Autonomy Community of Interest (COI) Test and Evaluation, Verification and Validation (TEVV) Working Group Technology Investment Strategy 2015–2018. Accessed on August 1, 2022 at https://apps.dtic.mil/dtic/tr/fulltext/u2/1010194.pdf.

[14] J. Rogin, State Department cables warned of safety issues at Wuhan lab studying bat coronaviruses, Washington Post 4 (2020) 16.

[15] F. Lentzos, G. Koblenz, Fifty-nine labs around world handle the deadliest pathogens–only a quarter score high on safety, The Conversation 14 (2021).

[16] O. Dyer, Covid-19: China stymies investigation into pandemic's origins, Br. Med. J. 374 (2021).

[17] J.J. Zhong, Recent advances in bioreactor engineering, Korean J. Chem. Eng. 27 (4) (2010) 1035–1041.

[18] M. Sokolov, Decision making and risk management in biopharmaceutical engineering—opportunities in the age of covid-19 and digitalization, Ind. Eng. Chem. Res. 59 (40) (2020) 17587–17592.

[19] J. Jumper, R. Evans, A. Pritzel, T. Green, M. Figurnov, O. Ronneberger, D. Hassabis, Highly accurate protein structure prediction with AlphaFold, Nature 596 (7873) (2021) 583–589.

[20] T. Edgar, D. Manz, Research Methods for Cyber Security, Syngress, 2017.

[21] C. Buck, C. Olenberger, A. Schweizer, F. Völter, T. Eymann, Never trust, always verify: a multivocal literature review on current knowledge and research gaps of zero-trust, Comput. Secur. 110 (2021) 102436.

[22] B. Horowitz, Introduction of the Life Cycle-Ready AI Concept, SERC Workshop: Model Centric Engineering, Georgetown University, Washington, DC, 2019. April 16 & 17, 2019.

[23] K. Wong, China Claims Successful Test of Hypersonic Waverider, Jane's 360, 2018. from https://www.janes.com/article/82295/china-claims-successful-test-of-hypersonic-waverider.

[24] W.F. Lawless, J. Llinas, D.A. Sofge, R. Mittu, Engineering Artificially Intelligent Systems, Springer International Publishing, 2021.

[25] M. Mylrea, M. Nielsen, J. John, M. Abbaszadeh, Digital twin industrial immune system: AI-driven cybersecurity for critical infrastructures, in: Systems Engineering and Artificial Intelligence, Springer, Cham, 2021, pp. 197–212.

[26] W.F. Lawless, R. Mittu, D. Sofge, I.S. Moskowitz, Russell, S. (Eds.), Artificial Intelligence for the Internet of Everything, Elsevier, 2019.

[27] National Academies of Sciences, Engineering, and Medicine, Safeguarding the Bioeconomy, National Academies Press, 2020.

[28] W.F. Lawless, R. Mittu, D.A. Sofge, T. Shortell, T.A. McDermott, Introduction to "systems engineering and artificial intelligence" and the chapters, in: Systems Engineering and Artificial Intelligence, Springer, Cham, 2021, pp. 1–22.

[29] M. Mylrea, S.N.G. Gourisetti, Blockchain for supply chain cybersecurity, optimization and compliance, in: 2018 Resilience Week (RWS), IEEE, 2018, pp. 70–76.

[30] A. Jacovi, A. Marasović, T. Miller, Y. Goldberg, Formalizing trust in artificial intelligence: prerequisites, causes and goals of human trust in AI, in: Proceedings of the 2021 ACM Conference on Fairness, Accountability, and Transparency, 2021, March, pp. 624–635.

[31] K. Simonyan, A. Vedaldi, A. Zisserman, Deep inside convolutional networks: visualising image classification models and saliency maps, in: 2nd International Conference on Learning Representations ICLR, Proceedings, Workshop Track, 2014. https://arxiv.org/abs/1312.6034.

[32] O. Kovaleva, A. Romanov, A. Rogers, A. Rumshisky, Revealing the dark secrets of BERT, in: Proceedings of the 2019 Conference on Empirical Methods in Natural Language Processing and the 9th International Joint Conference on Natural Language Processing (EMNLP-IJCNLP), Association for Computational Linguistics, Hong Kong, China, 2019, pp. 4365–4374. https://doi.org/10.18653/v1/D19-1445.

[33] P.W. Koh, P. Liang, Understanding black-box predictions vi influence functions, in: D. Precup, Y.W. Teh (Eds.), Proceedings of Machine Learning Research, Vol. 70, PMLR, International Convention Centre, Sydney, Australia, 2017, pp. 1885–1894. http://proceedings.mlr.press/v70/koh17a.html.

[34] A. Ettinger, A. Elgohary, P. Resnik, Probing for semantic evidence of composition by means of simple classification tasks, in: Proceedings of the 1st Workshop on Evaluating Vector-Space Representations for NLP, Association for Computational Linguistics, Berlin, Germany, 2016, pp. 134–139.

[35] Y. Goyal, Z. Wu, J. Ernst, D. Batra, D. Parikh, Counterfactual visual explanations, in: International Conference on Machine Learning, PMLR, 2019, pp. 2376–2384.

[36] T. Miller, Contrastive explanation: a structural-model approach, CoRR (2018). abs/1811.03163 (2018). arXiv:1811.03163 http://arxiv.org/abs/1811.03163.

[37] A. Marasović, C. Bhagavatula, J.S. Park, R. Le Bras, N.A. Smith, Y. Choi, Natural language rationales with full-stack visual reasoning: from pixels to semantic frames to commonsense graphs, in: Findings of the Association for Computational Linguistics: EMNLP 2020. Association for Computational Linguistics, 2020, pp. 2810–2829. Online https://doi.org/10.18653/v1/2020. findings-emnlp.253.

[38] A. Ghorbani, J. Wexler, J.Y. Zou, B. Kim, Towards automatic concept-based explanations, in: H.M. Wallach, H. Larochelle, A. Beygelzimer, F. d'Alché-Buc, E.B. Fox, R. Garnett (Eds.), Advances in Neural Information Processing Systems 32: Annual Conference on Neural Information Processing Systems 2019, NeurIPS 2019, 2019, pp. 9273–9282. 8–14 December 2019, Vancouver, BC, Canada http://papers.nips.cc/paper/9126-towardsautomatic-concept-based-explanations.

[39] T.A. McDermott, M.R. Blackburn, P.A. Beling, Artificial intelligence and future of systems engineering, Sytems Engineering and Artificial Intelligence, Springer, 2021, pp. 47–59.

[40] T. Cody, S. Adams, P. Beling, Motivating a systems theory of AI, Insight 23 (1) (2020) 37–40.

[41] M. Russinovich, M. Costa, C. Fournet, D. Chisnall, A. Delignat-Lavaud, S. Clebsch, V. Bhatia, Toward confidential cloud computing: extending hardware-enforced cryptographic protection to data while in use, Queue 19 (1) (2021) 49–76.

[42] W. Du, M.J. Atallah, Secure multi-party computation problems and their applications: a review and open problems, in: Proceedings of the 2001 Workshop on New Security Paradigms, 2001, September, pp. 13–22.

12

A framework of human factors methods for safe, ethical, and usable artificial intelligence in defense

Paul M. Salmon[a], Brandon J. King[a], Scott McLean[a], Gemma J.M. Read[a], Christopher Shanahan[b], and Kate Devitt[c,d]

[a]CENTRE FOR HUMAN FACTORS AND SOCIOTECHNICAL SYSTEMS, UNIVERSITY OF THE SUNSHINE COAST, SUNSHINE COAST, QLD, AUSTRALIA [b]DEFENCE SCIENCE TECHNOLOGY GROUP, CANBERRA, NSW, AUSTRALIA [c]HUMAN-CENTRED COMPUTING, SCHOOL OF ELECTRICAL ENGINEERING AND COMPUTER SCIENCE, UNIVERSITY OF QUEENSLAND, BRISBANE, QLD, AUSTRALIA [d]DATA AND INFORMATION SERVICES, QUEENSLAND GOVERNMENT CUSTOMER AND DIGITAL GROUP, QUEENSLAND GOVERNMENT, BRISBANE, QLD, AUSTRALIA

1 Introduction

Artificial intelligence (AI) is becoming increasingly sophisticated and ubiquitous, with AI technologies now contributing to almost all aspects of everyday life. While defense has lagged behind some other domains in terms of AI development, the last two decades have seen increasing interest and rapid progress [1]. Applications include uncrewed combat aerial vehicles (UCAVs), autonomous underwater vehicles (AUVs), bomb disposal robots, decision and planning support, data mining, cybersecurity, intelligence, and logistics and maintenance, to name only a few. AI is now firmly on the defense agenda worldwide.

Though there are potentially widespread benefits, the risks associated with AI and autonomous agents have been discussed in many areas [2–7]. These risks are varied and range from safety and privacy risks to ethical and even existential risks [4,5]. Ensuring that AI technologies are designed and operate in an ethical manner, an area known as "machine ethics" [8], is one critical endeavor that is currently receiving much attention, both in defense and other areas such as healthcare, security, and law enforcement. In defense, various sets of principles for the responsible design and use of AI technologies have been proposed, including those recently outlined by NATO [9], and the UK [10], US [11], and Australian [12] defense forces. Though these principles support a strong commitment to ethical design and implementation of AI, there is little guidance available on how these principles can be incorporated and assessed during AI system design and operation. This represents a critical gap in the knowledge base around how to design and implement ethical AI technologies [6,13].

Putting AI in the Critical Loop. https://doi.org/10.1016/B978-0-443-15988-6.00002-9

The critical role of HFE methods in the design and evaluation of AI has been discussed by many, and in various contexts [2,6,14–18]. This has led to the specification of key HFE concepts that require consideration during AI design and operation. For example, Sujan et al. [18] outlined eight core HFE considerations when designing and using AI in healthcare: situation awareness, workload, automation bias, explanation and trust, human-AI teaming, training, relationships between staff and patients, and ethics. As with the ethical principles discussed earlier, there is relatively little guidance available on how aspects such as situation awareness, workload, trust, and human-AI teaming can be assessed and optimized throughout the AI system life cycle. This issue is compounded by the fact that, although there are many HFE methods available, few have been developed specifically for use in AI design and evaluation [6]. In the case of situation awareness, for example, there are close to 20 different measurement approaches described in the literature, including situation awareness requirements analysis, freeze probe recall, real-time probe, observer rating, systems analysis, and posttask subjective rating methods [19]. Though situation awareness has been identified as a critical consideration in AI design and operation (e.g., Ref. [18]), it is unclear what combination of methods should be applied during design and evaluation efforts to ensure that appropriate levels of situation awareness can be achieved. Guidance is therefore required on what HFE methods should be applied to support the design and evaluation of AI technologies. Especially within defense, there is a critical need to explore the intersection between HFE principles and methods and principles for responsible AI development and use. Specifically, this includes how HFE methods can be applied to ensure that AI technologies are designed to operate ethically and support ethical behavior as well as meeting other critical requirements around usability and safety. The need for such guidance is supported by a series of catastrophic AI-related failures in other domains where HFE was not embedded in design processes [20,21].

The aim of this chapter is to present a prototype framework for safe, ethical, and usable AI in defense that could be used by stakeholders to ensure that critical ethical principles (e.g., [9]) are implemented during AI design and operation. The intention was to create a framework of methods that could be used to create safe and usable AI systems that operate ethically and support ethical behavior across the broader defense system.

2 Method

The prototype framework was developed based on a review of HFE methods and two workshops involving researchers experienced in the development and application of HFE methods. The first workshop involved the identification of different categories of HFE methods that could be applied to support the design, evaluation, and operation of AI throughout the Australian Defense Force (ADF) capability life cycle (Australian Defense [22]). The second workshop involved an assessment of 16 categories of HFE methods and their capacity to assess ethical principles derived from the UK, US, and Australian (AUKUS) ethical frameworks for defense and a modified version of NATO's principles of responsible AI use [9].

2.1 Integration of responsible AI principles

Various sets of principles for the responsible design and use of defense AI technologies have been proposed, including those recently outlined by NATO, and the AUKUS defense forces. For the purposes of developing our prototype framework, we initially adopted the following principles of responsible AI use from NATO's AI strategy [9]:

- **Lawfulness**: AI applications will be developed and used in accordance with national and international law, including international humanitarian law and human rights law, as applicable.
- **Responsibility and accountability**: AI applications will be developed and used with appropriate levels of judgment and care; clear human responsibility shall apply to ensure accountability.
- **Explainability and traceability**: AI applications will be appropriately understandable and transparent and include the use of review methodologies, sources, and procedures. This includes verification, assessment, and validation mechanisms at either a NATO and/or national level.
- **Reliability**: AI applications will have explicit, well-defined use cases. The safety, security, and robustness of such capabilities will be subject to testing and assurance within those use cases across their entire life cycle, including through established NATO and/or national certification procedures.
- **Governability**: AI applications will be developed and used according to their intended functions and will allow for: appropriate human-machine interaction; the ability to detect and avoid unintended consequences; and the ability to take steps, such as disengagement or deactivation of systems, when such systems demonstrate unintended behavior.
- **Bias mitigation**: Proactive steps will be taken to minimize any unintended bias in the development and use of AI applications and in datasets.

Prior to our methods review, we compared these NATO principles with those of the AUKUS nations to ensure that key principles would not be overlooked. This involved mapping the AUKUS principles onto the NATO principles where the authors felt that the AUKUS principles were adequately covered by the NATO principles. Only one of the principles from the AUKUS nations could not be directly mapped to the NATO principles: Human centricity, defined as "the impact of AI-enabled systems on humans must be assessed and considered, for a full range of effects both positive and negative across the entire system lifecycle" [10]. Though elements of human centricity were covered within the NATO principle of Governability, the authors felt that an explicit principle around the impacts on human behavior and health and well-being was an important inclusion for the prototype framework. Accordingly, human centricity was added to the principles to be considered when developing the prototype framework.

2.2 Human factors and ergonomics methods review

There are now well over 100 structured HFE methods available for designing and evaluating aspects of human, team, organization, and system performance (see [23,24]). These include methods to support the design of equipment and workspaces, and methods designed to understand, assess, and enhance critical aspects of behavior such as perception, decision-making, situation awareness, cognitive workload, teamwork, and human-machine interaction, to name only a few [6]. From a broader systems perspective, HFE methods also contribute to the design of policies, procedures, training, and education programs as well as risk and safety management via activities such as risk assessment, incident reporting, and accident analysis, and to the development of national and international regulatory frameworks and standards [6]. Ideally, HFE methods are applied throughout the system life cycle [24] to inform needs analysis and design requirements, to assist in the design and evaluation of concepts and prototypes, and to assess performance and inform necessary sociotechnical system refinements following implementation.

As discussed earlier, the potential utility of HFE methods in supporting the design of AI has been noted by many. Primarily this argument is based on three common forms of HFE application: 1. the use of HFE methods to help optimize interactions between humans and technologies (e.g., interface and workspace design); 2. the use of HFE methods to optimize human performance when using or working with technologies (e.g., designing to enhance situation awareness and optimize workload); and 3. the use of HFE methods to optimize key aspects of the broader sociotechnical system in which technologies are to be used (e.g., procedure and training design, risk assessment, accident analysis). Given the breadth of HFE methods available across these three application areas, an initial review was undertaken to identify categories of HFE methods that could potentially be applied throughout the ADF capability life cycle when considering AI technologies specifically. The review included relevant HFE methods textbooks [23,24] as well as relevant AI-related articles from the peer-reviewed HFE literature (e.g., [18]). Based on the review, 16 categories of HFE methods were identified (Table 12.1).

2.3 Workshop 1: Mapping HFE methods to AI-based ADF capability life cycle phases

A workshop involving the first four coauthors as participants was held to determine which categories of HFE method could be applied at different points in an AI-based ADF capability life cycle. Specifically, each category of HFE method was discussed with regard to its potential application throughout the ADF capability life cycle. This discussion involved considering what activities are undertaken during each life-cycle phase and then determining whether each form of HFE method could be usefully applied in defense AI applications. For example, when considering the Risk Mitigation and Requirement Setting phase of the life cycle, the workshop participants discussed how HFE methods could be used to inform the "development and progression of capability options through the

Table 12.1   HFE methods.

Category of method	Commonly applied methods
Physical HFE methods and standards	Anthropometric data
	Workspace design guidelines
Task analysis	Hierarchical Task Analysis (HTA [25])
Cognitive task analysis	Applied Cognitive Task Analysis (ACTA [26])
	Concurrent verbal protocol analysis [27]
	Critical Decision Method [28]
Process charting	Operator Sequence Diagrams (OSDs [24])
Human error identification	Systematic Human Error Reduction and Prediction Approach [29]
Situation awareness assessment	Situation Awareness Global Assessment Technique (SAGAT [30])
	Situation Awareness Present Assessment (SPAM [31])
Trust assessment	Checklist for Trust Between People and Automation (CTPA [32])
	Human-Centered Artificial Intelligence (HCAI) trustworthiness scale [13]
Mental workload assessment	NASA Task Load Index (NASA-TLX [33])
	Primary and secondary task performance measures; Psycho-physiological measures
Teamwork assessment	Coordination Demands Analysis [34]
Interface analysis	HCI Checklist [35]
	Link analysis [36]
	Layout analysis [37]
Usability evaluation	System Usability Scale (SUS [38])
Performance time prediction	Critical Path Analysis (CPA [39])
	Keystroke Level Model (KLM [40])
Design	Sociotechnical Systems Design Toolkit (STS-DT [41])
Systems analysis	Cognitive Work Analysis (CWA [42])
	Event Analysis of Systemic Teamwork (EAST [21])
Risk assessment	Systems Theoretic Accident Model and Process – Systems Theory Process Analysis (STAMP-STPA [43])
	Networked Hazard Analysis and Risk Management System (Net-HARMS [44])
Accident analysis	Systems Theoretic Accident Model and Process – Causal Analysis based on Systems Theory (STAMP-CAST [43])
	Accident Mapping (AcciMap [45])

Adapted from P.M. Salmon, T. Carden, P.A. Hancock, Putting the humanity into inhuman systems: how human factors and ergonomics can be used to manage the risks associated with artificial general intelligence, Hum. Factors Ergon. Manuf. Ind. 31(2) (2021) 223–236, https://doi.org/10.1002/hfm.20883.

investment approval process" (ADF [22]) and identified key analysis activities such as risk assessment and requirements analysis. Categories of HFE methods identified by participants that could provide input into these activities included risk assessment and accident analysis methods, design methods, physical HFE methods, task analysis methods, cognitive task analysis methods, situation awareness assessment, and trust assessment methods. The discussion was based on the workshop participants' knowledge of HFE methods and their experiences in applying methods from within each category. Collectively, the four workshop participants have extensive experience in applying a wide range of HFE methods, including coauthorship of various methods textbooks covering well over

100 HFE methods (e.g., Refs. [23,24,46,47]) and multiple applications in areas such as defense, transport, workplace safety, sport and outdoor recreation, cybersecurity, and disaster management. Where a category of methods was deemed suitable for application during a stage in the AI life cycle, specific methods from each category were discussed and suitable methods were selected for consideration for inclusion in the prototype framework.

2.4 Workshop 2: Mapping of HFE methods to modified NATO principles of responsible AI use

A second workshop was held with the participants from Workshop 1 to identify which HFE methods could be applied to assess aspects of each of the modified NATO principles of responsible AI use. Specifically, each HFE method was discussed with regard to its potential use in assessing whether ethical principles are fulfilled during the design and operation of AI technologies. This discussion involved considering each principle and associated definition and systematically working through each HFE method to determine whether it could be applied to attain or assess aspects of each principle. For example, with the principle of human centricity, the workshop participants discussed how HFE methods could be used to support assessment of the principle that "the impact of AI-enabled systems on humans must be assessed and considered, for a full range of effects both positive and negative across the entire system lifecycle" (Australian Defense [22]). In the case of human centricity, the participants concluded that all 16 categories of HFE methods could be applied as they could be used to assess positive and negative impacts on human operator performance (e.g., physical HFE, situation awareness, decision-making), team performance (e.g., teamwork assessment), and system performance (e.g., risk assessment, systems analysis, accident analysis).

The outputs of both workshops were subsequently used to develop a prototype framework for safe, ethical, and usable AI. This involved the first author identifying specific HFE methods for application throughout the AI-based capability life cycle, and then identifying which of the modified NATO principles for responsible AI use could be assessed during the different life-cycle phases and activities. The prototype framework was subsequently reviewed by the remaining five coauthors and refined based on their feedback.

3 Results

3.1 Applicability of HFE methods across the defense capability life cycle

The potential use of each category of HFE method across the defense capability life-cycle phases [22] is presented in Table 12.2. The full list of specific methods within each category is presented in Appendix A.

As shown in Table 12.2, when specifically considering AI systems, all 16 categories of HFE method can be applied to support at least one stage of the Defense Capability Life

Table 12.2 Potential use of HFE methods across the defense capability life cycle.

Category of method	Defence Capability Lifecycle phase			
	Strategy and concepts	Risk mitigation and requirement setting	Acquisition	In-service and disposal
Physical HFE methods	✓*	✓	✓	✓
Task analysis	✓*	✓	✓	✓
Cognitive task analysis	✓*	✓	✓	✓
Process charting	✓*		✓	✓
Human error identification	✓*		✓	✓
Situation awareness assessment	✓*	✓	✓	✓
Trust assessment	✓*	✓	✓	✓
Mental workload assessment	✓*		✓	✓
Teamwork assessment	✓*		✓	✓
Interface analysis	✓*		✓	✓
Usability evaluation	✓*		✓	✓
Performance time prediction	✓*		✓	✓
Design	✓*	✓	✓	
Systems analysis	✓*	✓	✓	✓
Risk assessment	✓*	✓	✓	✓
Accident analysis	✓*	✓		✓

*Denotes methods that can be applied to the analysis of existing Defence systems in order to identify capability needs

Cycle. Most phases are well supported by the different categories of HFE method. The Risk Mitigation and Requirement Setting phase, including the development and progression of capability options, had the least categories of applicable HFE methods, with nine applicable categories.

Methods that were applicable across all phases of the capability life cycle were also identified. From the 16 categories of methods, seven were considered to be applicable to all phases of the defense capability life cycle, including physical HFE, task analysis, cognitive task analysis, situation awareness assessment, trust assessment, systems analysis, and risk assessment methods.

3.2 Suitability of HFE methods for assessing principles of responsible AI use

The mapping of each category of HFE method to the modified NATO principles of responsible AI use is presented in Table 12.3. The mapping of the full list of specific methods within each category to the principles is presented in Appendix B.

As shown in Table 12.3, all 16 categories of HFE methods can be applied to assess at least one of the modified NATO principles for responsible AI use. Table 12.3 also shows

Table 12.3 Mapping of HFE method categories to the modified NATO principles of responsible AI use.

Category of method	Modified NATO principles of responsible AI use						
	Lawfulness	Responsibility / accountability	Explainability / traceability	Reliability	Governability	Bias mitigation	Human centricity
Physical HFE methods				✓	✓		✓
Task analysis		✓	✓	✓	✓	✓	✓
Cognitive task analysis		✓	✓	✓	✓	✓	✓
Process charting		✓	✓	✓	✓		✓
Human error identification	✓			✓	✓		✓
Situation awareness assessment	✓	✓	✓	✓	✓	✓	✓
Trust assessment	✓	✓	✓	✓		✓	✓
Mental workload assessment		✓	✓	✓			✓
Teamwork assessment		✓	✓	✓	✓		✓
Interface analysis		✓	✓	✓	✓		✓
Usability evaluation		✓	✓	✓	✓		✓
Performance time prediction				✓			✓
Design	✓	✓	✓	✓	✓	✓	✓
Systems analysis	✓	✓	✓	✓	✓	✓	✓
Risk assessment	✓	✓	✓	✓	✓	✓	✓
Accident analysis	✓	✓	✓	✓	✓	✓	✓

that six categories of HFE methods can be used to assess aspects of all of the modified principles, specifically: situation awareness assessment, trust assessment, design, systems analysis, risk assessment, and accident analysis methods. The mapping exercise also demonstrates, however, that coverage of the principles of lawfulness and bias mitigation is less comprehensive, with only seven categories of methods deemed to be applicable to assessments of lawfulness, and eight categories of methods deemed to be applicable to bias mitigation. Further work should explore the extent to which these and other legal and ethical assessment methods can be used to assess lawfulness and bias mitigation, or whether there is a need to develop new methods for this purpose.

The finding that multiple categories of methods were found to be applicable to each principle suggests that an integrated toolkit approach could be adopted whereby various methods are applied to comprehensively assess the extent to which each principle is achieved. For example, when assessing human centricity, various HFE methods could be used to understand the impacts on different facets of human performance, such as decision-making, situation awareness, workload, and errors. Further, this integrated toolkit could be multidisciplinary in nature and incorporate methods from other areas such as law and ethics. The wide applicability of HFE methods does, however, raise concerns around potential difficulties in ensuring a consistent approach to the assessment of AI technologies. Guidelines on appropriate methods to use in relation to each principle are therefore critical and future research could explore the development of a standardized approach.

3.3 Prototype framework of human factors and ergonomics methods for safe, ethical, and usable AI

The prototype HFE framework for safe, ethical, and usable AI is presented in Fig. 12.1. Included in Fig. 12.1 are example HFE methods that could be applied during each phase of the defense capability life cycle.

As shown in Fig. 12.1, it is recommended that HFE methods be applied to assess AI technology ethics, emergent risks, and usability during multiple activities across the defense capability life cycle. These activities include:

1. **Identification of capability need**. The identification of a capability need can involve a range of analysis methods and can include ongoing analyses of performance or current system state. This can be conducted through the application of methods such as risk assessment and accident analysis, or targeted gap analyses using task and systems analysis methods.

2. **Assessment of the sociotechnical system in which the AI will be deployed**. The importance of assessing the broader sociotechnical system in which AI technologies will operate has been emphasized (e.g., Ref. [17]). This activity involves the use of systems analysis methods to help understand what users and existing technologies, processes, and social and organizational structures the AI will need to interact with. Models of the existing system that describe its components and interactions are also critical.

3. **Prospective risk assessment.** Formal risk assessment involves the use of structured methods to prospectively identify potential hazards that may create adverse outcomes during use of the AI technology [48]. Various forms of qualitative and quantitative methods exist, enabling the identification of hazards and associated risks and/or an estimation of their likelihood of occurrence [49].

4. **Design**. Design activities involve the use of structured design methods to develop and refine AI design concepts. Design methods should focus not only on the technology itself, but also on the interaction between the AI technology and human users, and its integration into the broader sociotechnical system.

5. **Desktop evaluation**. Desktop evaluations play a critical role in the refinement of design concepts and can be undertaken with early design concepts through to final prototypes. A wide range of evaluation methods are available to support interface evaluations, task and allocation of functions analysis, usability assessments, and user trials.

6. **Experimentation/user trials.** Experimentation and user trials enable various aspects of operator, team, technology, and sociotechnical system performance to be assessed in a controlled setting, including situation awareness, trust, workload, decision-making, potential for design-induced error, and teamwork and coordination. At this stage the findings can be used to refine the AI technology or the aspects of the sociotechnical system in which it will be used, including standard operating procedures, training programs, other technologies, maintenance activities, and incident reporting processes.

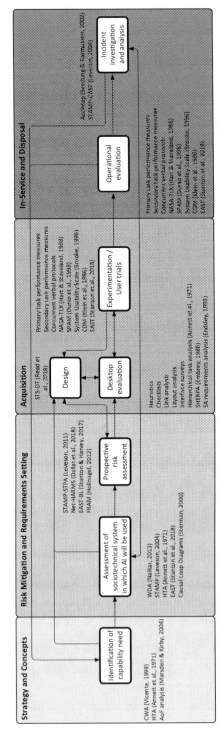

FIG. 12.1 Prototype framework for safe, ethical, and usable AI including example methods to apply at each stage of the life cycle.

7. **Operational evaluation**. Operational evaluation enables evaluation of the AI technology once implemented and used during operational activities.
8. **Incident investigation and analysis.** Incident investigation and analysis should occur following adverse events and near misses involving the AI technology, with the intention being to support the development of interventions that prevent or manage future occurrences. Various methods are available to support incident investigation and analysis, with state-of-the-art methods avoiding a focus on individual actions ("human error") and instead supporting the identification of interrelated contributory factors across the overall sociotechnical system.

In Fig. 12.2, the modified NATO principles for responsible AI use are mapped onto the framework to show where each principle can be assessed during the ADF capability life cycle.

As shown in Fig. 12.2, all seven principles can be assessed via HFE methods across the defense capability life cycle. According to the mapping exercise, aspects of all seven principles can be assessed during the Risk Mitigation and Requirements Setting, Acquisition, and In-Service and Disposal phases; however, only lawfulness, bias mitigation, and human centricity can be considered during the Strategy and Concepts phase using current HFE methods. This suggests that there is a critical gap around HFE methods to assess ethical principles when identifying capability needs, acknowledging that this may be due to the fact that the AI technology has not yet been designed or acquired. Regardless, further work should explore the development of processes that could be used to prospectively consider ethical principles when identifying a particular capability need. This will enable ethical requirements to drive design and acquisition processes.

The mapping exercise emphasizes the importance of applying HFE methods early in the life cycle to ensure the design of safe, ethical, and usable AI. In particular, this entails understanding the capability need, the broader sociotechnical system, prospective risk assessment around ethical principles, and the design and evaluation of early prototypes. It is therefore recommended that all seven principles be assessed in some way prior to the development of prototype AI system designs. This can be achieved through the use of prospective risk assessment methods during the Risk Mitigation and Requirements Setting phase, and the use of a range of HFE methods to evaluate and refine early design concepts. A failure to consider ethical principles during the early phases of the capability life cycle could result in the acquisition of AI systems that do not satisfy the seven principles of responsible AI use.

4 Discussion

Ensuring that AI technologies enable ethical decision-making is arguably one of the most important contemporary challenges in defense research. As the discipline that focuses on human health and wellbeing and the interactions between humans and other system components, the critical role of HFE in this endeavor has been emphasized [6]. The

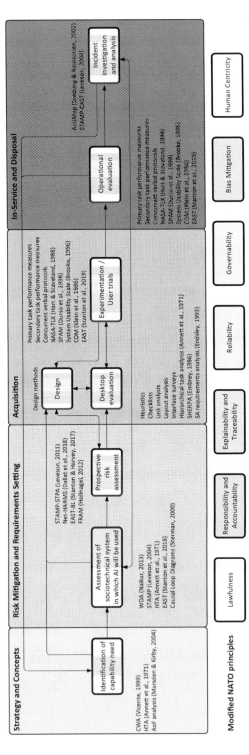

FIG. 12.2 Modified NATO principles mapped onto prototype framework for safe, ethical, and usable AI including example methods to apply at Each Stage of the Life cycle.

aim of this chapter was to present a prototype framework of HFE methods that could be used by stakeholders to ensure that critical ethical principles (e.g., Ref. [9]) are implemented during AI design and operation.

The review of HFE methods revealed that all 16 categories can be applied to AI design and/or evaluation during at least one phase of the defense capability cycle, with most phases being well supported by the different HFE method categories. From the 16 categories of methods, seven were considered to be applicable to all phases of the defense capability life cycle, including physical HFE, task analysis, cognitive task analysis, situation awareness assessment, trust assessment, systems analysis, and risk assessment methods. For example, risk assessment methods such as the Networked Hazard Analysis and Risk Management System (Net-HARMS; [44]) or the Systems Theoretic Process Analysis (STPA; [50]) could be applied initially during the Strategy and Concepts phase to identify risks associated with current technologies and processes to support the identification of a new capability need. During the Risk Mitigation and Requirements Setting phase, the same methods could be used to prospectively identify the safety and ethical risks associated with different capability options to inform decision-making around acquisition. During the Acquisition phase, a more detailed assessment of risks could then be undertaken with the capability itself, and likewise risks could be identified and analyzed using the same methods during the In-Service and Disposal phase. Overall, the applicability of the methods reviewed provides further support to the notion that HFE methods have a key role to play in AI design, implementation, and operation [3,6,18].

The mapping of HFE methods to the modified set of NATO principles for responsible AI use revealed that all 16 categories of HFE method can be applied to assess aspects of at least one of the seven modified principles. The categories of method that can be used to assess aspects of all principles include situation awareness assessment, trust assessment, design, systems analysis, risk assessment, and accident analysis methods. This finding suggests that these six categories of HFE methods will be critical for ensuring responsible AI use principles are implemented. It is therefore recommended that methods from these six categories should represent the minimal level of HFE input during AI design and evaluation activities.

Within HFE there is a long legacy of using toolkits of integrated HFE methods when tackling complex issues (e.g., Refs. [24,51]). The aim is to enhance comprehensiveness by using different methods to assess different features of performance that cannot be covered by one method alone. Given the range of HFE methods found to be applicable, it is concluded that an integrated toolkit approach could be adopted to ensure comprehensive assessments of each principle. For example, when assessing human centricity various HFE methods could be used to understand the impacts on different facets of human performance such as decision-making, situation awareness, workload, and errors.

A prototype framework for safe, ethical and usable AI was proposed based on the methods review and mapping exercise. The prototype framework describes a range of activities that should be undertaken throughout the defense capability life cycle to ensure the design and operation of ethical defense AI technologies, along with a range of HFE methods that could be applied during each phase. The prototype framework also shows where in the ADF capability life cycle each of the modified NATO principles should be considered.

According to the prototype framework, there are multiple opportunities to assess each principle across the defense capability life cycle. This is encouraging and provides further evidence to support the use of HFE methods in AI design, implementation, and evaluation. The framework also confirms the importance of applying HFE methods early in the life cycle to ensure the design of safe, ethical, and usable AI [3,6,14]. In particular, the need to consider ethical aspects of AI when identifying a capability need and undertaking risk mitigation and requirements analyses is emphasized. It is therefore recommended that all seven principles be assessed prior to the development or acquisition of prototype AI technologies. This can be achieved through the use of prospective risk assessment methods during the Risk Mitigation and Requirement Setting phase, and the use of a range of HFE methods to evaluate and refine early design concepts during the Acquisition phase. A failure to consider ethical principles during the early phases of the capability life cycle could result in the implementation of AI systems that do not satisfy the seven principles of responsible AI use.

The prototype framework also suggests that there is an absence of HFE methods available specifically for assessing ethical aspects of potential AI technologies during the Strategy and Concepts phase. It is therefore recommended that future research should be undertaken to explore the development of methods that could be used to prospectively consider ethical principles when identifying a particular capability need. This will ensure that ethical requirements drive the initial AI system design and acquisition process.

4.1 Future areas of research

Though there are various critical areas of future research around the design of safe, ethical, and usable AI, the most pressing in relation to this chapter involve testing and further refinement of the prototype framework. It is therefore recommended that the framework be applied to current and future defense AI acquisition projects. Further development and refinement of the framework should also involve providing step-by-step practical guidance on how to apply the different HFE methods during each phase. Beyond this, recommendations on toolkits of methods to be applied for different forms of AI technology should be made. It may be, for example, that different HFE methods are required for different AI technologies such as physical systems (e.g., UCAVs and bomb disposal robots) versus data-based systems (e.g., decision support and data mining tools).

To close, we again emphasize that HFE has a critical role to play in ensuring that AI technologies can be deployed in a safe and ethical manner [2,6,14–18]. While the potential benefits of AI are significant, in defense as well as other areas such as transport, healthcare, and disaster management, there are also many potential risks. We encourage further development and use of the framework presented in this chapter, and hope that it provides a useful approach for those involved in the design and evaluation of AI technologies.

Appendix A Mapping of methods to the ADF capability life cycle phases.

HFE method	ADF capability lifecycle phases			
	Strategy and concepts	Risk mitigation and requirement setting	Acquisition	In-service and disposal
STAMP - Causal Analysis based on STAMP (STAMP-CAST)	✓			✓
Accident Mapping (AcciMap; including aggregate AcciMaps)	✓			✓
Functional Resonance Analysis Method (FRAM)	✓	✓		✓
Critical Decision Method (CDM)	✓	✓		✓
Applied Cognitive Task Analysis (ACTA)	✓	✓		✓
Schema World Action Research Method (SWARM)	✓	✓		✓
Malvern Capacity Estimate technique (MACE)	✓			✓
NASA-Task Load Index (NASA-TLX)	✓			✓
DRA Workload Scales (DRAWS)	✓			✓
Subjective Workload Assessment Technique (SWAT)	✓			✓
Workload profile technique	✓			✓
Situation Awareness requirements analysis	✓	✓		
Scenario-based design	✓	✓	✓	
Task-centred system design	✓	✓	✓	
Allocation of functions analysis	✓	✓	✓	
Sociotechnical Systems Design Toolkit (STS-DT)	✓	✓	✓	✓
Competency assessment				✓
Primary task performance measures			✓	✓
Secondary task performance measures			✓	✓
Line Operation Safety Audit (LOSA)			✓	✓
Systematic Human Error Reduction and Prediction Approach (SHERPA)	✓	✓	✓	
Human Error Assessment and Reduction Technique (HEART)	✓	✓	✓	
PreventiMaps	✓			
Eye tracking				✓
Operation Sequence Diagrams (OSD)	✓		✓	✓
Networked Hazard Analysis and Risk Management System (Net-HARMS)	✓		✓	✓
STAMP-Systems-Theoretic Process Analysis (STAMP-STPA)	✓		✓	✓
Event Analysis of Systemic Teamwork- Broken Links (EAST-BL)	✓		✓	✓
Situation Present Assessment Method (SPAM)	✓			✓
Situation Awareness Global Assessment Technique (SAGAT)	✓			✓
Situation Awareness Rating Technique (SART)	✓			✓
Physical ergonomics standards	✓	✓	✓	✓
Alarm design standards and guidelines	✓	✓	✓	✓
Event Analysis of Systemic Teamwork (EAST)	✓		✓	✓
Rich Pictures	✓			
Causal loop diagrams	✓			✓
Cognitive Work Analysis (CWA)	✓	✓		✓
Hierarchical Task Analysis (HTA)	✓	✓	✓	✓
Concurrent Verbal Protocol Analysis (VPA)	✓		✓	✓
Task decomposition	✓	✓	✓	✓
Team task analysis	✓	✓	✓	✓
Coordination Demand Analysis (CDA)	✓		✓	✓
Targeted Acceptable Responses to Generated Events or Tasks (TARGETs)	✓		✓	✓
Team Cognitive Task Analysis (TCTA)	✓		✓	✓
Groupware Task Analysis (GTA)	✓		✓	✓
System Usability Scale (SUS)	✓		✓	✓
Software Usability Measurement Inventory (SUMI)	✓		✓	✓
Supervisory Control and Data Acquisition (SCADA)	✓		✓	✓

Appendix B Mapping of methods to each of the modified NATO principles of responsible use of AI.

Method	Modified NATO principles of responsible AI use						
	Lawfulness	Responsibility / accountability	Explainability / traceability	Reliability	Governability	Bias mitigation	Human centricity
STAMP – Causal Analysis based on STAMP (CAST-STAMP)	✓	✓	✓	✓		✓	✓
Accident Mapping (AcciMap; including aggregate AcciMaps)	✓	✓	✓	✓			✓
Functional Resonance Analysis Method (FRAM)	✓	✓	✓	✓			✓
Critical Decision Method (CDM)		✓	✓	✓			✓
Applied Cognitive Task Analysis (ACTA)		✓	✓	✓			✓
Schema World Action Research Method (SWARM)		✓	✓	✓			✓
Malvern Capacity Estimate technique (MACE)		✓	✓				✓
NASA-Task Load Index (NASA-TLX)		✓	✓				✓
DRA Workload Scales (DRAWS)		✓	✓				✓
Subjective Workload Assessment Technique (SWAT)		✓	✓				✓
Workload profile technique		✓	✓				✓
Situation Awareness requirements analysis	✓	✓	✓		✓		✓
Scenario-based design	✓	✓	✓		✓		✓
Task-centred system design		✓	✓		✓		✓
Allocation of functions analysis		✓	✓	✓	✓		✓
Sociotechnical Systems Design Toolkit (STS-DT)	✓	✓	✓	✓	✓		✓
Competency assessment		✓	✓	✓	✓		✓
Primary task performance measures		✓	✓				✓
Secondary task performance measures		✓	✓				✓
Line Operation Safety Audit (LOSA)					✓		✓
Systematic Human Error Reduction and Prediction Approach (SHERPA)	✓			✓	✓		✓
Human Error Assessment and Reduction Technique (HEART)	✓			✓	✓		✓
PreventiMaps	✓			✓	✓		✓
Eye tracking							✓
Operation Sequence Diagrams (OSD)		✓	✓	✓	✓		✓
Networked Hazard Analysis and Risk Management System (Net-HARMS)	✓	✓	✓	✓	✓		✓
STAMP-Systems-Theoretic Process Analysis (STAMP-STPA)	✓	✓	✓	✓	✓		✓
Event Analysis of Systemic Teamwork- Broken Links (EAST-BL)	✓	✓	✓	✓	✓		✓
Situation Present Assessment Method (SPAM)	✓	✓	✓	✓			✓
Situation Awareness Global Assessment Technique (SAGAT)	✓	✓	✓	✓			✓
Situation Awareness Rating Technique (SART)	✓	✓	✓	✓			✓
Physical ergonomics standards				✓			✓
Event Analysis of Systemic Teamwork (EAST)	✓	✓	✓	✓	✓		✓
Rich Pictures	✓	✓	✓	✓	✓		✓
Causal loop diagrams	✓	✓	✓	✓	✓		✓
Cognitive Work Analysis (CWA)	✓	✓	✓	✓	✓	✓	✓
Hierarchical Task Analysis (HTA)	✓	✓		✓	✓		✓
Concurrent Verbal Protocol Analysis (VPA)	✓	✓	✓	✓	✓		✓
Task decomposition	✓	✓		✓	✓		✓
Team task analysis	✓	✓	✓	✓	✓		✓
Coordination Demand Analysis (CDA)	✓	✓	✓	✓	✓		✓
Targeted Acceptable Responses to Generated Events or Tasks (TARGETs)		✓	✓	✓			✓
Team Cognitive Task Analysis (TCTA)		✓	✓	✓			✓
Groupware Task Analysis (GTA)		✓	✓	✓			✓
System Usability Scale (SUS)		✓	✓				✓
Software Usability Measurement Inventory (SUMI)		✓	✓				✓
Supervisory Control and Data Acquisition (SCADA)		✓	✓				✓

References

[1] F.E. Morgan, B. Boudreaux, A.J. Lohn, M. Ashby, C. Curriden, K. Klima, D. Grossman, Military Applications of Artificial Intelligence: Ethical Concerns in an Uncertain World, RAND Corporation, 2020. https://www.rand.org/pubs/research_reports/RR3139-1.html.

[2] P.A. Hancock, Imposing limits on autonomous systems, Ergonomics 60 (2) (2017) 284–291, https://doi.org/10.1080/00140139.2016.1190035.

[3] P.A. Hancock, Avoiding adverse autonomous agent actions, Hum. Comput. Interact. 37 (3) (2022) 211–236, https://doi.org/10.1080/07370024.2021.1970556.

[4] S. McLean, G.J. Read, J. Thompson, C. Baber, N.A. Stanton, P.M. Salmon, The risks associated with artificial general intelligence: a systematic review, J. Exp. Theor. Artif. Intell. (2021) 1–15, https://doi.org/10.1080/0952813X.2021.1964003.

[5] V.C. Müller (Ed.), Risks of Artificial Intelligence, CRC Press, 2016.

[6] P.M. Salmon, T. Carden, P.A. Hancock, Putting the humanity into inhuman systems: how human factors and ergonomics can be used to manage the risks associated with artificial general intelligence, Hum. Factors Ergon. Manuf. Ind. 31 (2) (2021) 223–236, https://doi.org/10.1002/hfm.20883.

[7] S. Omohundro, Autonomous technology and the greater human good, J. Exp. Theor. Artif. Intell. 26 (3) (2014) 303–315.

[8] M. Brundage, Limitations and risks of machine ethics, in: V.C. Müller (Ed.), Risks of Artificial Intelligence, CRC Press, 2016, p. 291.

[9] NATO, Summary of the NATO Artificial Intelligence Strategy, 2021. https://www.nato.int/cps/en/natohq/official_texts_187617.htm?selectedLocale=en.

[10] Ministry of Defence, Ambitious, Safe, Responsible: Our Approach to the Delivery of AI Enabled Capability in Defence, 2022. https://www.gov.uk/government/publications/ambitious-safe-responsible-our-approach-to-the-delivery-of-ai-enabled-capability-in-defence.

[11] U.S. Department of Defence, DOD Adopts Ethical Principles for Artificial Intelligence, 2020. https://www.defense.gov/News/Releases/Release/Article/2091996/dod-adopts-ethical-principles-for-artificial-intelligence/.

[12] Department of Defence, A Method for Ethical AI in Defence [Technical Report], 2021. https://www.dst.defence.gov.au/publication/ethical-ai.

[13] B. Schneiderman, Human-Centered AI, Oxford University Press, 2022.

[14] P.A. Hancock, Some pitfalls in the promises of automated and autonomous vehicles, Ergonomics 62 (4) (2019) 479–495, https://doi.org/10.1080/00140139.2018.1498136.

[15] D. Petrat, Artificial intelligence in human factors and ergonomics: an overview of the current state of research, Discover Artif. Intell. 1 (1) (2021) 1–10, https://doi.org/10.1007/s44163-021-00001-5.

[16] P.M. Salmon, The horse has bolted! Why human factors and ergonomics has to catch up with autonomous vehicles (and other advanced forms of automation): commentary on Hancock (2019): some pitfalls in the promises of automated and autonomous vehicles, Ergonomics 62 (4) (2019) 502–504, https://doi.org/10.1080/00140139.2018.1563333.

[17] P.M. Salmon, Commentary: controlling the demon: autonomous agents and the urgent need for controls, Hum. Comput. Interact. 37 (3) (2022) 246–247, https://doi.org/10.1080/07370024.2021.1977127.

[18] M. Sujan, R. Pool, P. Salmon, Eight human factors and ergonomics principles for healthcare artificial intelligence, BMJ Health Care Inform. 29 (1) (2022), https://doi.org/10.1136/bmjhci-2021-100516.

[19] N.A. Stanton, P.M. Salmon, G.H. Walker, E. Salas, P.A. Hancock, State-of-science: situation awareness in individuals, teams and systems, Ergonomics 60 (4) (2017) 449–466, https://doi.org/10.1080/00140139.2017.1278796.

[20] G.J. Read, A. O'Brien, N.A. Stanton, P.M. Salmon, Learning lessons for automated vehicle design: using systems thinking to analyse and compare automation-related accidents across transport domains, Saf. Sci. 153 (2022) 105822.

[21] N.A. Stanton, P.M. Salmon, G. Walker, M. Stanton, Models and methods for collision analysis: a comparison study based on the Uber collision with a pedestrian, Saf. Sci. 120 (2019) 117–128.

[22] Australian Defence Force, Capability Life Cycle Manual (Version 2.1), 2021. https://www.dica.org.au/wp-content/uploads/2020/10/Capability-Life-Cycle-Manual-v2-1.pdf.

[23] P.M. Salmon, N.A. Stanton, G.H. Walker, A. Hulme, N. Goode, J. Thompson, G.J. Read, Handbook of Systems Thinking Methods, CRC Press, 2022.

[24] N.A. Stanton, P.M. Salmon, L.A. Rafferty, G.H. Walker, C. Baber, D.P. Jenkins, Human Factors Methods: A Practical Guide for Engineering and Design, CRC Press, 2013.

[25] J. Annett, K.D. Duncan, R.B. Stammers, M.J. Gray. Task, Analysis. Department of Employment Training Information Paper 6, HMSO, London, 1971.

[26] L.G. Militello, R.J. Hutton, Applied cognitive task analysis (ACTA): a practitioner's toolkit for understanding cognitive task demands, Ergonomics 41 (11) (1998) 1618–1641.

[27] G. Walker, Verbal protocol analysis, in: Handbook of Human Factors and Ergonomics Methods, CRC Press, 2004, pp. 327–337.

[28] G.A. Klein, R. Calderwood, D. Macgregor, Critical decision method for eliciting knowledge, IEEE Trans. Syst. Man Cybern. 19 (3) (1989) 462–472.

[29] D.E. Embrey, SHERPA: a systematic human error reduction and prediction approach [paper presentation], in: International Meeting on Advances in Human Factors in Nuclear Power Systems, Knoxville, Tennessee, April 21–24, 1986. https://inis.iaea.org/search/search.aspx?orig_q=RN:18074340.

[30] M.R. Endsley, Measurement of situation awareness in dynamic systems, Hum. Factors 37 (1) (1995) 65–84, https://doi.org/10.1518/001872095779049499.

[31] F.T. Durso, C.A. Hackworth, T.R. Truitt, J. Crutchfield, D. Nikolic, C.A. Manning, Situation awareness as a predictor of performance for en route air traffic controllers, Air Traffic Control Q. 6 (1) (1998) 1–20, https://doi.org/10.2514/atcq.6.1.1.

[32] J.Y. Jian, A.M. Bisantz, C.G. Drury, Foundations for an empirically determined scale of trust in automated systems, Int. J. Cognit. Ergon. 4 (1) (2000) 53–71.

[33] S.G. Hart, L.E. Staveland, Development of NASA-TLX (task load index): results of empirical and theoretical research, Adv. Psychol. 52 (1988) 139–183, https://doi.org/10.1016/S0166-4115(08)62386-9.

[34] S.C. Burke, Team task analysis, in: N.A. Stanton, A. Hedge, K. Brookhuis, E. Salas, H. Hendrick (Eds.), Handbook of Human Factors and Ergonomics Methods, CRC Press, 2004, pp. 56:1–56:8.

[35] S. Ravden, G. Johnson, Evaluating Usability of Human-Computer Interfaces: A Practical Method, Halsted Press, 1989.

[36] N.A. Stanton, M.S. Young, A Guide to Methodology in Ergonomic: Designing for Human Use, Taylor & Francis, 1999.

[37] R. Easterby, H. Zwaga, Information Design, Wiley, 1984.

[38] Brooke, SUS: a 'quick and dirty' usability scale, in: P.W. Jordan, B. Thomas, B.A. Weerdmeester, I. McClelland (Eds.), Usability Evaluation in Industry, Taylor & Francis, 1996, pp. 189–194. https://hell.meiert.org/core/pdf/sus.pdf.

[39] K.G. Lockyer, J. Gordon, Critical Path Analysis and Other Project Network Techniques, Beekman Books Incorporated, 1991.

[40] S.K. Card, T.P. Moran, A. Newell, The Psychology of Human-Computer Interaction, RCR Press, 1983. https://doi.org/10.1201/9780203736166.

[41] G.J.M. Read, P.M. Salmon, N. Goode, M.G. Lenné, A sociotechnical design toolkit for bridging the gap between systems-based analysis and system design, Hum. Factors Ergon. Manuf. Ind. 28 (6) (2018) 327–341.

[42] K.J. Vicente, Cognitive Work Analysis: Toward Safe, Productive, and Healthy Computer-Based Work, Lawrence Erlbaum, 1999.

[43] N. Leveson, A new accident model for engineering safer systems, Saf. Sci. 42 (4) (2004) 237–270, https://doi.org/10.1016/S0925-7535(03)00047-X.

[44] C. Dallat, P.M. Salmon, N. Goode, Identifying risks and emergent risks across sociotechnical systems: the NETworked Hazard analysis and risk management system (NET-HARMS), Theor. Issues Ergon. Sci. 19 (4) (2018) 456–482, https://doi.org/10.1080/1463922X.2017.1381197.

[45] I. Svedung, J. Rasmussen, Graphic representation of accident scenarios: mapping system structure and the causation of accidents, Saf. Sci. 40 (5) (2002) 397–417, https://doi.org/10.1016/S0925-7535(00)00036-9.

[46] P.M. Salmon, N.A. Stanton, M. Lenné, D.P. Jenkins, L. Rafferty, G.H. Walker, Human Factors Methods and Accident Analysis: Practical Guidance and Case Study Applications, CRC Press, 2011.

[47] N.A. Stanton, P.M. Salmon, G.H. Walker, C. Baber, D. Jenkins, Human Factors Methods- a Practical Guide for Engineering and Design, Ashgate, 2005.

[48] P. Chemweno, L. Pintelon, P.N. Muchiri, A. Van Horenbeek, Risk assessment methodologies in maintenance decision making: a review of dependability modelling approaches, Reliab. Eng. Syst. Saf. 173 (2018) 64–77, https://doi.org/10.1016/j.ress.2018.01.011.

[49] C. Dallat, P.M. Salmon, N. Goode, Risky systems versus risky people: to what extent do risk assessment methods consider the systems approach to accident causation? A review of the literature, Saf. Sci. 119 (2019) 266–279, https://doi.org/10.1016/j.ssci.2017.03.012.

[50] N.G. Leveson, Applying systems thinking to analyze and learn from events, Saf. Sci. 49 (1) (2011) 55–64, https://doi.org/10.1016/j.ssci.2009.12.021.

[51] B. Kirwan, Human error identification techniques for risk assessment of high risk systems—part 2: towards a framework approach, Appl. Ergon. 29 (5) (1998) 299–318.

13

A schema for harms-sensitive reasoning, and an approach to populate its ontology by human annotation

Ariel M. Greenberg

JOHNS HOPKINS UNIVERSITY APPLIED PHYSICS LABORATORY, LAUREL, MD, UNITED STATES

1 Introduction: Chess bot incident begs for harms reasoning licensure

During a chess match held on July 19, 2022, between children and robots at the Moscow Open, a robot, in the course of a move, grabbed and broke his 7-year-old opponent's finger. Russian officials claim the child violated safety protocols,[a] but victim blaming is arguably the wrong precedent to set. While roboticists acknowledge that the gripper strength was far overpowered (Robotiq two-finger gripper with a grip force of 4.5–50 pounds[b]) for this application, using weaker or padded grippers, or holding children responsible, are bandages that miss the point—that a robot's capacity to appreciate the potential harmful consequences of its actions must be brought to be commensurate with its capability to perform the actions.

At that point, carefully constructed licensure would permit the machine to operate in a specific context, in this case, to physically interact with children, bound by the principle of nonmaleficence, to do no harm. Autonomous systems, due in large part to their complexity and stochasticity, are fundamentally challenging to test, evaluate, verify, and validate. Licensure approaches may be a more appropriate way to assure performance, similarly to how we permit drivers to operate motor vehicles after candidates pass exams that test

---

[a]"There are certain safety rules and the child, apparently, violated them. When he made his move, he did not realize he first had to wait," Sergey Smagin, Vice President of the Russian Chess Federation said. "This is an extremely rare case, the first I can recall," he added. "Apparently, children need to be warned. It happens." And Sergey Lazarev, President of the Moscow Chess Federation, said, "The robot broke the child's finger. This is of course bad."

[b]Robotiq 2F-85 and 2F-140 Grippers.

knowledge and skill. Once the system under study is subjected to and passes a battery of testing, use or deployment for a particular purpose is certifiable.

2 Generating values-driven behavior

One way to implement effective nonmaleficence is to assert policies for the machine to abide by. In passing scissors to a person, the robot may be instructed to do so with the handle out, or be hard coded with constraints that prevent it from taking the action any other way. This approach, however, is brittle, as the machine is likely to encounter edge cases or circumstantial quirks that make the policy untriggered or less appropriate. Instead, we have adopted a values-driven approach to generating behavior, that seeks to derive appropriate behavior from the first principles of values versus first deriving behavior such that it solves the inverse problem of objective satisfaction. In this approach, the machine is built to appreciate that the reason we pass scissors with the handle out is because the pointy tips and sharp blades of scissors are potentially hazardous to skin. This conclusion is drawn by the following reasoning steps in decomposition:

> A part of the scissors is sharp (ATTRIBUTE OF OBJECT) →
>   Sharps lacerate skin (DANGER OF INJURY). →
>   Lacerated skin is painful (AFFECT OF INJURY) →
>   Pain is undesirable (VALUE).

The ultimate application for this ontology is to enable machines to draw in harms consideration earlier in the process of generating behavior to achieve an objective, and to positively reframe the objective in terms of the values that it serves. This approach diverges from post hoc governance that restricts behaviors from a repertoire should they violate a principle or lead to a forbidden outcome. In the positivist framing, the machine can discover that the safe way to pass scissors, for example, is with the handle out. By traversing the knowledge graph representing the ontology, the machine is able to recognize that scissors have a pointy end (attribute of object) that can puncture skin (an insult, in the medical sense of the cause of injury), and that flesh is vulnerable to contusion (injury) by contact with pointy ends.

The advantages of this approach are myriad. Insofar as the behavior is derived through knowledge and reason from principle (as opposed to by mimicry of policy), it is necessarily more generalizable and transferrable, and arguably more trustworthy.

3 Moral-scene assessment: Minds, and affordances to them

In our previous work on moral-scene assessment, we proposed that sizing up a scene in moral terms could begin by placing each entity within an image in one of two spaces: Minds, with axes of *agency* (minds that instantiate intentional mental states to perform actions) and its counterpart *patiency* (minds subject to the experience of suffering), and objects, according to the hazards and/or *affordances* (the possible uses of objects, further described in Section 6) they furnish. A description of sensing approaches by which the uncertainty ellipse representing covariance error of a candidate mind may be placed within the axes of its agency-patiency space is given in Ref. [1]. In that chapter, we also outlined some of the percepts to engage harms and affordance recognition, to distinguish weapons from toys, food from the inedible, and skin and vegetation from other materials. In addition, we distinguished harm as negative impact to experiencing minds, from damage as negative impact to objects. This chapter aims to backfill production and use of these percepts with a knowledge schema and inference specification.

4 Injury: How physical harms come to be

4.1 Injury modeling and classification

In seeking to extend the initial observation described in Section 2 about scissor-passing, we searched case law for circumstances in which injuries occurred around and in interaction with household objects. In particular, we were curious to abstract upon discussions concerning mechanisms of harm. We came across many (too many) incidents of children swallowing balloons.

One case, *Landrine v. Mego Corp* (https://casetext.com/case/landrine-v-mego-corporation), provided in the judgment a crisp insight: "Balloons in and of themselves are not dangerous." The implication is that it is not the object that presents the hazard, but the intended or accidental use of it; the age of the potentially harmed individual serves as a proxy to assess susceptibility to harms from that use. These factors are listed in Table 13.1.

When a person brings a (deflated) balloon to their mouth, the typical intended action by an adult is to inflate it. However, an adult may also intend to swallow the balloon, in the rare cases wherein concealing the object within their body serves as a means to smuggle a substance within the balloon. In either of these cases, the adult has the prerogative to take these intended actions. A young child, on the other hand, in bringing a balloon to their lips, may intend to explore the object with their mouth, and in so doing, attempt to chew and swallow it. In this case, the balloon presents a gastrointestinal hazard that upon ingestion leads to sustaining the injury of strangulated bowels, or the balloon presents a choking hazard by blocking the air passageway.

Table 13.1 Action—Bring balloon to mouth.

Intended use	Age group	Anticipated/accidental use
Typical	Adult	Inflate
Explore	Child	(Accidental) Swallow
Smuggle	Adult	(Intentional) Swallow

To systematize this analysis, we sought classifications for how injuries occur, formalized into a coding system that captures the ways in which harms come to be. Ideally, we wanted a description of injury, including mechanism of harm, to the level of detail in which a human vulnerability is causally connected to a hazardous attribute. The World Health Organization's International Classification of External Causes of Injury (ICECI, https://www.who.int/standards/classifications/other-classifications/international-classification-of-external-causes-of-injury) is just such a resource.

ICECI [2] puts forth a model for the epidemiology of injury that connects the elements of environment, host (the injured party or patient), vector (the implicated object), with agent (the force or energy sustained by the body of the injured). For machines to generalize to novel circumstances, we ask: What are the attributes of each of these model elements that allow harm to come to the host? Specifically, we are interested in the attributes that enable the agent (again, in this context, the harmful force or energy) to be of harm to the host, and in relationship to the affordances of the vector.

4.2 Data sources

The ICECI classification system is adapted and applied in the US Center for Disease Control and Prevention's Injury Surveillance Guidelines. From these guidelines, the US Consumer Product Safety Commission developed the National Electronic Injury Surveillance System [3]. This data source codes incidents and includes a description of how the injuries have come to be. For each patient admitted to an emergency department, staff collect a brief narrative about how the injury occurred, along with the victim's demographics. This narrative for a typical line item, indicating that an 80-year-old male was diagnosed as having swallowed a foreign body, appears as:

> *"80 YOM. SWALLOWED HEARING AID BATTERIES AFTER MISTOOK THE BATTERIES [FOR] THE PILLS HE MEANT TO TAKE. DX: SWALLOWED FB".*

5 Knowledge representation

5.1 Extending methods to recognize affordances to recognize hazards

The concept of *affordances* closely resembles what we seek in enabling hazard recognition:

> *"The affordances of the environment are what it offers the animal, what it provides or furnishes, either **for good or ill**… [The word affordance] implies the complementarity of the animal and the environment."*
>
> — *Gibson [4, p. 127]*

Whereas most research into visual affordance and function understanding is concerned with positive object offerings (for good), for example, recognizing in an image that a guitar is playable or that a bike is rideable, we are interested in recognizing negative object offerings (for ill). We call these detrimental object offerings dis-affordances, or "deffordances".

There are many effective strategies to deduce object affordances from visual input [5]. One particular method [6] combines computer vision with knowledge engineering techniques. This method gathers evidence from data sources including images and online text. Through web parsing and feature extraction, evidence of different types is collected: physical, categorical, and visual attributes, along with human-object interactions where available. In a Markov Logic Network, a knowledge base is learned. This knowledge base is represented as a knowledge graph wherein objects are nodes connected by different edge types indicating relationships between the objects, such as common attributes, affordances, or categories. In this way, the knowledge is encoded that both basketballs and apples (object nodes) are round (attribute edge) and rollable (affordance edge), but only the apple is edible (affordance edge). Next, these edges are broken out with their type represented by node to reformulate the graph such that edges now may also link disparate and negative relationships: The basketball is *made of* **plastic**, which is *not* **edible** [*relationship,* **node**]. By this construction, novel inferences may be performed, for example, so that other plastic objects may be recognized as inedible.

5.2 Recognizing detriments

In a general sense, harms might be formulated as a product of the collision of an affordance and a dis-affordance (or "deffordance"). Consider the diagnosis "*ingestion of a foreign body*" from the example given from NEISS in Section 6. That foreign body (i.e., button battery) has some physical size and shape properties (small, flat, round) that make it ingestible (affordance, by the heuristic that the physical characteristics of the object are such that it can fit through a toilet paper roll tube, a rough model of the pharynx), but in contrast to edible objects with those same properties (e.g., pill), those objects with the material property of *inorganic* are inedible (deffordance). Whereas a pill is commonly composed of excipient (the inactive starchy material binding the active substance), batteries are ordinarily encased in metal. Size characteristics are attainable through RGB(+D) computer vision techniques leveraging fiducial markers or distance and focal length, and the material differences are discernable through the spectral signatures available through hyperspectal imaging, as described in the reference in Section 3. The conclusion of detriment at the coincidence of *ingestible* but not *edible* is possible to reason to from object attributes. Were an ingestible but inedible object to be swallowed, the swallowed foreign body would be an insult to host, who would sustain an injury consequent to that insult, such as gastrointestinal distress.

This reasoning is depicted in Fig. 13.1, adopting the coloring conventions in Ref. [6] and extending for deffordances, and for insults and injuries, with those extensions given in brackets. The dashed-outline (purple in the online version) category is used here for precise values of physical characteristics gathered through direct measurements or from manufacturer specifications (0.05 m), whereas the vertical and horizontal-hatched (green

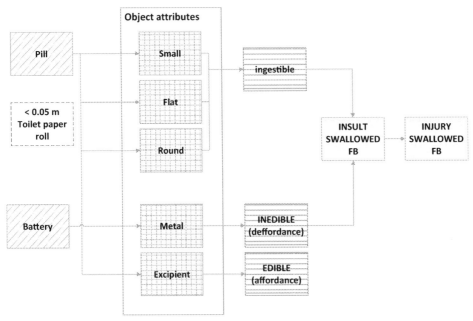

FIG. 13.1 Deffordance collision.

in the online version) category includes a qualitative description of those physical characteristics (small).

- Box with diagonal hatching (blue in online version) = isA
- Box with horizontal hatching (orange in online version) = hasAffordance [+hasDeffordance]
- Box with vertical and horizontal hatching (green in online version) = hasVisualAttribute
- Box with dashed outline (purple in online version) = hasSize, hasWeight
- [Filled box (yellow in online version) = hasInsult, hasInjury]

5.3 Ontology schema and the inference of dangerousness

Building on this example presented to anticipate (and ultimately preempt) injury from attributes of the object, environment, and person, we construct an ontology to enable reasoning about the danger presented in situations of affordance collision. As per Landrine v. Mego Corp., dangerousness is not an intrinsic property of an entity, but rather an inference (*isDangerousTo*) relating an entity attribute to a vulnerability, and a human's particular susceptibility to be harmed by exposure to that attribute. The ontology will be represented as a knowledge graph similar to those for visual affordances [5,6], augmented with detriments and the abstracted entities and relationships needed to infer object dangerousness. Following are the beginnings of a notational depiction, linking to the classification scheme offered by WHO's International Classification of External Causes of Injury (ICECI):

- **Object O** (ICECI Code C3 for object/substance producing injury),
  - with attributes $\{A_o\}$ that present hazards $\{H_O^A\}$

isDangerousTo (O, P)

- **Person P** with attributes $\{A_p\}$ that confer differential susceptibility impinging upon vulnerabilities $\{V_P^A\}$
  - (in Context C, including intent and behavior represented by ICECI Codes C1 Intent, C4 Place of Occurrence detailed in code P, C5 Activity when injured)

Because: A_o *presentsHazardtoVulnerability*(H_O^A, V_P^A) via Mechanism of injury (ICECI Codes C2, M1).

Susceptibility, the liability for a general human vulnerability to be experienced by a particular person, is not captured explicitly in ICECI, but can be estimated from statistics performed on the demographics of incident reports. In this way, the ontology may recognize a child's greater susceptibility to choking on a balloon compared to adults, pertaining to the general human vulnerability of choking on elastics, but accentuated for children with narrow throats and strong oral reflexes.

To furnish the concepts needed to perform this inference of dangerousness in a general sense, the ontology's schema includes the items in Table 13.2, depicted graphically in Fig. 13.2. The boxes shown represent relationships traversed in the knowledge graph to relate object, environmental, and personal attributes, ultimately to injuries.

Attributes of the object with the scene, and those ambient within the environment may be discerned through computer vision techniques involving machine learning, and activate the corresponding visual attribute nodes within the knowledge graph. These attributes are connected in the graph to the hazards such attributes present, and those hazards are connected to insults they induce in people affected by such hazards. In this way, the battery from our example before, with its attributes of small and round, connects to the hazard of (unintended) swallowing, and then to the insult of a swallowed foreign body.

Table 13.2 Harm ontology elements.

	Object	Person
Opaque	• **Object and environment attributes** $\{A_{o,E}\}$, which may be provided to the system opaquely by means of ML/CV	• **Personal attributes** $\{A_p\}$, which may be provided to the system by declaration
Legible	• **Hazards** $\{H\}$, and how *object and environment attributes* amount to the existence of particular hazards	• **Susceptibility**, and how *personal attributes* confer differential susceptibility to particular persons
	• **Insults**, and how *hazards* lead to those insults	• **Vulnerability** $\{V_P^A\}$, and how *differential susceptibility* leads particular persons to be differentially vulnerable to a particular *insult*
	• **Injury,** and how *insults* impinge on *vulnerabilities* such that the person would come to harm	

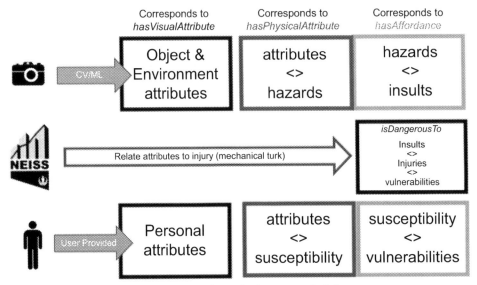

FIG. 13.2 Graphical depiction of the notional KG Schema for *isDangerousTo* inference.

Attributes of the person within the scene, which may be provided to the ontology by designer, prescriber, or user themselves, are connected to the differential susceptibility those personal attributes confer, and those susceptibilities are connected to the vulnerabilities made more likely by such susceptibility. In this way, the personal attribute of (high, or low) age confers an increased susceptibility to the vulnerability of having swallowed a foreign body.

When injuries impinging on those vulnerabilities are connected to the injuries sustained upon insult involving the scene's objects and environment, the *isDangerousTo* inference is complete. Thus, a button battery *isDangerousTo* an elder, at risk of the injury of poisoning.

To implement this schema, we will adopt modeling patterns from resources such as the OBO Foundry (https://obofoundry.org/ [7]), in particular to include relevant established features such as *disposition due to material basis in.*

5.4 Legibility of harms reasoning

As compared to machine learning approaches that render black boxes which, even if effective, are still difficult to understand and therefore to adjust, a critical feature of this harm ontology design is its availability to be introduced to the evaluative standards of the society in which it operates. These normative considerations may have hard limits shaped by that society's values, that ought to circumscribe the conclusions drawn by the system. We will craft as legible (as opposed to opaque) from the start those steps in inference to receive injection of normative considerations, and to be auditable in those terms.

The filled arrows in Fig. 13.2 indicate a "don't care condition" for the legibility of the step. In this way, opaque machine learning based computer vision techniques may be employed for producing the attributes of object and environmental attributes, and personal attributes may be provided instead of derived. The demands of two initial steps are lesser than the rest, insofar as we balance ease of collection with explainability of determination. That is, in terms of developing auditable harms sensitivity, we care less about how pixels are interpreted to become attributes, than about how hazards are expected to lead to insults.

6 Population of the ontology

For machines to generalize this nascent harms-reasoning capability to novel circumstances, we must establish in a knowledge base the attributes of each element within the WHO's epidemiological model of injury that allow harm to come to the host. Though such knowledge is not readily attainable by machines,[c] humans are naturally adept at providing responses to queries about salience, such that abstracting the attributes of an object relevant and important to a downstream injury due to that object is elicitable.

Thus, we intend to construct microtasks wherein human participants perform structured decomposition of incident reports in NEISS to annotate with attributes the elements of object (PROD, as notated in NEISS), person (PERS), and environment (ENVI) implicated in the injury. Importantly, with this reasoning so captured, novel inferences will be legible in terms of attributes, as compared to black box approaches, which are necessarily not so.

The microtask questions (bold) and responses (italics) for the sample NEISS entry from earlier are given (Fig. 13.3).

Although we are not the first to suggest that some information may be crowdsourced in order to develop and scale machines responsive to moral considerations, we take this approach to be novel. In the MIT Moral Machines project, for instance, people respond to a moral dilemma, such as how a trolley problem ought to be adjudicated. This is a democratic way of collecting impressions of appropriate machine action, but we argue that democracy at that level of examination is ripe for moral error. Instead, we prefer to derive machine behavior from first principles, and to draw people into the effort by asking them what is salient about a scene, to enter as evidence to apply those principles. In the crowdsourcing approach here described, respondents are asked not to perform high-level adjudication, but to provide low-level relevant factors. In this way, crowdsourcing is used to populate common-sense knowledge, stopping short of its use to make decisions based on that knowledge.

Substantial design effort is needed to ensure that the human annotation tasks ask for what is needed, are formulated to elicit useful responses, and validate the knowledge

---

[c]Since the writing of this chapter, we devised a new approach that uses large language models to populate the ontology, wherein the human elicitation component described in the chapter becomes a validation step. Detailed in the forthcoming publication, J. Jebari, A.M. Greenberg, Machine understanding of harms: theory and implementation (2024).

- **List the likely attributes of the PROD that enabled this to happen**

    - *Batteries resemble pills*

    - *PROD attributes: Small, round, flat*

    - *Affordances conferred by attributes: swallowable*

- **If any, which attributes of the PERS make them particularly susceptible to the injury**

    - *Age degrades vision*

    - *Personal attributes: AGE [NOT SEX, NOT RACE], ETOH NOT CONSUMED*

- **If any, which attributes of the ENVI make them particularly susceptible to the injury**

    - *Age increases ambient abundance of pills and hearing aid batteries*

- **Determination (presented to respondent for validation)*:***

    - Deffordance collision for *Ingestible foreign body*:

    - *SWALLOWABLE (affordance) BUT NOT EDIBLE (affordance)*

FIG. 13.3 Microtask and responses for a sample NEISS entry.

graph as it grows. The microtasks will conform to common multimodal annotation standards, and may be hosted on commercial platforms like Amazon's Mechanical Turk or government service platforms like APL's Natural Language Technologies Center for Excellence TURKLE (https://hltcoe.turkle.org/). Amazon's offering is available to international audiences, an essential feature for exploring cultural variation in responses.

Beyond initial human annotation to initialize the ontology scaffold, online learning can occur similarly to APL's Agent Spark (https://pai.agent.jhuapl.edu/) by scraping news feeds. In particular, this system would attend to reports in the Artificial Intelligence Incident Database (https://incidentdatabase.ai). As is available, we are eager to join this KG with those addressing related concerns, such the Consortium on the Landscape of AI Safety's Knowledge Graph System (CKGS, https://www.clais.org/the-landscape) or those embedded in general-purpose large language model systems like GPT-3 and its visual correlate DALL-E.

7 Parameterizations

Among other factors in reasoning about harms, time horizon, operating context, and cultural norms are prominently impactful. These factors must be parameterized for them to influence the operation of the inference engine.

Time horizon: We often incur short-term expense for a long-term benefit ("no pain, no gain"). Routinely in medicine, we trade an immediate detriment for an anticipated payoff. Allergy shots and mammograms are each unpleasant in the moment but stave off potential future greater discomfort. Compared to the time-consistent delay discounting seen in finance, wherein valuation decreases by a constant factor per unit of delay (e.g., exponential discounting), humans tend to engage in time-inconsistent delay discounting (e.g., hyperbolic discounting; see Ref. [8]). In present-based hyperbolic (vice exponential) discounting, benefits, if not immediate, are similarly valued in the future, regardless of how far out they are accrued. In this way, small rewards that occur sooner are preferred over larger rewards that occur later. Whether machines acting on our behalf should delay-discount like we do as individuals, or like we do as institutions that serve individuals (e.g., banks), remains an outstanding question.

Operating context: Actionable conclusions drawn about harm vary dramatically by situation. Though sharing much of the same core, reasoning about harm in manufacturing contexts prioritizes different concerns from those in eldercare or urban search and rescue contexts. Priority variation by context must be elicited and encoded.

Cultural norms: Norms indicate what is valued in a society or a local community, and the exchange rates among those values are not held as sacred (sacred values allow no trade-offs). Body modification, like tattoos or ritual scarification, is an example of an action in which physical harm is traded for a status benefit. In these cases, a machine ought not intervene to prohibit the body modification, at risk of rendering the individual "saved" from such harm excommunicated. Norms regarding respect for community elders (balancing among other factors, the elders' autonomy, fragility, and dignity) should dictate the course of intervention. Fighter pilots, as an example in contrast, operate according to very different norms. The Automatic Ground Collision Avoidance System avoids deadly crashes by pulling up when the aircraft is headed for surface impact, but its automatic engagement in seizing control over the aircraft may at times infringe upon the pilot's autonomy, especially in cases that interfere with the pilot's intent, such as attempting a daring combat maneuver. Cross-cultural variation must be captured for systems that abide by norms to be portable between contexts.

8 Handling harms of various types

8.1 Extension to nonphysical harms

Physical harms are the most tractable, so it was our intention to develop the implementation to address these first, and then extend the approach to nonphysical harms.[d] Nonphysical harms include the following:

---

[d]Since the writing of this chapter, we conceived of a new conceptualization for extending the method to nonphysical harm types, by generalizing harms as setbacks to interests to wellbeing. Detailed in the forthcoming publication, J. Jebari, A.M. Greenberg, Machine understanding of harms: theory and implementation (2024).

Financial harm negatively impacts on an individual's economic situation. Measured in dollars (or other local currency), this type of harm may be the result of property damage, loss of funds, or damage to earning potential. Financial harms most resemble physical harms in terms of their tractability.

Psychological harm negatively impacts an individual's psychological state, transiently or persistently. Emotional assaults that amount to an individual's feelings being hurt, or chronically such that that individual may be traumatized are both of concern. Disinformational assaults that distort an individual's beliefs and detrimentally influence decision-making also qualify. These harms, which to some extent appear in tort law as intentional infliction of emotional distress, do not quite comport to the schema developed for physical harms. In future work, we will examine the translation of insults and injuries to assaults and distress.

Dignitary and reputational harms negatively impact an individual's perceived social status. As a trivial example, should an eldercare bot apply make-up to a senior-living resident such that the individual is embarrassed, that individual's dignity has been infringed. Denial of an earned body modification constitutes a reputational harm.

Aesthetic, cultural/historical harms are those that impact individuals who place value on objects of social significance. When those objects are defaced or destroyed, the affected community is harmed. The Monuments Men in World War II traded risk to their physical wellbeing to protect artifacts from destruction, and humanity from the consequent harm. This is an indirect form of harm in that a damaged object leads to people suffering. Related is deprivation harm, in which the destruction or denial of food and shelter that satisfies basic needs results in human suffering.

Moralistic or handoff harm is negative impact imposed on those who receive a judgment or decision to make, without the appropriate information, time, or consideration of receptivity needed to do so with appropriate deliberation. In these cases, an individual bears the suffering from their sense of responsibility for the poor call, but has effectively been set up for failure. This type of harm is particularly prone to person-machine contexts in which a decision to be made by person is escalated to them by a machine with insufficient time, information, or attention to address it. See Ref. [9] for further discussion on this type.

8.2 Balancing the incommensurable

Initially, this ontology will be designed to anticipate harms physical in nature. Then, we will look to test extension of the schema in representing harms of other sorts described in the previous section, and seek applicable corpora from which to draw in populating the knowledge graph. Table 13.3 is a worksheet toward this end.

We acknowledge that, for example, physical injury is incommensurable to dignitary harm, but in practice trade-offs are inevitable and hard choices must be made. Perhaps there is a Pareto frontier per circumstance that could be empirically determined to draw from to guide machine decision-making (this, in contrast, might indeed be democratically determined). Such an examination would be the focus of our next phase of work.

Table 13.3 Harm types and data sources.

Type (incl. vulnerability impinged, harm experienced)	Scale	Applicable corpora	Annotation microtask	Schema extensions required	Priority criteria
Physical	Severity of injury diagnostic code, quality-adjusted life-year	NEISS	Attribute identification	None	TBD
Financial	$, as in impact to earning potential	Life insurance (wrt SES, problematic)	TBD	TBD	TBD
Psychological, emotional	TBD	Tort law of IIED (intentional infliction of emotional harm)	TBD	TBD	TBD
Dignitary, social, reputational	TBD	TBD	TBD	TBD	TBD
Aesthetic, cultural	TBD	TBD	TBD	TBD	TBD

9 Conclusion

In this chapter, we described an ontology to enable machines to recognize and reason about potential harms to humans. We adapted the concept of affordances to include detriments and inform on hazards, and outlined a knowledge graph schema linking the concepts of object and environmental attributes, hazards, insults, susceptibility, vulnerabilities, and ultimately injuries, so that an object's dangerousness to a particular person may be inferred. We then offered an approach to populate this knowledge graph by presenting incident reports from relevant corpora to human for them to annotate salient factors in the course of annotation microtasks. These methods, along with various parameterizations necessary to formalize for implementation, were developed to address sensitivity to physical harms. In future work, we will attend to nonphysical harms and the challenge of balancing across the different types for particular contexts.

Acknowledgments

Thanks to the JHU Discovery program for their grant supporting this investigation *Enabling Machines to Reason about Potential Harms to Humans*, and to the team including (alphabetically listed) David Handelman (APL), Brian Hutler (Temple), Debra Mathews, and Travis Rieder (JHU Berman Institute of Bioethics).

References

[1] A.M. Greenberg, Deciding Machines: Moral-Scene Assessment for Intelligent Systems. Human-Machine Shared Contexts, Elsevier Inc., 2020, https://doi.org/10.1016/B978-0-12-820543-3.00006-7.

[2] Y. Holder, M. Peden, E. Krug, J. Lund, G. Gururaj, O. Kobusingye, Injury surveillance guidelines, in: O.K.Y. Holder, M. Peden, E. Krug, J. Lund, G. Gururaj (Eds.), Centers for Disease Control and Prevention, World Health Organization, Atlanta, USA, 2001. https://apps.who.int/iris/bitstream/handle/10665/42451/9241591331.pdf?sequence=1.

[3] The national electronic injury surveillance system a tool for researchers division of hazard and injury data systems U. S. Consumer Product Safety Commission, System 2000 (2000). https://www.cpsc.gov/s3fs-public/pdfs/blk_media_2000d015.pdf.

[4] J.J. Gibson, The Ecological Approach to Visual Perception, Lawrence Erlbaum Associates, Hillsdale, NJ, 1986.

[5] M. Hassanin, S. Khan, M. Tahtali, Visual affordance and function understanding, ACM Comput. Surv. 54 (3) (2018) 1–26, https://doi.org/10.1145/3446370.

[6] Y. Zhu, A. Fathi, L. Fei-Fei, Reasoning about object affordances in a knowledge base representation, in: Lecture Notes in Computer Science, 2014, pp. 408–424. *Eccv*, no. Chapter 27 http://dblp.org/rec/conf/eccv/ZhuFF14%0Ahttp://link.springer.com.accesdistant.upmc.fr/content/pdf/10.1007%2F978-3-319-10605-2_27.pdf.

[7] R. Jackson, N. Matentzoglu, J.A. Overton, R. Vita, J.P. Balhoff, P.L. Buttigieg, S. Carbon, et al., OBO foundry in 2021: operationalizing open data principles to evaluate ontologies, Database 2021 (October) (2021) 1–9, https://doi.org/10.1093/database/baab069.

[8] A.M. Haith, T.R. Reppert, R. Shadmehr, Evidence for hyperbolic temporal discounting of reward in control of movements, J. Neurosci. 32 (34) (2012) 11727–11736, https://doi.org/10.1523/JNEUROSCI.0424-12.2012.

[9] A.M. Greenberg, J.L. Marble, Foundational concepts in person-machine teaming, Front. Phys. 10 (2023) 1–16, https://doi.org/10.3389/fphy.2022.1080132.

Index

Note: Page numbers followed by *f* indicate figures and *t* indicate tables.

Printed in the United States
by Baker & Taylor Publisher Services